BookWare Companion Series™

Digital Signal Processing
Using Matlab®

Books in the BookWare Companion Series™

Djaferis	*Automatic Control: The Power of Feedback Using MATLAB®*
Frederick/Chow	*Feedback Control Problems Using MATLAB® and the Control System Toolbox*
Ingle/Proakis	*Digital Signal Processing Using MATLAB®*
Pfeiffer	*Basic Probability Topics Using MATLAB®*
Proakis/Salehi	*Contemporary Communications Systems Using MATLAB®*
Rashid	*Electronics Circuit Design Using ELECTRONICS WORKBENCH*
Stonick/Bradley	*Labs for Signals and Systems Using MATLAB®*
Strum/Kirk	*Contemporary Linear Systems Using MATLAB®*

BookWare Companion Series™

DIGITAL SIGNAL PROCESSING USING MATLAB®

Vinay K. Ingle
John G. Proakis
Northeastern University

Brooks/Cole
Thomson Learning™

Pacific Grove • Albany • Belmont • Boston • Cincinnati • Johannesburg • London • Madrid • Melbourne
Mexico City • New York • Scottsdale • Singapore • Tokyo • Toronto

Publisher: *Bill Stenquist*
Marketing Team: *Christina DeVeto and Nathan Wilbur*
Editorial Assistant: *Shelley Gesicki*
Production Editor: *Mary Vezilich*

Production Service: *Greg Hubit Bookworks*
Cover Design: *Denise Davidson*
Print Buyer: *Barbara Stephan*
Typesetting: *Integre Technical Publishing Co., Inc.*
Printing and Binding: *Webcom Ltd.*

Library of Congress Cataloging-in-Publication Data
Ingle, Vinay K.
    Digital signal processing using MATLAB / Vinay K. Ingle, John G. Proakis.
        p.   cm.
    Includes bibliographical references
    ISBN 0-534-37174-4
    1. Signal processing—Digital techniques—Mathematics.   2. MATLAB.
I. Proakis, John G.   II. Title.
TK5102.9.I534   2000
621.382′2′078553042—dc21   99-28962

ABOUT THE SERIES

"The purpose of computing is insight, not numbers."
—R. W. Hamming, *Numerical Methods for Engineers and Scientists*,
McGraw-Hill, Inc.

It is with this spirit in mind that we present the BookWare Companion Series.™

Increasingly, the latest technologies and modern methods are crammed into courses already dense with important theory. As a result, many instructors now ask, "Are we simply teaching students the latest technology, or are we teaching them to reason?" We believe that these two alternatives need not be mutually exclusive. In fact, this series was founded on the belief that computer solutions and theory can be mutually reinforcing. Properly applied, computing can illuminate theory and help students to think, analyze, and reason in meaningful ways. It can also help them to understand the relationships and connections between new information and existing knowledge and to cultivate problem-solving skills, intuition, and critical thinking. The BookWare Companion Series was developed in response to this mission.

Specifically, the series is designed for educators who want to integrate computer-based learning tools into their courses, and for students who want to go further than their textbook alone allows. The former will find in the series the means by which to use powerful software tools to support their course activities without having to customize the applications themselves. The latter will find relevant problems and examples quickly and easily available and will have electronic access to them. Important for both educators and students is the premise on which the series is based: that students learn best when they are actively involved in their own learning. The BookWare Companion Series will engage them, provide a taste of real-life issues, demonstrate clear techniques for solving real problems, and challenge them to understand and apply these techniques on their own.

To serve your needs better, we are continually looking for ways to improve the series. Toward that end, please join us at our BookWare Companion Resource Center website:

http://www.brookscole.com/engineering/ee/bookware.html

You can recommend ways to make the series even better, share your ideas about using technology in the classroom with your colleagues, suggest a specific problem or example for the next edition, or just let us know what's on your mind. We look forward to hearing from you, and we thank you for your continuing support.

Bill Stenquist	Publisher	bill.stenquist@brookscole.com
Shelley Gesicki	Editorial Assistant	shelley.gesicki@brookscole.com
Nathan Wilbur	Marketing Manager	nathan.wilbur@brookscole.com
Christina DeVeto	Marketing Assistant	christina.deveto@brookscole.com

CONTENTS

PREFACE   xi

1   INTRODUCTION   1

Overview of Digital Signal Processing   2
A Few Words about MATLAB®   5

2   DISCRETE-TIME SIGNALS AND SYSTEMS   7

Discrete-time Signals   7
Discrete Systems   20
Convolution   22
Difference Equations   29
Problems   35

3   THE DISCRETE-TIME FOURIER ANALYSIS   40

The Discrete-time Fourier Transform (DTFT)   40
The Properties of the DTFT   47

MATLAB is a registered trademark of The MathWorks, Inc.

The Frequency Domain Representation of LTI Systems   53
Sampling and Reconstruction of Analog Signals   60
Problems   74

4   THE z-TRANSFORM   80

———————————■———————————

The Bilateral z-Transform   80
Important Properties of the z-Transform   84
Inversion of the z-Transform   89
System Representation in the z-Domain   95
Solutions of the Difference Equations   105
Problems   111

5   THE DISCRETE FOURIER TRANSFORM   116

———————————■———————————

The Discrete Fourier Series   117
Sampling and Reconstruction in the z-Domain   124
The Discrete Fourier Transform   129
Properties of the Discrete Fourier Transform   139
Linear Convolution using the DFT   154
The Fast Fourier Transform   160
Problems   172

6   DIGITAL FILTER STRUCTURES   182

———————————■———————————

Basic Elements   183
IIR Filter Structures   183
FIR Filter Structures   197

Lattice Filter Structures 208

Problems 219

7 FIR FILTER DESIGN 224

───────────────■───────────────

Preliminaries 224

Properties of Linear-phase FIR Filters 228

Window Design Techniques 243

Frequency Sampling Design Techniques 264

Optimal Equiripple Design Technique 277

Problems 294

8 IIR FILTER DESIGN 301

───────────────■───────────────

Some Preliminaries 302

Characteristics of Prototype Analog Filters 305

Analog-to-Digital Filter Transformations 327

Lowpass Filter Design Using MATLAB 345

Frequency-band Transformations 350

Comparison of FIR vs. IIR Filters 363

Problems 364

9 APPLICATIONS IN ADAPTIVE FILTERING 371

───────────────■───────────────

LMS Algorithm for Coefficient Adjustment 373

System Identification or System Modeling 376

Suppression of Narrowband Interference in a Wideband Signal 377

Adaptive Line Enhancement 380

Adaptive Channel Equalization   380

Summary   383

10   APPLICATIONS IN COMMUNICATIONS   384

Pulse-Code Modulation   384

Differential PCM (DPCM)   388

Adaptive PCM and DPCM (ADPCM)   392

Delta Modulation (DM)   396

Linear Predictive Coding (LPC) of Speech   399

Dual-tone Multifrequency (DTMF) Signals   403

Binary Digital Communications   408

Spread-Spectrum Communications   409

Summary   411

BIBLIOGRAPHY   412

INDEX   413

PREFACE

From the beginning of the 1980s we have witnessed a revolution in computer technology and an explosion in user-friendly applications. This revolution is still continuing today with low-cost personal computer systems that rival the performance of expensive workstations. This technological prowess should be brought to bear on the educational process and, in particular, on effective teaching that can result in enhanced learning. This companion book on digital signal processing (DSP) makes a small contribution toward that goal.

The teaching methods in signal processing have changed over the years from the simple "lecture-only" format to a more integrated "lecture-laboratory" environment in which practical hands-on issues are taught using DSP hardware. However, for effective teaching of DSP the lecture component must also make extensive use of computer-based explanations, examples, and exercises. For the last several years, the MATLAB software developed by *The MathWorks, Inc.* has established itself as the de facto standard for numerical computation in the signal-processing community and as a platform of choice for algorithm development. There are several reasons for this development, but one most important reason is that MATLAB is available on practically all computing platforms. For several years the Professional Version of MATLAB was the only version available on the market. The advent of a Student Edition has now made it possible to use it in classrooms. Recently, several textbooks in DSP have appeared that generally provide exercises that can be done using MATLAB. However, for students (and for practicing engineers interested in DSP) there are no "how-to" references for effective use of MATLAB in DSP. In this book we have made an attempt at integrating MATLAB with traditional topics in DSP so that it can be used to explore difficult topics and solve problems to gain insight. Many problems or design algorithms in DSP require considerable computation. It is for these that MATLAB provides a convenient tool so that many scenarios can be tried with ease. Such an approach can enhance the learning process.

SCOPE OF THE BOOK

This book is primarily intended for use as a supplement in junior- or senior-level undergraduate courses on DSP. We assume that the student (or user) is familiar with the fundamentals of MATLAB. Those topics are not covered since several tutorial books and manuals on MATLAB are available. Similarly, this book is not written as a textbook in DSP because of the availability of excellent textbooks. What we have tried to do is to provide enough depth to the material augmented by MATLAB functions and examples so that the presentation is consistent, logical, and enjoyable. Therefore this book can also be used as a self-study guide by anyone interested in DSP.

ORGANIZATION OF THE BOOK

The first eight chapters of this book discuss traditional material covered in an introductory course on DSP. The last two chapters are presented as applications in DSP with emphasis on MATLAB-based projects. The following is a list of chapters and a brief description of their contents:

Chapter 1, Introduction: This chapter introduces readers to the discipline of signal processing and discusses the advantages of DSP over analog signal processing. A brief introduction to MATLAB is also provided.

Chapter 2, Discrete-time Signals and Systems: This chapter provides a brief review of discrete-time signals and systems in the time domain. Appropriate use of MATLAB functions is demonstrated.

Chapter 3, The Discrete-time Fourier Analysis: This chapter discusses discrete-time signal and system representation in the frequency domain. Sampling and reconstruction of analog signals are also presented.

Chapter 4, The z-Transform: This chapter provides signal and system description in the complex frequency domain. MATLAB techniques are introduced to analyze z-transforms and to compute inverse z-transforms. Solutions of difference equations using the z-transform and MATLAB are provided.

Chapter 5, The Discrete Fourier Transform: This chapter is devoted to the computation of the Fourier transform and to its efficient implementation. The discrete Fourier series is used to introduce the discrete Fourier transform, and several of its properties are demonstrated using MATLAB. Topics such as fast convolution and fast Fourier transform are thoroughly discussed.

Chapter 6, Digital Filter Structures: This chapter discusses several structures for the implementation of digital filters. Several useful MATLAB functions are developed for the determination and implementation of these structures. Lattice and ladder filters are also introduced and discussed.

Chapter 7, FIR Filter Design: This chapter and the next introduce the important topic of digital filter design. Three important design techniques for FIR filters — namely, window design, frequency sampling design, and the equiripple filter design — are discussed. Several design examples are provided using MATLAB.

Chapter 8, IIR Filter Design: Included in this chapter are techniques in IIR filter design. It begins with analog filter design and introduces such topics as filter transformations and filter-band transformation. Once again several design examples using MATLAB are provided.

Chapter 9, Applications in Adaptive Filtering: This chapter is the first of two chapters on projects using MATLAB. Included is an introduction to the theory and implementation of adaptive FIR filters with projects in system identification, interference suppression, narrowband frequency enhancement, and adaptive equalization.

Chapter 10, Applications in Communications: This chapter focuses on several projects dealing with waveform representation and coding, and with digital communications. Included is a description of pulse-code modulation (PCM), differential PCM (DPCM) and adaptive DPCM (AD-PCM), delta modulation (DM) and adaptive DM (ADM), linear predictive coding (LPC), generation and detection of dual-tone multifrequency (DTMF) signals, and a description of signal detection applications in binary communications and spread-spectrum communications.

ABOUT THE SOFTWARE

This book is an outgrowth of our teaching of a MATLAB-based undergraduate DSP course over several years. Many MATLAB functions discussed in this book were developed in this course. These functions are available at the BookWare Companion Resource Center, online at http://www.brookscole.com/engineering/ee/bookware.html. The book also contains numerous MATLAB scripts in many examples. These scripts are also available at the Resource Center and are kept in individual directories created for each chapter. In addition, many figures were produced as MATLAB plots, and their scripts are available in the `figures` directory. Students should study these scripts to gain insight into the MATLAB procedures. We will appreciate any comments, corrections, or compact coding of these programs and scripts. Solutions to problems and the associated script files will be made available to instructors in the near future.

Further information about MATLAB and related publications may be obtained from:

The MathWorks, Inc.
24 Prime Park Way
Natick, MA 01760-1500
Phone: (508) 647-7000   Fax: (508) 647-7101
E-mail: info@mathworks.com
http://www.mathworks.com

ACKNOWLEDGMENTS

We are indebted to our numerous students in our ECE-1456 course at Northeastern University who provided us a forum to test teaching ideas using MATLAB and who endured our constant emphasis on MATLAB. Some efficient MATLAB functions are due to these students. We are also indebted to our reviewers, whose constructive criticism resulted in a better presentation of the material: Abeer Alwan, University of California, Los Angeles; Steven Chin, Catholic University; and Joel Trussel, North Carolina State University. We also thank all those readers who provided numerous corrections and refinements that have enhanced the present edition. In particular, we thank Prof. Huaichen of Xidian University, P. R. China.

We would like to thank Tom Robbins, former editor at PWS Publishing Company, for his initiative in creating the BookWare Companion Series and for his enthusiastic support of MATLAB in classroom teaching, especially in DSP. We sincerely appreciate the support exended by Ms. Naomi Bulock of MathWorks, Inc. who was always helpful in providing the newest versions of MATLAB. Finally, we would like to thank Brooks/Cole and Bookworks for the final preparation of the manuscript.

Vinay K. Ingle
John G. Proakis
Boston, Massachusetts

1 *INTRODUCTION*

Over the past several decades the field of digital signal processing (DSP) has grown to be important both theoretically and technologically. A major reason for its success in industry is due to the development and use of low-cost software and hardware. New technologies and applications in various fields are now poised to take advantage of DSP algorithms. This will lead to a greater demand for electrical engineers with background in DSP. Therefore it is necessary to make DSP an integral part of any electrical engineering curriculum.

Not long ago an introductory course on DSP was given mainly at the graduate level. It was supplemented by computer exercises on filter design, spectrum estimation, and related topics using mainframe (or mini) computers. However, considerable advances in personal computers and software over the past decade made it possible to introduce a DSP course to undergraduates. Since DSP applications are primarily algorithms that are implemented either on a DSP processor [11] or in software, a fair amount of programming is required. Using interactive software, such as MATLAB, it is now possible to place more emphasis on learning new and difficult concepts than on programming algorithms. Interesting practical examples can be discussed, and useful problems can be explored.

With this philosophy in mind, we have developed this book as a *companion book* (to traditional textbooks like [16, 19]) in which MATLAB is an integral part in the discussion of topics and concepts. We have chosen MATLAB as the programming tool primarily because of its wide availability on computing platforms in many universities across the country. Furthermore, a student edition of MATLAB has been available for several years, placing it among the least expensive software for educational purposes. We have treated MATLAB as a computational and programming toolbox containing several tools (sort of a super calculator with several keys) that can be used to explore and solve problems and, thereby, enhance the learning process.

This book is written at an introductory level in order to introduce undergraduate students to an exciting and practical field of DSP. We emphasize that this is not a textbook in the traditional sense but a companion book in which more attention is given to problem solving and hands-on experience with MATLAB. Similarly, it is not a tutorial book in MATLAB. We assume that the student is familiar with MATLAB and is currently taking a course in DSP. The book provides basic analytical tools needed to process real-world signals (a.k.a. analog signals) using digital techniques. We deal mostly with discrete-time signals and systems, which are analyzed in both the time and the frequency domains. The analysis and design of processing structures called *filters* and *spectrum analyzers* is one of the most important aspects of DSP and is treated in great detail in this book. Many advanced topics in DSP (which are generally covered in a graduate course) are not treated in this book, but it is hoped that the experience gained in this book will allow students to tackle advanced topics with greater ease and understanding.

In this chapter we provide a brief overview of both DSP and MATLAB.

OVERVIEW OF DIGITAL SIGNAL PROCESSING
■

In this modern world we are surrounded by all kinds of signals in various forms. Some of the signals are natural, but most of the signals are manmade. Some signals are necessary (speech), some are pleasant (music), while many are unwanted or unnecessary in a given situation. In an engineering context, signals are carriers of information, both useful and unwanted. Therefore extracting or enhancing the useful information from a mix of conflicting information is a simplest form of signal processing. More generally, signal processing is an operation designed for extracting, enhancing, storing, and transmitting useful information. The distinction between useful and unwanted information is often subjective as well as objective. Hence signal processing tends to be application dependent.

HOW ARE SIGNALS PROCESSED?   The signals that we encounter in practice are mostly analog signals. These signals, which vary continuously in time and amplitude, are processed using electrical networks containing active and passive circuit elements. This approach is known as analog signal processing (ASP)—for example, radio and television receivers.

$$\text{Analog signal:} \quad x_a(t) \quad \longrightarrow \quad \boxed{\text{Analog signal processor}} \quad \longrightarrow \quad y_a(t) \quad \text{:Analog signal}$$

They can also be processed using digital hardware containing adders, multipliers, and logic elements or using special-purpose microprocessors. However, one needs to convert analog signals into a form suitable for digital hardware. This form of the signal is called a digital signal. It takes one of the finite number of values at specific instances in time, and hence it can be represented by binary numbers, or bits. The processing of digital signals is called DSP; in block diagram form it is represented by

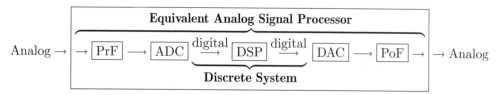

where the various block elements are discussed below.

PrF: This is a prefilter or an antialiasing filter, which conditions the analog signal to prevent aliasing.

ADC: This is called an analog-to-digital converter, which produces a stream of binary numbers from analog signals.

Digital signal processor: This is the heart of DSP and can represent a general-purpose computer or a special-purpose processor, or digital hardware, and so on.

DAC: This is the inverse operation to the ADC, called a digital-to-analog converter, which produces a staircase waveform from a sequence of binary numbers, a first step towards producing an analog signal.

PoF: This is a postfilter to smooth out staircase waveform into the desired analog signal.

It appears from the above two approaches to signal processing, analog and digital, that the DSP approach is the more complicated, containing more components than the "simpler looking" ASP. Therefore one might ask a question: Why process signals digitally? The answer lies in many advantages offered by DSP.

ADVANTAGES OF DSP OVER ASP

A major drawback of ASP is its limited scope for performing complicated signal processing applications. This translates into nonflexibility in processing and complexity in system designs. All of these generally lead to expensive products. On the other hand, using a DSP approach, it is possible to convert an inexpensive personal computer into a powerful signal processor. Some important advantages of DSP are these:

1. Systems using the DSP approach can be developed using software running on a general-purpose computer. Therefore DSP is relatively convenient to develop and test, and the software is portable.

2. DSP operations are based solely on additions and multiplications, leading to extremely stable processing capability—for example, stability independent of temperature.

3. DSP operations can easily be modified in real time, often by simple programming changes, or by reloading of registers.

4. DSP has lower cost due to VLSI technology, which reduces costs of memories, gates, microprocessors, and so forth.

The principal disadvantage of DSP is the speed of operations, especially at very high frequencies. Primarily due to the above advantages, DSP is now becoming a first choice in many technologies and applications, such as consumer electronics, communications, wireless telephones, and medical imaging.

TWO IMPORTANT CATEGORIES OF DSP

Most DSP operations can be categorized as being either signal *analysis* tasks or signal *filtering* tasks as shown below.

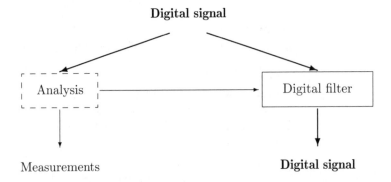

Signal analysis This task deals with the measurement of signal properties. It is generally a frequency-domain operation. Some of its applications are

- spectrum (frequency and/or phase) analysis
- speech recognition
- speaker verification
- target detection

Signal filtering This task is characterized by the "signal in–signal out" situation. The systems that perform this task are generally called *filters*. It is usually (but not always) a time-domain operation. Some of the applications are

- removal of unwanted background noise
- removal of interference
- separation of frequency bands
- shaping of the signal spectrum

In some applications, such as voice synthesis, a signal is first analyzed to study its characteristics, which are then used in digital filtering to generate a synthetic voice.

In the first half of this book we will deal with the signal-analysis aspect of DSP. In Chapter 2 we will begin with basic descriptions of discrete-time signals and systems. These signals and systems are analyzed in the frequency domain in Chapter 3. A generalization of the frequency-domain description, called the z-transform, is introduced in Chapter 4. The practical algorithms for computing the Fourier transform are discussed in Chapter 5 in the form of the discrete Fourier transform and the fast Fourier transform.

The second half of this book is devoted to the signal-filtering aspect of DSP. In Chapter 6 we describe various implementations and structures of digital filters. In Chapter 7 we provide design techniques and algorithms for designing one type of digital filter called finite-duration impulse response (or FIR) filters, while in Chapter 8 we provide a similar treatment for another type of filter called infinite-duration impulse response (or IIR) filters. In both chapters we discuss only the simpler but practically useful techniques of filter design. More advanced techniques are not covered. Finally, the last two chapters provide some practical applications in the form of projects that can be done using material learned in the first eight chapters. In Chapter 9 concepts in adaptive filtering are introduced, and simple projects in system identification, interference suppression, adaptive line enhancement, and so forth are discussed. In Chapter 10 a brief introduction to digital communications is presented with projects in such topics as PCM, DPCM, and LPC being outlined.

In all these chapters the central theme is the generous use and adequate demonstration of MATLAB tools. Most of the existing MATLAB functions for DSP are described in detail, and their correct use is demonstrated in many examples. Furthermore, many new MATLAB functions are developed to provide insights into the working of many algorithms. We believe that this "hand-holding" approach will enable students to dispel fears about DSP and will provide an enriching learning experience.

A FEW WORDS ABOUT MATLAB®

MATLAB is an interactive, matrix-based system for scientific and engineering numeric computation and visualization. Its strength lies in the fact that complex numerical problems can be solved easily and in a fraction of the time required with a programming language such as Fortran or C. It is also powerful in the sense that by using its relatively simple programming capability, MATLAB can be easily extended to create new commands and functions.

Linux? *

MATLAB is available on a number of computing environments: PCs running DOS, Win95, and Win98, Apple Macintosh, UNIX workstations, and several parallel machines. The basic MATLAB program is further enhanced by the availability of numerous toolboxes (a collection of specialized functions in a specific topic) over the years. The information in this book generally applies to all these environments. Although functions described in the first edition were developed under MATLAB Version 4.2, these functions have been made compatible with the present version of MATLAB. Furthermore, through the services of the BookWare Companion Resource Center (www.brookscole.com/engineering/ee/bookware.htm), every effort will be made to preserve this compatibility under future versions of MATLAB.

The scope and power of MATLAB go far beyond the few words given in this section. It is senseless to provide a concise information or tutorial on MATLAB when excellent books and guides are available on this topic. Students should consult the MATLAB User's Guide [2] and Reference Guide [1]. Similarly, students should attempt the tutorial given in [3]. The information given in all these references, along with the online facility, usually is sufficient for students to use this book.

DISCRETE-TIME SIGNALS AND SYSTEMS

We begin with the concepts of signals and systems in discrete time. A number of important types of signals and their operations are introduced. Linear and shift-invariant systems are discussed mostly because they are easier to analyze and implement. The convolution and the difference equation representations are given special attention because of their importance in digital signal processing and in MATLAB. The emphasis in this chapter is on the representations and implementation of signals and systems using MATLAB.

DISCRETE-TIME SIGNALS

Signals are broadly classified into analog and discrete signals. An analog signal will be denoted by $x_a(t)$, in which the variable t can represent any physical quantity, but we will assume that it represents time in seconds. A discrete signal will be denoted by $x(n)$, in which the variable n is integer-valued and represents discrete instances in time. Therefore it is also called a discrete-time signal, which is a *number sequence* and will be denoted by one of the following notations:

$$x(n) = \{x(n)\} = \{\cdots, x(-1), x(0), x(1), \cdots\}$$
$$\uparrow$$

where the *up-arrow* indicates the sample at $n = 0$.

In MATLAB we can represent a *finite-duration* sequence by a *row vector* of appropriate values. However, such a vector does not have any information about sample position n. Therefore a correct representation of $x(n)$ would require two vectors, one each for x and n. For example, a sequence $x(n) = \{2, 1, -1, 0, 1, 4, 3, 7\}$ can be represented in MATLAB by

$$\;\uparrow$$

```
>> n=[-3,-2,-1,0,1,2,3,4];  x=[2,1,-1,0,1,4,3,7];
```

Generally, we will use the **x**-vector representation alone when the sample position information is not required or when such information is trivial (e.g. when the sequence begins at $n = 0$). An arbitrary *infinite-duration* sequence cannot be represented in MATLAB due to the finite memory limitations.

<table>
<tr><td>TYPES OF
SEQUENCES</td><td>We use several elementary sequences in digital signal processing for analysis purposes. Their definitions and MATLAB representations are given below.</td></tr>
</table>

    1. *Unit sample sequence*:

$$\delta(n) = \begin{cases} 1, & n = 0 \\ 0, & n \neq 0 \end{cases} = \left\{ \cdots, 0, 0, \underset{\uparrow}{1}, 0, 0, \cdots \right\}$$

In MATLAB the function `zeros(1,N)` generates a row vector of N zeros, which can be used to implement $\delta(n)$ over a finite interval. However, the logical relation `n==0` is an elegant way of implementing $\delta(n)$. For example, to implement

$$\delta(n - n_0) = \begin{cases} 1, & n = n_0 \\ 0, & n \neq n_0 \end{cases}$$

over the $n_1 \leq n_0 \leq n_2$ interval, we will use the following MATLAB function.

```
function [x,n] = impseq(n0,n1,n2)
% Generates x(n) = delta(n-n0); n1 <= n <= n2
% ------------------------------------------------
% [x,n] = impseq(n0,n1,n2)
%
n = [n1:n2]; x = [(n-n0) == 0];
```

    2. *Unit step sequence*:

$$u(n) = \begin{cases} 1, & n \geq 0 \\ 0, & n < 0 \end{cases} = \left\{ \cdots, 0, 0, \underset{\uparrow}{1}, 1, 1, \cdots \right\}$$

In MATLAB the function `ones(1,N)` generates a row vector of N ones. It can be used to generate $u(n)$ over a finite interval. Once again an elegant approach is to use the logical relation n>=0. To implement

$$u(n - n_0) = \begin{cases} 1, & n \geq n_0 \\ 0, & n < n_0 \end{cases}$$

over the $n_1 \leq n_0 \leq n_2$ interval, we will use the following MATLAB function.

```
function [x,n] = stepseq(n0,n1,n2)
% Generates x(n) = u(n-n0); n1 <= n <= n2
% ------------------------------------------
% [x,n] = stepseq(n0,n1,n2)
%
n = [n1:n2]; x = [(n-n0) >= 0];
```

3. *Real-valued exponential sequence*:

$$x(n) = a^n, \forall n; \; a \in \mathbb{R}$$

In MATLAB an array operator ".^" is required to implement a real exponential sequence. For example, to generate $x(n) = (0.9)^n$, $0 \leq n \leq 10$, we will need the following MATLAB script:

```
>> n = [0:10]; x = (0.9).^n;
```

4. *Complex-valued exponential sequence*:

$$x(n) = e^{(\sigma + j\omega_0)n}, \forall n$$

where σ is called an attenuation and ω_0 is the frequency in radians. A MATLAB function exp is used to generate exponential sequences. For example, to generate $x(n) = \exp[(2 + j3)n]$, $0 \leq n \leq 10$, we will need the following MATLAB script:

```
>> n = [0:10]; x = exp((2+3j)*n);
```

5. *Sinusoidal sequence*:

$$x(n) = \cos(\omega_0 n + \theta), \forall n$$

where θ is the phase in radians. A MATLAB function cos (or sin) is used to generate sinusoidal sequences. For example, to generate $x(n) = 3\cos(0.1\pi n + \pi/3) + 2\sin(0.5\pi n)$, $0 \leq n \leq 10$, we will need the following MATLAB script:

```
>> n = [0:10]; x = 3*cos(0.1*pi*n+pi/3) + 2*sin(0.5*pi*n);
```

6. *Random sequences*: Many practical sequences cannot be described by mathematical expressions like those above. These sequences are called random (or stochastic) sequences and are characterized by parameters of the associated probability density functions or their statistical moments. In MATLAB two types of (pseudo-) random sequences are available. The `rand(1,N)` generates a length N random sequence whose elements are uniformly distributed between $[0, 1]$. The `randn(1,N)` generates a length N Gaussian random sequence with mean 0 and variance 1. Other random sequences can be generated using transformations of the above functions.

7. *Periodic sequence*: A sequence $x(n)$ is periodic if $x(n) = x(n+N)$, $\forall n$. The smallest integer N that satisfies the above relation is called the *fundamental* period. We will use $\tilde{x}(n)$ to denote a periodic sequence. To generate P periods of $\tilde{x}(n)$ from one period $\{x(n), \quad 0 \le n \le N-1\}$, we can copy $x(n)$ P times:

```
>> xtilde = [x,x,...,x];
```

But an elegant approach is to use MATLAB's powerful indexing capabilities. First we generate a matrix containing P rows of $x(n)$ values. Then we can concatenate P rows into a long row vector using the construct (:). However, this construct works only on columns. Hence we will have to use the matrix transposition operator ' to provide the same effect on rows.

```
>> xtilde = x' * ones(1,P);   % P columns of x; x is a row vector
>> xtilde = xtilde(:);        % long column vector
>> xtilde = xtilde';          % long row vector
```

Note that the last two lines can be combined into one for compact coding. This is shown in Example 2.1.

OPERATIONS ON SEQUENCES

Here we briefly describe basic sequence operations and their MATLAB equivalents.

1. *Signal addition*: This is a sample-by-sample addition given by

$$\{x_1(n)\} + \{x_2(n)\} = \{x_1(n) + x_2(n)\}$$

It is implemented in MATLAB by the arithmetic operator "+". However, the lengths of $x_1(n)$ and $x_2(n)$ must be the same. If sequences are of unequal lengths, or if the sample positions are different for equal-length sequences, then we cannot directly use the operator +. We have to first augment $x_1(n)$ and $x_2(n)$ so that they have the same position vector **n** (and hence the same length). This requires careful attention to MATLAB's indexing operations. In particular, logical operation of intersection "&",

relational operations like "<=" and "==", and the find function are required to make $x_1(n)$ and $x_2(n)$ of equal length. The following function, called the sigadd function, demonstrates these operations.

```
function [y,n] = sigadd(x1,n1,x2,n2)
% implements y(n) = x1(n)+x2(n)
% --------------------------
% [y,n] = sigadd(x1,n1,x2,n2)
%   y = sum sequence over n, which includes n1 and n2
%   x1 = first sequence over n1
%   x2 = second sequence over n2 (n2 can be different from n1)
%
n = min(min(n1),min(n2)):max(max(n1),max(n2));  % duration of y(n)
y1 = zeros(1,length(n)); y2 = y1;               % initialization
y1(find((n>=min(n1))&(n<=max(n1))==1))=x1;       % x1 with duration of y
y2(find((n>=min(n2))&(n<=max(n2))==1))=x2;       % x2 with duration of y
y = y1+y2;                                        % sequence addition
```

Its use is illustrated in Example 2.2.

2. *Signal multiplication*: This is a sample-by-sample multiplication (or "dot" multiplication) given by

$$\{x_1(n)\} \cdot \{x_2(n)\} = \{x_1(n)x_2(n)\}$$

It is implemented in MATLAB by the array operator ".*". Once again the similar restrictions apply for the .* operator as for the + operator. Therefore we have developed the sigmult function, which is similar to the sigadd function.

```
function [y,n] = sigmult(x1,n1,x2,n2)
% implements y(n) = x1(n)*x2(n)
% --------------------------
% [y,n] = sigmult(x1,n1,x2,n2)
%   y = product sequence over n, which includes n1 and n2
%   x1 = first sequence over n1
%   x2 = second sequence over n2 (n2 can be different from n1)
%
n = min(min(n1),min(n2)):max(max(n1),max(n2));  % duration of y(n)
y1 = zeros(1,length(n)); y2 = y1;               %
y1(find((n>=min(n1))&(n<=max(n1))==1))=x1;       % x1 with duration of y
y2(find((n>=min(n2))&(n<=max(n2))==1))=x2;       % x2 with duration of y
y = y1 .* y2;                                     % sequence multiplication
```

Its use is also given in Example 2.2.

3. *Scaling*: In this operation each sample is multiplied by a scalar α.

$$\alpha\{x(n)\} = \{\alpha x(n)\}$$

An arithmetic operator "∗" is used to implement the scaling operation in MATLAB.

4. *Shifting*: In this operation each sample of $x(n)$ is shifted by an amount k to obtain a shifted sequence $y(n)$.

$$y(n) = \{x(n-k)\}$$

If we let $m = n - k$, then $n = m + k$ and the above operation is given by

$$y(m+k) = \{x(m)\}$$

Hence this operation has no effect on the vector **x**, but the vector **n** is changed by adding k to each element. This is shown in the function sigshift.

```
function [y,n] = sigshift(x,m,n0)
% implements y(n) = x(n-n0)
% -------------------------
% [y,n] = sigshift(x,m,n0)
%
n = m+n0; y = x;
```

Its use is given in Example 2.2.

5. *Folding*: In this operation each sample of $x(n)$ is flipped around $n = 0$ to obtain a folded sequence $y(n)$.

$$y(n) = \{x(-n)\}$$

In MATLAB this operation is implemented by fliplr(x) function for sample values and by -fliplr(n) function for sample positions as shown in the sigfold function.

```
function [y,n] = sigfold(x,n)
% implements y(n) = x(-n)
% -----------------------
% [y,n] = sigfold(x,n)
%
y = fliplr(x); n = -fliplr(n);
```

6. *Sample summation*: This operation differs from signal addition operation. It adds all sample values of $x(n)$ between n_1 and n_2.

$$\sum_{n=n_1}^{n_2} x(n) = x(n_1) + \cdots + x(n_2)$$

It is implemented by the sum(x(n1:n2)) function.

7. *Sample products*: This operation also differs from signal multiplication operation. It multiplies all sample values of $x(n)$ between n_1 and n_2.

$$\prod_{n_1}^{n_2} x(n) = x(n_1) \times \cdots \times x(n_2)$$

It is implemented by the `prod(x(n1:n2))` function.

8. *Signal energy*: The energy of a sequence $x(n)$ is given by

$$\mathcal{E}_x = \sum_{-\infty}^{\infty} x(n)x^*(n) = \sum_{-\infty}^{\infty} |x(n)|^2$$

where superscript * denotes the operation of complex conjugation[1]. The energy of a finite-duration sequence $x(n)$ can be computed in MATLAB using

```
>> Ex = sum(x .* conj(x)); % one approach
>> Ex = sum(abs(x) .^ 2); % another approach
```

9. *Signal power*: The average power of a periodic sequence with fundamental period N is given by

$$\mathcal{P}_x = \frac{1}{N} \sum_{0}^{N-1} |x(n)|^2$$

□ **EXAMPLE 2.1**   Generate and plot each of the following sequences over the indicated interval.

    a. $x(n) = 2\delta(n+2) - \delta(n-4)$, $-5 \leq n \leq 5$.
    b. $x(n) = n\left[u(n) - u(n-10)\right] + 10e^{-0.3(n-10)}\left[u(n-10) - u(n-20)\right]$, $0 \leq n \leq 20$.
    c. $x(n) = \cos(0.04\pi n) + 0.2w(n)$, $0 \leq n \leq 50$, where $w(n)$ is a Gaussian random sequence with zero mean and unit variance.
    d. $\tilde{x}(n) = \{..., 5, 4, 3, 2, 1, 5, 4, 3, 2, 1, 5, 4, 3, 2, 1, ...\}$; $-10 \leq n \leq 9$.

Solution   **a.** $x(n) = 2\delta(n+2) - \delta(n-4)$, $-5 \leq n \leq 5$.

```
>> n = [-5:5];
>> x = 2*impseq(-2,-5,5) - impseq(4,-5,5);
>> stem(n,x); title('Sequence in Problem 2.1a')
>> xlabel('n'); ylabel('x(n)');
```

The plot of the sequence is shown in Figure 2.1a.

---

[1]The symbol * denotes many operations in digital signal processing. Its font (roman or computer) and its position (normal or superscript) will distinguish each operation.

b. $x(n) = n\left[u(n) - u(n-10)\right] + 10e^{-0.3(n-10)}\left[u(n-10) - u(n-20)\right]$, $0 \le n \le 20$.

```
>> n = [0:20];
>> x1 = n.*(stepseq(0,0,20)-stepseq(10,0,20));
>> x2 = 10*exp(-0.3*(n-10)).*(stepseq(10,0,20)-stepseq(20,0,20));
>> x = x1+x2;
>> subplot(2,2,3); stem(n,x); title('Sequence in Problem 2.1b')
>> xlabel('n'); ylabel('x(n)');
```

The plot of the sequence is shown in Figure 2.1b.

c. $x(n) = \cos(0.04\pi n) + 0.2w(n), \quad 0 \le n \le 50$.

```
>> n = [0:50];
>> x = cos(0.04*pi*n)+0.2*randn(size(n));
>> subplot(2,2,2); stem(n,x); title('Sequence in Problem 2.1c')
>> xlabel('n'); ylabel('x(n)');
```

The plot of the sequence is shown in Figure 2.1c.

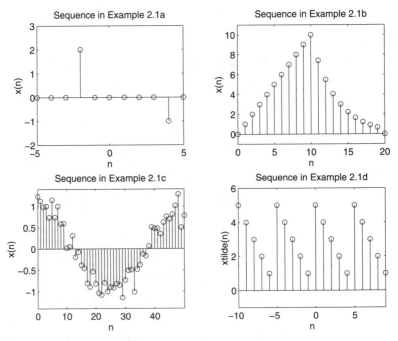

FIGURE 2.1 *Sequences in Example 2.1*

d. $\tilde{x}(n) = \{..., 5, 4, 3, 2, 1, 5, 4, 3, 2, 1, 5, 4, 3, 2, 1, ...\}; \; -10 \leq n \leq 9.$

Note that over the given interval, the sequence $\tilde{x}(n)$ has four periods.

```
>> n = [-10:9]; x = [5,4,3,2,1];
>> xtilde = x' * ones(1,4);
>> xtilde = (xtilde(:))';
>> subplot(2,2,4); stem(n,xtilde); title('Sequence in Problem 2.1d')
>> xlabel('n'); ylabel('xtilde(n)');
```

The plot of the sequence is shown in Figure 2.1d. □

□ **EXAMPLE 2.2** Let $x(n) = \{1, 2, 3, 4, 5, 6, 7, 6, 5, 4, 3, 2, 1\}$. Determine and plot the following sequences.

a. $x_1(n) = 2x(n-5) - 3x(n+4)$
b. $x_2(n) = x(3-n) + x(n)x(n-2)$

Solution

The sequence $x(n)$ is nonzero over $-2 \leq n \leq 10$. Hence

```
>> n = -2:10; x = [1:7,6:-1:1];
```

will generate $x(n)$.

a. $x_1(n) = 2x(n-5) - 3x(n+4)$.
The first part is obtained by shifting $x(n)$ by 5 and the second part by shifting $x(n)$ by -4. This shifting and the addition can be easily done using the **sigshift** and the **sigadd** functions.

```
>> [x11,n11] = sigshift(x,n,5); [x12,n12] = sigshift(x,n,-4);
>> [x1,n1] = sigadd(2*x11,n11,-3*x12,n12);
>> subplot(2,1,1); stem(n1,x1); title('Sequence in Example 2.2a')
>> xlabel('n'); ylabel('x1(n)');
```

The plot of $x_1(n)$ is shown in Figure 2.2a.
b. $x_2(n) = x(3-n) + x(n)x(n-2)$.
The first term can be written as $x(-(n-3))$. Hence it is obtained by first folding $x(n)$ and then shifting the result by 3. The second part is a multiplication of $x(n)$ and $x(n-2)$, both of which have the same length but different support (or sample positions). These operations can be easily done using the **sigfold** and the **sigmult** functions.

```
>> [x21,n21] = sigfold(x,n); [x21,n21] = sigshift(x21,n21,3);
>> [x22,n22] = sigshift(x,n,2); [x22,n22] = sigmult(x,n,x22,n22);
>> [x2,n2] = sigadd(x21,n21,x22,n22);
>> subplot(2,1,2); stem(n2,x2); title('Sequence in Example 2.2b')
>> xlabel('n'); ylabel('x2(n)');
```

The plot of $x_2(n)$ is shown in Figure 2.2b. □

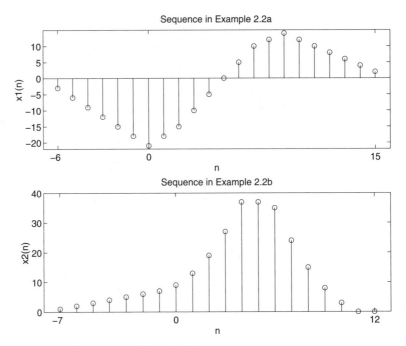

FIGURE 2.2 *Sequences in Example 2.2*

This example shows that the four `sig*` functions developed in this section provide a convenient approach for sequence manipulations.

☐   **EXAMPLE 2.3**   Generate the complex-valued signal

$$x(n) = e^{(-0.1+j0.3)n}, \quad -10 \le n \le 10$$

and plot its magnitude, phase, the real part, and the imaginary part in four separate subplots.

Solution   **MATLAB Script** _____

```
>> n = [-10:1:10]; alpha = -0.1+0.3j;
>> x = exp(alpha*n);
>> subplot(2,2,1); stem(n,real(x));title('real part');xlabel('n')
>> subplot(2,2,2); stem(n,imag(x));title('imaginary part');xlabel('n')
>> subplot(2,2,3); stem(n,abs(x));title('magnitude part');xlabel('n')
>> subplot(2,2,4); stem(n,(180/pi)*angle(x));title('phase part');xlabel('n')
```

The plot of the sequence is shown in Figure 2.3.   ☐

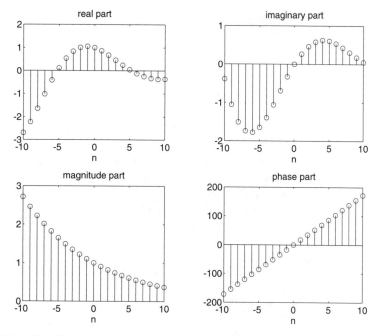

FIGURE 2.3 *Complex-valued sequence plots in Example 2.3*

SOME USEFUL
RESULTS

There are several important results in discrete-time signal theory. We will discuss some that are useful in digital signal processing.

Unit sample synthesis Any arbitrary sequence $x(n)$ can be synthesized as a weighted sum of delayed and scaled unit sample sequences, such as

$$x(n) = \sum_{k=-\infty}^{\infty} x(k)\delta(n-k) \tag{2.1}$$

We will use this result in the next section.

Even and odd synthesis A real-valued sequence $x_e(n)$ is called even (symmetric) if

$$x_e(-n) = x_e(n)$$

Similarly, a real-valued sequence $x_o(n)$ is called odd (antisymmetric) if

$$x_o(-n) = -x_o(n)$$

Then any arbitrary real-valued sequence $x(n)$ can be decomposed into its even and odd components

$$x(n) = x_e(n) + x_o(n) \qquad (2.2)$$

where the even and odd parts are given by

$$x_e(n) = \frac{1}{2}[x(n) + x(-n)] \quad \text{and} \quad x_o(n) = \frac{1}{2}[x(n) - x(-n)] \qquad (2.3)$$

respectively. We will use this decomposition in studying properties of the Fourier transform. Therefore it is a good exercise to develop a simple MATLAB function to decompose a given sequence into its even and odd components. Using MATLAB operations discussed so far, we can obtain the following evenodd function.

```
function [xe, xo, m] = evenodd(x,n)
% Real signal decomposition into even and odd parts
% -----------------------------------------------
% [xe, xo, m] = evenodd(x,n)
%
if any(imag(x) ~= 0)
        error('x is not a real sequence')
end
m = -fliplr(n);
m1 = min([m,n]); m2 = max([m,n]); m = m1:m2;
nm = n(1)-m(1); n1 = 1:length(n);
x1 = zeros(1,length(m));
x1(n1+nm) = x; x = x1;
xe = 0.5*(x + fliplr(x));
xo = 0.5*(x - fliplr(x));
```

The sequence and its support are supplied in x and n arrays, respectively. It first checks if the given sequence is real and determines the support of the even and odd components in m array. It then implements (2.3) with special attention to the MATLAB indexing operation. The resulting components are stored in xe and xo arrays.

□ **EXAMPLE 2.4**   Let $x(n) = u(n) - u(n-10)$. Decompose $x(n)$ into even and odd components.

Solution   The sequence $x(n)$, which is nonzero over $0 \le n \le 9$, is called a *rectangular pulse*. We will use MATLAB to determine and plot its even and odd parts.

```
>> n = [0:10]; x = stepseq(0,0,10)-stepseq(10,0,10);
>> [xe,xo,m] = evenodd(x,n);
>> figure(1); clf
>> subplot(2,2,1); stem(n,x); title('Rectangular pulse')
```

```
>> xlabel('n'); ylabel('x(n)'); axis([-10,10,0,1.2])
>> subplot(2,2,2); stem(m,xe); title('Even Part')
>> xlabel('n'); ylabel('xe(n)'); axis([-10,10,0,1.2])
>> subplot(2,2,4); stem(m,xo); title('Odd Part')
>> xlabel('n'); ylabel('xo(n)'); axis([-10,10,-0.6,0.6])
```

The plots shown in Figure 2.4 clearly demonstrate the decomposition. □

A similar decomposition for complex-valued sequences is explored in Problem 2.5.

The geometric series A one-sided exponential sequence of the form $\{\alpha^n, \quad n \geq 0\}$, where α is an arbitrary constant, is called a geometric series. In digital signal processing, the convergence and expression for the sum of this series are used in many applications. The series converges for $|\alpha| < 1$, while the sum of its components converges to

$$\sum_{n=0}^{\infty} \alpha^n \longrightarrow \frac{1}{1-\alpha}, \quad \text{for } |\alpha| < 1 \tag{2.4}$$

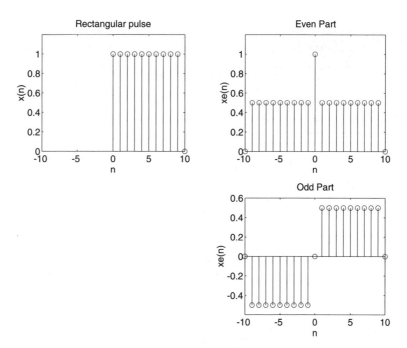

FIGURE 2.4 *Even-odd decomposition in Example 2.4*

We will also need an expression for the sum of any finite number of terms of the series given by

$$\sum_{n=0}^{N-1} \alpha^n = \frac{1 - \alpha^N}{1 - \alpha}, \ \forall \alpha \tag{2.5}$$

These two results will be used throughout this book.

Correlations of sequences   Correlation is an operation used in many applications in digital signal processing. It is a measure of the degree to which two sequences are similar. Given two real-valued sequences $x(n)$ and $y(n)$ of finite energy, the *crosscorrelation* of $x(n)$ and $y(n)$ is a sequence $r_{xy}(\ell)$ defined as

$$r_{x,y}(\ell) = \sum_{n=-\infty}^{\infty} x(n)y(n - \ell) \tag{2.6}$$

The index ℓ is called the shift or lag parameter. The special case of (2.6) when $y(n) = x(n)$ is called *autocorrelation* and is defined by

$$r_{xx}(\ell) = \sum_{n=-\infty}^{\infty} x(n)x(n - \ell) \tag{2.7}$$

It provides a measure of self-similarity between different alignments of the sequence. MATLAB functions to compute auto- and crosscorrelations are discussed later in the chapter.

DISCRETE SYSTEMS

Mathematically, a discrete-time system (or *discrete system* for short) is described as an operator $T[\cdot]$ that takes a sequence $x(n)$ (called *excitation*) and transforms it into another sequence $y(n)$ (called *response*). That is,

$$y(n) = T[x(n)]$$

In DSP we will say that the system processes an *input* signal into an *output* signal. Discrete systems are broadly classified into *linear* and *nonlinear* systems. We will deal mostly with linear systems.

LINEAR SYSTEMS

A discrete system $T[\cdot]$ is a linear operator $L[\cdot]$ if and only if $L[\cdot]$ satisfies the principle of superposition, namely,

$$L[a_1 x_1(n) + a_2 x_2(n)] = a_1 L[x_1(n)] + a_2 L[x_2(n)], \forall a_1, a_2, x_1(n), x_2(n)$$

(2.8)

Using (2.1) and (2.8), the output $y(n)$ of a linear system to an arbitrary input $x(n)$ is given by

$$y(n) = L[x(n)] = L\left[\sum_{n=-\infty}^{\infty} x(k)\delta(n-k)\right] = \sum_{n=-\infty}^{\infty} x(k) L[\delta(n-k)]$$

The response $L[\delta(n-k)]$ can be interpreted as the response of a linear system at time n due to a unit sample (a well-known sequence) at time k. It is called an *impulse response* and is denoted by $h(n,k)$. The output then is given by the *superposition summation*

$$y(n) = \sum_{n=-\infty}^{\infty} x(k)h(n,k)$$

(2.9)

The computation of (2.9) requires the *time-varying* impulse response $h(n,k)$, which in practice is not very convenient. Therefore time-invariant systems are widely used in DSP.

Linear time-invariant (LTI) system A linear system in which an input-output pair, $x(n)$ and $y(n)$, is invariant to a shift n in time is called a linear time-invariant system. For an LTI system the $L[\cdot]$ and the shifting operators are reversible as shown below.

$$x(n) \longrightarrow \boxed{L[\cdot]} \longrightarrow y(n) \longrightarrow \boxed{\text{Shift by } k} \longrightarrow y(n-k)$$

$$x(n) \longrightarrow \boxed{\text{Shift by } k} \longrightarrow x(n-k) \longrightarrow \boxed{L[\cdot]} \longrightarrow y(n-k)$$

We will denote an LTI system by the operator $LTI[\cdot]$. Let $x(n)$ and $y(n)$ be the input-output pair of an LTI system. Then the time-varying function $h(n,k)$ becomes a time-invariant function $h(n-k)$, and the output from (2.9) is given by

$$y(n) = LTI[x(n)] = \sum_{k=-\infty}^{\infty} x(k)h(n-k)$$

(2.10)

The impulse response of an LTI system is given by $h(n)$. The mathematical operation in (2.10) is called a *linear convolution sum* and is denoted by

$$y(n) \overset{\triangle}{=} x(n) * h(n)$$

(2.11)

Hence an LTI system is completely characterized in the time domain by the impulse response $h(n)$ as shown below.

$$x(n) \longrightarrow \boxed{h(n)} \longrightarrow y(n) = x(n) * h(n)$$

We will explore several properties of the convolution in Problem 2.12.

Stability   This is a very important concept in linear system theory. The primary reason for considering stability is to avoid building harmful systems or to avoid burnout or saturation in the system operation. A system is said to be *bounded-input bounded-output (BIBO) stable* if every bounded input produces a bounded output.

$$|x(n)| < \infty \Rightarrow |y(n)| < \infty, \forall x, y$$

An LTI system is BIBO stable if and only if its impulse response is *absolutely summable.*

$$\text{BIBO Stability} \iff \sum_{-\infty}^{\infty} |h(n)| < \infty \qquad (2.12)$$

Causality   This important concept is necessary to make sure that systems can be built. A system is said to be causal if the output at index n_0 depends only on the input up to and including the index n_0; that is, the output does not depend on the future values of the input. An LTI system is causal if and only if the impulse response

$$h(n) = 0, \quad n < 0 \qquad (2.13)$$

Such a sequence is termed a *causal sequence.* In signal processing, unless otherwise stated, we will always assume that the system is causal.

CONVOLUTION

We introduced the convolution operation (2.11) to describe the response of an LTI system. In DSP it is an important operation and has many other uses that we will see throughout this book. Convolution can be evaluated in many different ways. If the sequences are mathematical functions (of finite or infinite duration), then we can analytically evaluate (2.11) for all n to obtain a functional form of $y(n)$.

□ **EXAMPLE 2.5**   Let the rectangular pulse $x(n) = u(n) - u(n-10)$ of Example 2.4 be an input to an LTI system with impulse response

$$h(n) = (0.9)^n \, u(n)$$

Determine the output $y(n)$.

Solution   The input $x(n)$ and the impulse response $h(n)$ are shown in Figure 2.5. From (2.11)

$$y(n) = \sum_{k=0}^{9} (1) \, (0.9)^{(n-k)} \, u(n-k) = (0.9)^n \sum_{k=0}^{9} (0.9)^{-k} \, u(n-k) \qquad \textbf{(2.14)}$$

The sum in 2.14 is almost a geometric series sum except that the term $u(n-k)$ takes different values depending on n and k. There are three different conditions under which $u(n-k)$ can be evaluated.

CASE i $n < 0$: Then $u(n-k) = 0$, $0 \le k \le 9$. Hence from (2.14)

$$y(n) = 0 \qquad \textbf{(2.15)}$$

In this case the nonzero values of $x(n)$ and $h(n)$ *do not overlap.*

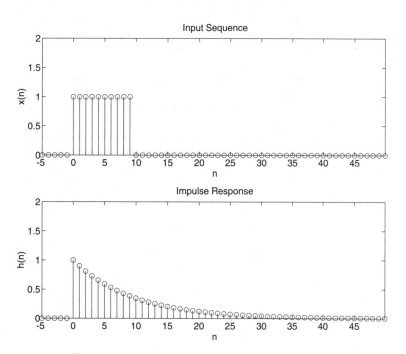

FIGURE 2.5   *The input sequence and the impulse response in Example 2.5*

CASE ii $0 \leq n < 9$: Then $u(n-k) = 1, \quad 0 \leq k \leq n$. Hence from (2.14)

$$y(n) = (0.9)^n \sum_{k=0}^{n} (0.9)^{-k} = (0.9)^n \sum_{k=0}^{n} \left[(0.9)^{-1}\right]^k \qquad (2.16)$$

$$= (0.9)^n \frac{1 - (0.9)^{-(n+1)}}{1 - (0.9)^{-1}} = 10 \left[1 - (0.9)^{n+1}\right], \quad 0 \leq n < 9$$

In this case the impulse response $h(n)$ *partially overlaps* the input $x(n)$.

CASE iii $n \geq 9$: Then $u(n-k) = 1, 0 \leq k \leq 9$ and from (2.14)

$$y(n) = (0.9)^n \sum_{k=0}^{9} (0.9)^{-k} \qquad (2.17)$$

$$= (0.9)^n \frac{1 - (0.9)^{-10}}{1 - (0.9)^{-1}} = 10 \, (0.9)^{n-9} \left[1 - (0.9)^{10}\right], \quad n \geq 9$$

In this last case $h(n)$ *completely overlaps* $x(n)$.

The complete response is given by (2.15), (2.16), and (2.17). It is shown in Figure 2.6 which depicts the distortion of the input pulse. □

The above example can also be done using a method called graphical convolution, in which (2.11) is given a graphical interpretation. In this method $h(n - k)$ is interpreted as a *folded-and-shifted* version of $h(k)$. The output $y(n)$ is obtained as a sample sum under the overlap of $x(k)$ and $h(n - k)$. We use an example to illustrate this.

□   **EXAMPLE 2.6**   Given the following two sequences

$$x(n) = \left[3, 11, 7, 0, -1, 4, 2\right], \quad -3 \leq n \leq 3; \qquad h(n) = \left[2, 3, 0, -5, 2, 1\right], \quad -1 \leq n \leq 4$$

determine the convolution $y(n) = x(n) * h(n)$.

Solution   In Figure 2.7 we show four plots. The top-left plot shows $x(k)$ and $h(k)$, the original sequences. The top-right plot shows $x(k)$ and $h(-k)$, the folded version

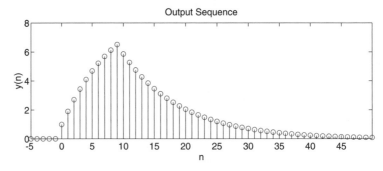

FIGURE 2.6   *The output sequence in Example 2.5*

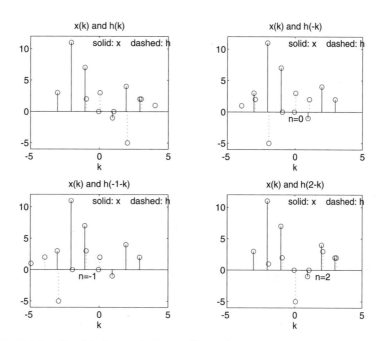

FIGURE 2.7 *Graphical convolution in Example 2.6*

of $h(k)$. The bottom-left plot shows $x(k)$ and $h(-1-k)$, the folded-and-shifted-by--1 version of $h(k)$. Then

$$\sum_k x(k)h(-1-k) = 3 \times (-5) + 11 \times 0 + 7 \times 3 + 0 \times 2 = 6 = y(-1)$$

The bottom-right plot shows $x(k)$ and $h(2-k)$, the folded-and-shifted-by-2 version of $h(k)$, which gives

$$\sum_k x(k)h(2-k) = 11 \times 1 + 7 \times 2 + 0 \times (-5) + (-1) \times 0 + 4 \times 3 + 2 \times 2 = 41 = y(2)$$

Thus we have obtained two values of $y(n)$. Similar graphical calculations can be done for other remaining values of $y(n)$. Note that the beginning point (first nonzero sample) of $y(n)$ is given by $n = -3 + (-1) = -4$, while the end point (the last nonzero sample) is given by $n = 3 + 4 = 7$. The complete output is given by

$$y(n) = \left\{ 6, 31, 47, 6, -51, -5, 41, 18, -22, -3, 8, 2 \right\}$$

Students are strongly encouraged to verify the above result. Note that the resulting sequence $y(n)$ has a *longer length* than both the $x(n)$ and $h(n)$ sequences. □

MATLAB
IMPLEMEN-
TATION

If arbitrary sequences are of infinite duration, then MATLAB cannot be used *directly* to compute the convolution. MATLAB does provide a built-in function called `conv` that computes the convolution between two finite-duration sequences. The `conv` function assumes that the two sequences begin at $n = 0$ and is invoked by

```
>> y = conv(x,h);
```

For example, to do the convolution in Example 2.5, we could use

```
>> x = [3, 11, 7, 0, -1, 4, 2];
>> h = [2, 3, 0, -5, 2, 1];
>> y = conv(x,h)
y =
      6    31    47     6   -51    -5    41    18   -22    -3     8     2
```

to obtain the correct $y(n)$ values. However, the `conv` function neither provides nor accepts any timing information if the sequences have arbitrary support. What is needed is a beginning point and an end point of $y(n)$. Given finite duration $x(n)$ and $h(n)$, it is easy to determine these points. Let

$$\{x(n); \ n_{xb} \leq n \leq n_{xe}\} \quad \text{and} \quad \{h(n); \ n_{hb} \leq n \leq n_{he}\}$$

be two finite-duration sequences. Then referring to Example 2.6 we observe that the beginning and end points of $y(n)$ are

$$n_{yb} = n_{xb} + n_{hb} \quad \text{and} \quad n_{ye} = n_{xe} + n_{he}$$

respectively. A simple extension of the `conv` function, called `conv_m`, which performs the convolution of arbitrary support sequences can now be designed.

```
function [y,ny] = conv_m(x,nx,h,nh)
% Modified convolution routine for signal processing
% ----------------------------------------------------
% [y,ny] = conv_m(x,nx,h,nh)
%   [y,ny] = convolution result
%   [x,nx] = first signal
%   [h,nh] = second signal
%
nyb = nx(1)+nh(1); nye = nx(length(x)) + nh(length(h));
ny = [nyb:nye];
y = conv(x,h);
```

☐ **EXAMPLE 2.7**   Perform the convolution in Example 2.6 using the `conv_m` function.

Solution   **MATLAB Script**
```
>> x = [3, 11, 7, 0, -1, 4, 2]; nx = [-3:3];
>> h = [2, 3, 0, -5, 2, 1]; ny = [-1:4];
```

```
>> [y,ny] = conv_m(x,nx,h,nh)
y =
    6    31    47     6   -51    -5    41    18   -22    -3     8     2
ny =
   -4    -3    -2    -1     0     1     2     3     4     5     6     7
```

Hence

$$y(n) = \left\{ 6, 31, 47, 6, -51, -5, 41, 18, -22, -3, 8, 2 \right\}$$

as in Example 2.6.   □

An alternate method in MATLAB can be used to perform the convolution. This method uses a matrix-vector multiplication approach, which we will explore in Problem 2.13.

SEQUENCE CORRE-LATIONS REVISITED

If we compare the convolution operation (2.11) with that of the crosscorrelation of two sequences defined in (2.6), we observe a close resemblance. The crosscorrelation $r_{yx}(\ell)$ can be put in the form

$$r_{yx}(\ell) = y(\ell) * x(-\ell)$$

with the autocorrelation $r_{xx}(\ell)$ in the form

$$r_{xx}(\ell) = x(\ell) * x(-\ell)$$

Therefore these correlations can be computed using the conv function if sequences are of finite duration.

□   EXAMPLE 2.8   In this example we will demonstrate one application of the crosscorrelation sequence. Let

$$x(n) = \left[3, 11, 7, 0, -1, 4, 2 \right]$$

be a prototype sequence, and let $y(n)$ be its noise-corrupted-and-shifted version

$$y(n) = x(n - 2) + w(n)$$

where $w(n)$ is Gaussian sequence with mean 0 and variance 1. Compute the crosscorrelation between $y(n)$ and $x(n)$.

Solution

From the construction of $y(n)$ it follows that $y(n)$ is "similar" to $x(n - 2)$ and hence their crosscorrelation would show the strongest similarity at $\ell = 2$. To test this out using MATLAB, let us compute the crosscorrelation using two different noise sequences.

```
% noise sequence 1
>> x = [3, 11, 7, 0, -1, 4, 2]; nx=[-3:3]; % given signal x(n)
>> [y,ny] = sigshift(x,nx,2);              % obtain x(n-2)
>> w = randn(1,length(y)); nw = ny;        % generate w(n)
>> [y,ny] = sigadd(y,ny,w,nw);             % obtain y(n) = x(n-2) + w(n)
>> [x,nx] = sigfold(x,nx);                 % obtain x(-n)
>> [rxy,nrxy] = conv_m(y,ny,x,nx);         % crosscorrelation
>> subplot(1,1,1), subplot(2,1,1);stem(nrxy,rxy)
>> axis([-4,8,-50,250]);xlabel('lag variable l')
>> ylabel('rxy');title('Crosscorrelation: noise sequence 1')
%
% noise sequence 2
>> x = [3, 11, 7, 0, -1, 4, 2]; nx=[-3:3]; % given signal x(n)
>> [y,ny] = sigshift(x,nx,2);              % obtain x(n-2)
>> w = randn(1,length(y)); nw = ny;        % generate w(n)
>> [y,ny] = sigadd(y,ny,w,nw);             % obtain y(n) = x(n-2) + w(n)
>> [x,nx] = sigfold(x,nx);                 % obtain x(-n)
>> [rxy,nrxy] = conv_m(y,ny,x,nx);         % crosscorrelation
>> subplot(2,1,2);stem(nrxy,rxy)
>> axis([-4,8,-50,250]);xlabel('lag variable l')
>> ylabel('rxy');title('Crosscorrelation: noise sequence 2')
```

From Figure 2.8 we observe that the crosscorrelation indeed peaks at $\ell = 2$, which implies that $y(n)$ is similar to $x(n)$ shifted by 2. This approach can be

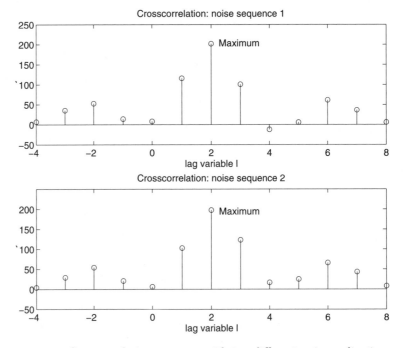

FIGURE 2.8 *Crosscorrelation sequence with two different noise realizations*

Chapter 2 ■ DISCRETE-TIME SIGNALS AND SYSTEMS

used in applications like radar signal processing in identifying and localizing targets. □

It should be noted that the signal-processing toolbox in MATLAB also provides a function called `xcorr` for sequence correlation computations. In its simplest form

```
>> xcorr(x,y)
```

computes the crosscorrelation between vectors x and y, while

```
>> xcorr(x)
```

computes the autocorrelation of vector x. This function is not available in the Student Edition of MATLAB. It generates results that are identical to the one obtained from the proper use of the `conv_m` function. However, the `xcorr` function cannot provide the timing (or lag) information (as done by the `conv_m` function), which then must be obtained by some other means. Therefore we will emphasize the use of the `conv_m` function.

DIFFERENCE EQUATIONS
■

An LTI discrete system can also be described by a linear constant coefficient difference equation of the form

$$\sum_{k=0}^{N} a_k y(n-k) = \sum_{m=0}^{M} b_m x(n-m), \quad \forall n \tag{2.18}$$

If $a_N \neq 0$, then the difference equation is of order N. This equation describes a recursive approach for computing the current output, given the input values and previously computed output values. In practice this equation is computed forward in time, from $n = -\infty$ to $n = \infty$. Therefore another form of this equation is

$$y(n) = \sum_{m=0}^{M} b_m x(n-m) - \sum_{k=1}^{N} a_k y(n-k) \tag{2.19}$$

A solution to this equation can be obtained in the form

$$y(n) = y_H(n) + y_P(n)$$

The *homogeneous part* of the solution, $y_H(n)$, is given by

$$y_H(n) = \sum_{k=1}^{N} c_k z_k^n$$

where $z_k, k = 1, \ldots, N$ are N roots (also called *natural frequencies*) of the characteristic equation

$$\sum_0^N a_k z^k = 0$$

This characteristic equation is important in determining the stability of systems. If the roots z_k satisfy the condition

$$|z_k| < 1, \ k = 1, \ldots, N \qquad (2.20)$$

then a causal system described by (2.19) is stable. The *particular part* of the solution, $y_P(n)$, is determined from the right-hand side of (2.18). In Chapter 4 we will discuss the analytical approach of solving difference equations using the z-transform.

MATLAB
IMPLEMEN-
TATION

A routine called `filter` is available to solve difference equations numerically, given the input and the difference equation coefficients. In its simplest form this routine is invoked by

```
y = filter(b,a,x)
```

where

```
b = [b0, b1, ..., bM]; a = [a0, a1, ..., aN];
```

are the coefficient arrays from the equation given in (2.18), and x is the input sequence array. The output y has the same length as input x. One must ensure that the coefficient a0 not be zero. We illustrate the use of this routine in the following example.

☐ **EXAMPLE 2.9**   Given the following difference equation

$$y(n) - y(n-1) + 0.9y(n-2) = x(n); \quad \forall n$$

a. Calculate and plot the impulse response $h(n)$ at $n = -20, \ldots, 100$.
b. Calculate and plot the unit step response $s(n)$ at $n = -20, \ldots, 100$.
c. Is the system specified by $h(n)$ stable?

Solution

From the given difference equation the coefficient arrays are

```
b = [1]; a=[1, -1, 0.9];
```

```
>> b = [1]; a = [1, -1, 0.9];
>> x = impseq(0,-20,120); n = [-20:120];
>> h = filter(b,a,x);
>> subplot(2,1,1); stem(n,h);
>> title('Impulse Response'); xlabel('n'); ylabel('h(n)')
```

The plot of the impulse response is shown in Figure 2.9.

b. MATLAB Script

```
>> x = stepseq(0,-20,120);
>> s = filter(b,a,x);
>> subplot(2,1,2); stem(n,s)
>> title('Step Response'); xlabel('n'); ylabel('s(n)')
```

The plot of the unit step response is shown in Figure 2.9.

c. To determine the stability of the system, we have to determine $h(n)$ for all n. Although we have not described a method to solve the difference equation, we can use the plot of the impulse response to observe that $h(n)$ is practically zero for $n > 120$. Hence the sum $\sum |h(n)|$ can be determined from MATLAB using

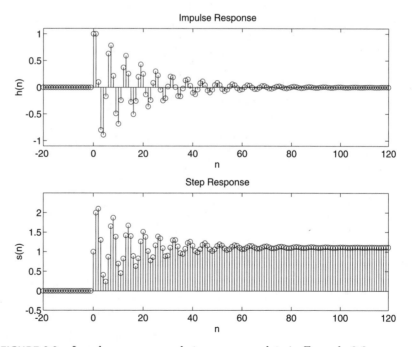

FIGURE 2.9 *Impulse response and step response plots in Example 2.9*

```
>> sum(abs(h))
ans = 14.8785
```

which implies that the system is stable. An alternate approach is to use the stability condition (2.20) using MATLAB's **roots** function.

```
>>z = roots(a);
>>magz = abs(z)
magz =   0.9487
         0.9487
```

Since the magnitudes of both roots are less than one, the system is stable. □

In the previous section we noted that if one or both sequences in the convolution are of infinite length, then the **conv** function cannot be used. If one of the sequences is of infinite length, then it is possible to use MATLAB for numerical evaluation of the convolution. This is done using the **filter** function as we will see in the following example.

□ **EXAMPLE 2.10** Let us consider the convolution given in Example 2.5. The input sequence is of finite duration

$$x(n) = u(n) - u(n - 10)$$

while the impulse response is of infinite duration

$$h(n) = (0.9)^n u(n)$$

Determine $y(n) = x(n) * h(n)$.

Solution If the LTI system, given by the impulse response $h(n)$, can be described by a difference equation, then $y(n)$ can be obtained from the **filter** function. From the $h(n)$ expression

$$(0.9) h(n - 1) = (0.9) (0.9)^{n-1} u(n - 1) = (0.9)^n u(n - 1)$$

or

$$h(n) - (0.9) h(n - 1) = (0.9)^n u(n) - (0.9)^n u(n - 1)$$

$$= (0.9)^n [u(n) - u(n - 1)] = (0.9)^n \delta(n)$$

$$= \delta(n)$$

The last step follows from the fact that $\delta(n)$ is nonzero only at $n = 0$. By definition $h(n)$ is the output of an LTI system when the input is $\delta(n)$. Hence substituting $x(n)$ for $\delta(n)$ and $y(n)$ for $h(n)$, the difference equation is

$$y(n) - 0.9y(n - 1) = x(n)$$

Now MATLAB's **filter** function can be used to compute the convolution indirectly.

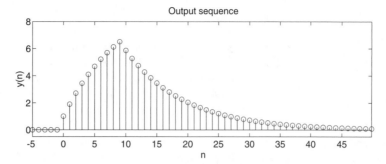

FIGURE 2.10 *Output sequence in Example 2.10*

```
>> b = [1]; a = [1,-0.9];
>> n = -5:50; x = stepseq(0,-5,50) - stepseq(10,-5,50);
>> y = filter(b,a,x);
>> subplot(1,1,1);
>> subplot(2,1,2); stem(n,y); title('Output sequence')
>> xlabel('n'); ylabel('y(n)'); axis([-5,50,-0.5,8])
```

The plot of the output is shown in Figure 2.10, which is exactly the same as that in Figure 2.6. □

In Example 2.10 the impulse response was a one-sided exponential sequence for which we could determine a difference equation representation. This means that not all infinite-length impulse responses can be converted into difference equations. The above analysis, however, can be extended to a linear combination of one-sided exponential sequences, which results in higher-order difference equations. We will discuss this topic of conversion from one representation to another one in Chapter 4.

ZERO-INPUT AND ZERO-STATE RESPONSES

In digital signal processing the difference equation is generally solved forward in time from $n = 0$. Therefore initial conditions on $x(n)$ and $y(n)$ are necessary to determine the output for $n \geq 0$. The difference equation is then given by

$$y(n) = \sum_{m=0}^{M} b_m x(n-m) - \sum_{k=1}^{N} a_k y(n-k); \ n \geq 0 \qquad (2.21)$$

subject to the initial conditions:

$$\{y(n); \ -N \leq n \leq -1\} \qquad \text{and} \qquad \{x(n); \ -M \leq n \leq -1\}$$

A solution to (2.21) can be obtained in the form

$$y(n) = y_{ZI}(n) + y_{ZS}(n)$$

where $y_{ZI}(n)$ is called the *zero-input* solution, which is a solution due to the initial conditions alone (assuming they exist), while the *zero-state* solution, $y_{ZS}(n)$, is a solution due to input $x(n)$ alone (or assuming that the initial conditions are zero). In MATLAB another form of the function `filter` can be used to solve for the difference equation, given its initial conditions. We will illustrate the use of this form in Chapter 4.

DIGITAL
FILTERS

Filter is a generic name that means a linear time-invariant system designed for a specific job of frequency selection or frequency discrimination. Hence discrete-time LTI systems are also called digital filters. There are two types of digital filters.

FIR filter If the unit impulse response of an LTI system is of finite duration, then the system is called a *finite-duration impulse response* (or FIR) filter. Hence for an FIR filter $h(n) = 0$ for $n < n_1$ and for $n > n_2$. The following part of the difference equation (2.18) describes a *causal* FIR filter:

$$y(n) = \sum_{m=0}^{M} b_m x(n-m) \qquad (2.22)$$

Furthermore, $h(0) = b_0$, $h(1) = b_1$, ..., $h(M) = b_M$, while all other $h(n)$'s are 0. FIR filters are also called *nonrecursive* or *moving average (MA)* filters. In MATLAB FIR filters are represented either as impulse response values $\{h(n)\}$ or as difference equation coefficients $\{b_m\}$ and $\{a_0 = 1\}$. Therefore to implement FIR filters, we can use either the `conv(x,h)` function (and its modifications that we discussed) or the `filter(b,1,x)` function. There is a difference in the outputs of these two implementations that should be noted. The output sequence from the `conv(x,h)` function has a *longer length* than both the $x(n)$ and $h(n)$ sequences. On the other hand, the output sequence from the `filter(b,1,x)` function has exactly the *same length* as the input $x(n)$ sequence. In practice (and especially for processing signals) the use of the `filter` function is encouraged.

IIR filter If the impulse response of an LTI system is of infinite duration, then the system is called an *infinite-duration impulse response* (or IIR) filter. The following part of the difference equation (2.18):

$$\sum_{k=0}^{N} a_k y(n-k) = x(n) \qquad (2.23)$$

describes a *recursive* filter in which the output $y(n)$ is recursively computed from its previously computed values and is called an *autoregressive (AR)* filter. The impulse response of such filter is of infinite duration and hence it represents an IIR filter. The general equation (2.18) also describes an IIR filter. It has two parts: an AR part and an MA part. Such an IIR filter is called an *autoregressive moving average*, or an ARMA, filter. In MATLAB IIR filters are described by the difference equation coefficients $\{b_m\}$ and $\{a_k\}$ and are implemented by the `filter(b,a,x)` function.

PROBLEMS

———————■———————

P2.1 Generate and plot the samples (use the `stem` function) of the following sequences using MATLAB.

a. $x_1(n) = \sum_{m=0}^{10} (m+1)\left[\delta(n-2m) - \delta(n-2m-1)\right], \quad 0 \le n \le 25$.

b. $x_2(n) = n^2\left[u(n+5) - u(n-6)\right] + 10\delta(n) + 20(0.5)^n\left[u(n-4) - u(n-10)\right]$.

c. $x_3(n) = (0.9)^n \cos(0.2\pi n + \pi/3), \quad 0 \le n \le 20$.

d. $x_4(n) = 10\cos(0.0008\pi n^2) + w(n), \quad 0 \le n \le 100$, where $w(n)$ is a random sequence uniformly distributed between $[-1,1]$. How do you characterize this sequence?

e. $\tilde{x}_5(n) = \{\ldots, 1, 2, 3, 2, 1, 2, 3, 2, 1, \ldots\}_{\text{PERIODIC}}$. Plot 5 periods.
$\quad\qquad\qquad\qquad\qquad\uparrow$

P2.2 Let $x(n) = \{1, -2, 4, 6, -5, 8, 10\}$. Generate and plot the samples (use the `stem` function) of
$\qquad\qquad\qquad\quad\uparrow$
the following sequences.

a. $x_1(n) = 3x(n+2) + x(n-4) - 2x(n)$

b. $x_2(n) = 5x(5+n) + 4x(n+4) + 3x(n)$

c. $x_3(n) = x(n+4)x(n-1) + x(2-n)x(n)$

d. $x_4(n) = 2e^{0.5n}x(n) + \cos(0.1\pi n)x(n+2), \quad -10 \le n \le 10$

e. $x_5(n) = \sum_{k=1}^{5} nx(n-k)$

P2.3 The complex exponential sequence $e^{j\omega_0 n}$ or the sinusoidal sequence $\cos(\omega_0 n)$ are periodic if the *normalized* frequency $f_0 \triangleq \dfrac{\omega_0}{2\pi}$ is a rational number; that is, $f_0 = \dfrac{K}{N}$, where K and N are integers.

a. Prove the above result.

b. Generate and plot $\cos(0.3\pi n), -20 \le n \le 20$. Is this sequence periodic? If it is, what is its fundamental period? From the examination of the plot what interpretation can you give to the integers K and N above?

c. Generate and plot $\cos(0.3n), -20 \le n \le 20$. Is this sequence periodic? What do you conclude from the plot? If necessary examine the values of the sequence in MATLAB to arrive at your answer.

P2.4 Decompose the sequences given in Problem 2.2 into their even and odd components. Plot these components using the `stem` function.

P2.5 A complex-valued sequence $x_e(n)$ is called *conjugate-symmetric* if

$$x_e(n) = x_e^*(-n)$$

———————————————————————————————————————

Similarly, a complex-valued sequence $x_o(n)$ is called *conjugate-antisymmetric* if

$$x_o(n) = -x_o^*(-n)$$

Then any arbitrary complex-valued sequence $x(n)$ can be decomposed into

$$x(n) = x_e(n) + x_o(n)$$

where $x_e(n)$ and $x_o(n)$ are given by

$$x_e(n) = \tfrac{1}{2}\left[x(n) + x^*(-n)\right] \qquad \text{and} \qquad x_o(n) = \tfrac{1}{2}\left[x(n) - x^*(-n)\right] \qquad \textbf{(2.24)}$$

respectively.

a. Modify the **evenodd** function discussed in the text so that it accepts an arbitrary sequence and decomposes it into its symmetric and antisymmetric components by implementing (2.24).

b. Decompose the following sequence:

$$x(n) = 10e^{-(0.4\pi n)}, \qquad 0 \le n \le 10$$

into its conjugate-symmetric and conjugate-antisymmetric components. Plot their real and imaginary parts to verify the decomposition. (Use the **subplot** function.)

P2.6 The operation of *signal dilation* (or *decimation* or *down-sampling*) is defined by

$$y(n) = x(nM)$$

in which the sequence $x(n)$ is down-sampled by an integer factor M. For example, if

$$x(n) = \{\ldots, -2, 4, 3, -6, 5, -1, 8, \ldots\}$$
$$\uparrow$$

then the down-sampled sequences by a factor 2 are given by

$$y(n) = \{\ldots, -2, 3, 5, 8, \ldots\}$$
$$\uparrow$$

a. Develop a MATLAB function **dnsample** that has the form

```
function y = dnsample(x,M)
```

to implement the above operation. Use the indexing mechanism of MATLAB with careful attention to the origin of the time axis $n = 0$.

b. Generate $x(n) = \sin(0.125\pi n)$, $-50 \le n \le 50$. Decimate $x(n)$ by a factor of 4 to generate $y(n)$. Plot both $x(n)$ and $y(n)$ using **subplot** and comment on the results.

c. Repeat the above using $x(n) = \sin(0.5\pi n)$, $-50 \le n \le 50$. Qualitatively discuss the effect of down-sampling on signals.

P2.7 Determine the autocorrelation sequence $r_{yx}(\ell)$ and the crosscorrelation sequence $r_{xy}(\ell)$ for the following sequences.

$$x(n) = (0.9)^n, \quad 0 \le n \le 20; \qquad y(n) = (0.8)^{-n}, \quad -20 \le n \le 0$$

What is your observation?

P2.8 In a certain concert hall, echoes of the original audio signal $x(n)$ are generated due to the reflections at the walls and ceiling. The audio signal experienced by the listener $y(n)$ is a combination of $x(n)$ and its echoes. Let

$$y(n) = x(n) + \alpha x(n - k)$$

where k is the amount of delay in samples and α is its relative strength. We want to estimate the delay using the correlation analysis.

a. Determine analytically the autocorrelation $r_{yy}(\ell)$ in terms of the autocorrelation $r_{xx}(\ell)$.

b. Let $x(n) = \cos(0.2\pi n) + 0.5\cos(0.6\pi n)$, $\alpha = 0.1$, and $k = 50$. Generate 200 samples of $y(n)$ and determine its autocorrelation. Can you obtain α and k by observing $r_{yy}(\ell)$?

P2.9 Three systems are given below.

$$T_1[x(n)] = 2^{x(n)}; \qquad T_2[x(n)] = 3x(n) + 4; \qquad T_3[x(n)] = x(n) + 2x(n-1) - x(n-2)$$

a. Use (2.8) to determine analytically whether the above systems are linear.

b. Let $x_1(n)$ be a uniformly distributed random sequence between $[0, 1]$ over $0 \leq n \leq 100$, and let $x_2(n)$ be a Gaussian random sequence with mean 0 and variance 10 over $0 \leq n \leq 100$. Using these sequences, test the linearity of the above systems. Choose any values for constants a_1 and a_2 in (2.8). You should use several realizations of the above sequences to arrive at your answers.

P2.10 Three systems are given below.

$$T_1[x(n)] = \sum_{0}^{n} x(k); \qquad T_2[x(n)] = \sum_{n-10}^{n+10} x(k); \qquad T_3[x(n)] = x(-n)$$

a. Use (2.9) to determine analytically whether the above systems are time-invariant.

b. Let $x(n)$ be a Gaussian random sequence with mean 0 and variance 10 over $0 \leq n \leq 100$. Using this sequence, test the time invariance of the above systems. Choose any values for sample shift k in (2.9). You should use several realizations of the above sequence to arrive at your answers.

P2.11 For the systems given in Problems 2.9 and 2.10 determine analytically their stability and causality.

P2.12 The linear convolution defined in (2.11) has several properties:

$$x_1(n) * x_2(n) = x_1(n) * x_2(n) \qquad\qquad : \text{Commutation}$$
$$[x_1(n) * x_2(n)] * x_3(n) = x_1(n) * [x_2(n) * x_3(n)] \qquad : \text{Association}$$
$$x_1(n) * [x_2(n) + x_3(n)] = x_1(n) * x_2(n) + x_1(n) * x_3(n) \qquad : \text{Distribution}$$
$$x(n) * \delta(n - n_0) = x(n - n_0) \qquad\qquad : \text{Identity}$$

(2.25)

a. Analytically prove these properties.

b. Using the following three sequences, verify the above properties.

$$x_1(n) = n[u(n + 10) - u(n - 20)]$$
$$x_2(n) = \cos(0.1\pi n)[u(n) - u(n - 30)]$$
$$x_3(n) = (1.2)^n[u(n + 5) - u(n - 10)]$$

Use the **conv_m** function.

P2.13 When the sequences $x(n)$ and $h(n)$ are of finite duration N_x and N_h, respectively, then their linear convolution (2.10) can also be implemented using *matrix-vector multiplication*. If elements of $y(n)$ and $x(n)$ are arranged in column vectors $\mathbf{x}$ and $\mathbf{y}$ respectively, then from (2.10) we obtain

$$y = Hx$$

where linear shifts in $h(n-k)$ for $n = 0, \ldots, N_h - 1$ are arranged as rows in the matrix $\mathbf{H}$. This matrix has an interesting structure and is called a *Toeplitz* matrix. To investigate this matrix, consider the sequences

$$x(n) = \left\{ 1, 2, 3, 4 \atop \uparrow \right\} \quad \text{and} \quad h(n) = \left\{ 3, 2, 1 \atop \uparrow \right\}$$

a. Determine the linear convolution $y(n) = h(n) * x(n)$.

b. Express $x(n)$ as a 4×1 column vector $\mathbf{x}$ and $y(n)$ as a 6×1 column vector $\mathbf{y}$. Now determine the 6×4 matrix $\mathbf{H}$ so that $\mathbf{y} = \mathbf{Hx}$.

c. Characterize the matrix $\mathbf{H}$. From this characterization can you give a definition of a Toeplitz matrix? How does this definition compare with that of time invariance?

d. What can you say about the first column and the first row of $\mathbf{H}$?

P2.14 MATLAB provides a function called `toeplitz` to generate a Toeplitz matrix, given the first row and the first column.

a. Using this function and your answer to Problem 2.13 part d, develop an alternate MATLAB function to implement linear convolution. The format of the function should be

```
function [y,H]=conv_tp(h,x)
% Linear Convolution using Toeplitz Matrix
% ----------------------------------------
% [y,H] = conv_tp(h,x)
% y = output sequence in column vector form
% H = Toeplitz matrix corresponding to sequence h so that y = Hx
% h = Impulse response sequence in column vector form
% x = input sequence in column vector form
```

b. Verify your function on the sequences given in Problem 2.13.

P2.15 Let $x(n) = (0.8)^n u(n)$.

a. Determine $x(n) * x(n)$ analytically.

b. Using the `filter` function, determine the first 50 samples of $x(n) * x(n)$. Compare your results with those of part a.

P2.16 A particular linear and time-invariant system is described by the difference equation

$$y(n) - 0.5y(n-1) + 0.25y(n-2) = x(n) + 2x(n-1) + x(n-3)$$

a. Determine the stability of the system.

b. Determine and plot the impulse response of the system over $0 \leq n \leq 100$. Determine the stability from this impulse response.

c. If the input to this system is $x(n) = [5 + 3\cos(0.2\pi n) + 4\sin(0.6\pi n)] u(n)$, determine the response $y(n)$ over $0 \leq n \leq 200$.

P2.17 A "simple" *digital differentiator* is given by

$$y(n) = x(n) - x(n-1)$$

which computes a backward first-order difference of the input sequence. Implement this differentiator on the following sequences and plot the results. Comment on the appropriateness of this simple differentiator.

a. $x(n) = 5\,[u(n) - u(n-20)]$: a rectangular pulse

b. $x(n) = n\,[u(n) - u(n-10)] + (20-n)\,[u(n-10) - u(n-20)]$: a triangular pulse

c. $x(n) = \sin\left(\dfrac{\pi n}{25}\right) [u(n) - u(n-100)]$: a sinusoidal pulse

THE DISCRETE-TIME FOURIER ANALYSIS

We have seen how a linear and time-invariant system can be represented using its response to the unit sample sequence. This response, called the unit impulse response $h(n)$, allows us to compute the system response to any arbitrary input $x(n)$ using the linear convolution as shown below.

$$x(n) \longrightarrow \boxed{h(n)} \longrightarrow y(n) = h(n) * x(n)$$

This convolution representation is based on the fact that any signal can be represented by a linear combination of scaled and delayed unit samples. Similarly, we can also represent any arbitrary discrete signal as a linear combination of basis signals introduced in Chapter 2. Each basis signal set provides a new signal representation. Each representation has some advantages and some disadvantages depending upon the type of system under consideration. However, when the system is linear and time-invariant, only one representation stands out as the most useful. It is based on the complex exponential signal set $\left\{ e^{j\omega n} \right\}$ and is called the *Discrete-time Fourier Transform*.

THE DISCRETE-TIME FOURIER TRANSFORM (DTFT)

If $x(n)$ is absolutely summable, that is, $\sum_{-\infty}^{\infty} |x(n)| < \infty$, then its discrete-time Fourier transform is given by

$$X(e^{j\omega}) \triangleq \mathcal{F}[x(n)] = \sum_{n=-\infty}^{\infty} x(n) e^{-j\omega n} \qquad (3.1)$$

The inverse discrete-time Fourier transform (IDTFT) of $X(e^{j\omega})$ is given by

$$x(n) \triangleq \mathcal{F}^{-1}\left[X(e^{j\omega})\right] = \frac{1}{2\pi}\int_{-\pi}^{\pi} X(e^{j\omega})e^{j\omega n}\,d\omega \qquad (3.2)$$

The operator $\mathcal{F}[\cdot]$ transforms a discrete signal $x(n)$ into a complex-valued continuous function $X(e^{j\omega})$ of real variable ω, called a digital frequency, which is measured in radians.

☐ **EXAMPLE 3.1** Determine the discrete-time Fourier transform of $x(n) = (0.5)^n u(n)$.

Solution The sequence $x(n)$ is absolutely summable; therefore its discrete-time Fourier transform exists.

$$X(e^{j\omega}) = \sum_{-\infty}^{\infty} x(n)e^{-j\omega n} = \sum_{0}^{\infty} (0.5)^n e^{-j\omega n}$$

$$= \sum_{0}^{\infty} \left(0.5e^{-j\omega}\right)^n = \frac{1}{1 - 0.5e^{-j\omega}} = \frac{e^{j\omega}}{e^{j\omega} - 0.5} \qquad \square$$

☐ **EXAMPLE 3.2** Determine the discrete-time Fourier transform of the following finite-duration sequence:

$$x(n) = \{1, 2, 3, 4, 5\}$$
$$\uparrow$$

Solution Using definition (3.1),

$$X(e^{j\omega}) = \sum_{-\infty}^{\infty} x(n)e^{-j\omega n} = e^{j\omega} + 2 + 3e^{-j\omega} + 4e^{-j2\omega} + 5e^{-j3\omega}$$

Since $X(e^{j\omega})$ is a complex-valued function, we will have to plot its magnitude and its angle (or the real and the imaginary part) with respect to ω separately to visually describe $X(e^{j\omega})$. Now ω is a real variable between $-\infty$ and ∞, which would mean that we can plot only a part of the $X(e^{j\omega})$ function using MATLAB. Using two important properties of the discrete-time Fourier transform, we can reduce this domain to the $[0, \pi]$ interval for real-valued sequences. We will discuss other useful properties of $X(e^{j\omega})$ in the next section.

☐

TWO IMPORTANT PROPERTIES

We will state the following two properties without proof.

1. **Periodicity:** The discrete-time Fourier transform $X(e^{j\omega})$ is periodic in ω with period 2π.

$$X(e^{j\omega}) = X(e^{j[\omega + 2\pi]})$$

Implication: We need only one period of $X(e^{j\omega})$ (i.e., $\omega \in [0, 2\pi]$, or $[-\pi, \pi]$, etc.) for analysis and not the whole domain $-\infty < \omega < \infty$.

2. **Symmetry:** For real-valued $x(n)$, $X(e^{j\omega})$ is conjugate symmetric.

$$X(e^{-j\omega}) = X^*(e^{j\omega})$$

or

$$\begin{aligned}
\mathrm{Re}[X(e^{-j\omega})] &= \mathrm{Re}[X(e^{j\omega})] && \text{(even symmetry)} \\
\mathrm{Im}[X(e^{-j\omega})] &= -\mathrm{Im}[X(e^{j\omega})] && \text{(odd symmetry)} \\
\left|X(e^{-j\omega})\right| &= \left|X(e^{j\omega})\right| && \text{(even symmetry)} \\
\angle X(e^{-j\omega}) &= -\angle X(e^{j\omega}) && \text{(odd symmetry)}
\end{aligned}$$

Implication: To plot $X(e^{j\omega})$, we now need to consider only a half period of $X(e^{j\omega})$. Generally, in practice this period is chosen to be $\omega \in [0, \pi]$.

MATLAB
IMPLEMEN-
TATION

If $x(n)$ is of infinite duration, then MATLAB cannot be used directly to compute $X(e^{j\omega})$ from $x(n)$. However, we can use it to evaluate the expression $X(e^{j\omega})$ over $[0, \pi]$ frequencies and then plot its magnitude and angle (or real and imaginary parts).

☐ **EXAMPLE 3.3**   Evaluate $X(e^{j\omega})$ in Example 3.1 at 501 equispaced points between $[0, \pi]$ and plot its magnitude, angle, real, and imaginary parts.

Solution

MATLAB Script _____

```
>> w = [0:1:500]*pi/500;  % [0, pi] axis divided into 501 points.
>> X = exp(j*w) ./ (exp(j*w) - 0.5*ones(1,501));
>> magX = abs(X); angX = angle(X);
>> realX = real(X); imagX = imag(X);
>> subplot(2,2,1); plot(w/pi,magX); grid
>> xlabel('frequency in pi units'); title('Magnitude Part'); ylabel('Magnitude')
>> subplot(2,2,3); plot(w/pi,angX); grid
>> xlabel('frequency in pi units'); title('Angle Part'); ylabel('Radians')
>> subplot(2,2,2); plot(w/pi,realX); grid
>> xlabel('frequency in pi units'); title('Real Part'); ylabel('Real')
>> subplot(2,2,4); plot(w/pi,imagX); grid
>> xlabel('frequency in pi units'); title('Imaginary Part'); ylabel('Imaginary')
```
_____

The resulting plots are shown in Figure 3.1. Note that we divided the w array by pi before plotting so that the frequency axes are in the units of π and therefore easier to read. **This practice is strongly recommended.**   ☐

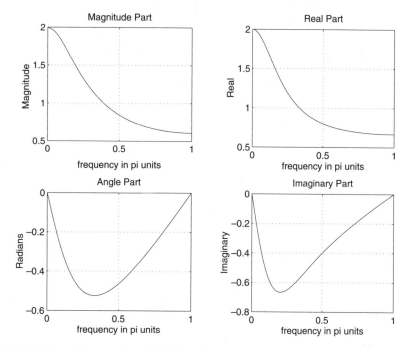

FIGURE 3.1 *Plots in Example 3.3*

If $x(n)$ is of finite duration, then MATLAB can be used to compute $X(e^{j\omega})$ numerically at any frequency ω. The approach is to implement (3.1) directly. If, in addition, we evaluate $X(e^{j\omega})$ at equispaced frequencies between $[0, \pi]$, then (3.1) can be implemented as a *matrix-vector multiplication* operation. To understand this, let us assume that the sequence $x(n)$ has N samples between $n_1 \leq n \leq n_N$ (i.e., not necessarily between $[0, N-1]$) and that we want to evaluate $X(e^{j\omega})$ at

$$\omega_k \overset{\triangle}{=} \frac{\pi}{M}k, \quad k = 0, 1, \ldots, M$$

which are $(M+1)$ equispaced frequencies between $[0, \pi]$. Then (3.1) can be written as

$$X(e^{j\omega_k}) = \sum_{\ell=1}^{N} e^{-j(\pi/M)kn_\ell}x(n_\ell), \quad k = 0, 1, \ldots, M$$

When $\{x(n_\ell)\}$ and $\{X(e^{j\omega_k})\}$ are arranged as *column* vectors $\mathbf{x}$ and $\mathbf{X}$, respectively, we have

$$\mathbf{X} = \mathbf{W}\mathbf{x} \tag{3.3}$$

where $\mathbf{W}$ is an $(M+1) \times N$ matrix given by

$$\mathbf{W} \triangleq \left\{ e^{-j(\pi/M)kn_\ell}; \ n_1 \leq n \leq n_N, \quad k = 0, 1, \ldots, M \right\}$$

In addition, if we arrange $\{k\}$ and $\{n_\ell\}$ as *row* vectors $\mathbf{k}$ and $\mathbf{n}$ respectively, then

$$\mathbf{W} = \left[\exp\left(-j\frac{\pi}{M}\mathbf{k}^T\mathbf{n} \right) \right]$$

In MATLAB we represent sequences and indices as row vectors; therefore taking the transpose of (3.3), we obtain

$$\mathbf{X}^T = \mathbf{x}^T \left[\exp\left(-j\frac{\pi}{M}\mathbf{n}^T\mathbf{k} \right) \right] \tag{3.4}$$

Note that $\mathbf{n}^T\mathbf{k}$ is an $N \times (M+1)$ matrix. Now (3.4) can be implemented in MATLAB as follows.

```
>> k = [0:M]; n = [n1:n2];
>> X = x * (exp(-j*pi/M)) .^ (n'*k);
```

☐   **EXAMPLE 3.4**   Numerically compute the discrete-time Fourier transform of the sequence $x(n)$ given in Example 3.2 at 501 equispaced frequencies between $[0, \pi]$.

Solution

MATLAB Script _____
```
>> n = -1:3; x = 1:5;
>> k = 0:500; w = (pi/500)*k;
>> X = x * (exp(-j*pi/500)) .^ (n'*k);
>> magX = abs(X); angX = angle(X);
>> realX = real(X); imagX = imag(X);
>> subplot(2,2,1); plot(k/500,magX);grid
>> xlabel('frequency in pi units'); title('Magnitude Part')
>> subplot(2,2,3); plot(k/500,angX/pi);grid
>> xlabel('frequency in pi units'); title('Angle Part')
>> subplot(2,2,2); plot(k/500,realX);grid
>> xlabel('frequency in pi units'); title('Real Part')
>> subplot(2,2,4); plot(k/500,imagX);grid
>> xlabel('frequency in pi units'); title('Imaginary Part')
```
_____

The frequency-domain plots are shown in Figure 3.2. Note that the angle plot is depicted as a discontinuous function between $-\pi$ and π. This is because the angle function in MATLAB computes the principal angle.   ☐

The procedure of the above example can be compiled into a MATLAB function, say a dtft function, for ease of implementation. This is explored in Problem 3.1. This numerical computation is based on definition (3.1).

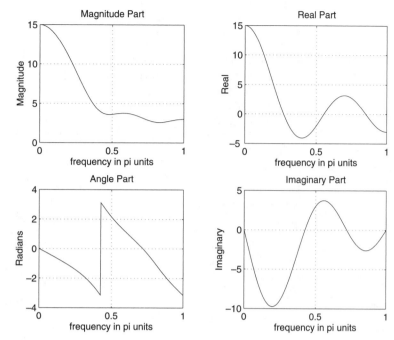

FIGURE 3.2   *Plots in Example 3.4*

It is not the most elegant way of numerically computing the discrete-time Fourier transform of a finite-duration sequence. Furthermore, it creates an $N \times (M + 1)$ matrix in (3.4) that may exceed the size limit in the Student Edition of MATLAB for large M and N. In Chapter 5 we will discuss in detail the topic of a computable transform called the discrete Fourier transform (DFT) and its efficient computation called the fast Fourier transform (FFT). Also there is an alternate approach based on the z-transform using the MATLAB function `freqz` for finite-duration sequences, which we will discuss in Chapter 4. In this chapter we will continue to use the approaches discussed so far for calculation as well as for investigation purposes.

In the next two examples we investigate the periodicity and symmetry properties using complex-valued and real-valued sequences.

□ **EXAMPLE 3.5**   Let $x(n) = (0.9 \exp(j\pi/3))^n$, $0 \le n \le 10$. Determine $X(e^{j\omega})$ and investigate its periodicity.

Solution   Since $x(n)$ is complex-valued, it satisfies only the periodicity property. Therefore it is uniquely defined over one period of 2π. However, we will evaluate and plot it at 401 frequencies over two periods between $[-2\pi, 2\pi]$ to observe its periodicity.

```
>> n = 0:10; x = (0.9*exp(j*pi/3)).^n;
>> k = -200:200; w = (pi/100)*k;
>> X = x * (exp(-j*pi/100)) .^ (n'*k);
>> magX = abs(X); angX =angle(X);
>> subplot(2,1,1); plot(w/pi,magX);grid
>> xlabel('frequency in units of pi'); ylabel('|X|')
>> title('Magnitude Part')
>> subplot(2,1,2); plot(w/pi,angX/pi);grid
>> xlabel('frequency in units of pi'); ylabel('radians/pi')
>> title('Angle Part')
```

From the plots in Figure 3.3 we observe that $X(e^{j\omega})$ is periodic in ω but is not conjugate-symmetric. □

□   **EXAMPLE 3.6**   Let $x(n) = (-0.9)^n$, $-10 \leq n \leq 10$. Investigate the conjugate-symmetry property of its discrete-time Fourier transform.

Solution   Once again we will compute and plot $X(e^{j\omega})$ over two periods to study its symmetry property.

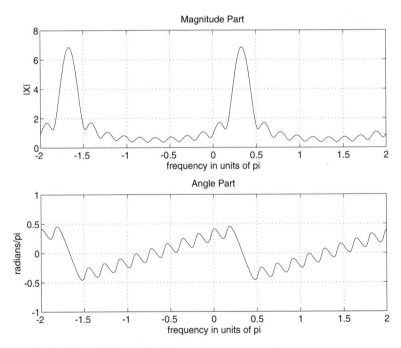

FIGURE 3.3   *Plots in Example 3.5*

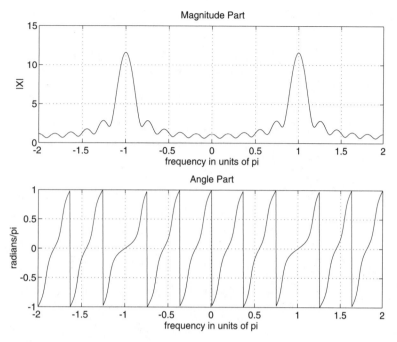

FIGURE 3.4 *Plots in Example 3.6*

```
subplot(1,1,1)
n = -5:5; x = (-0.9).^n;
k = -200:200; w = (pi/100)*k;
X = x * (exp(-j*pi/100)) .^ (n'*k);
magX = abs(X); angX =angle(X);
subplot(2,1,1); plot(w/pi,magX);grid
axis([-2,2,0,15])
xlabel('frequency in units of pi'); ylabel('|X|')
title('Magnitude Part')
subplot(2,1,2); plot(w/pi,angX/pi);grid
axis([-2,2,-1,1])
xlabel('frequency in units of pi'); ylabel('radians/pi')
title('Angle Part')
```

From the plots in Figure 3.4 we observe that $X(e^{j\omega})$ is not only periodic in ω but is also conjugate-symmetric. Therefore for real sequences we will plot their Fourier transform magnitude and angle responses from 0 to π. □

THE PROPERTIES OF THE DTFT

In the previous section we discussed two important properties that we needed for plotting purposes. We now discuss the remaining useful properties, which are given below without proof. Let $X(e^{j\omega})$ be the discrete-time Fourier transform of $x(n)$.

1. **Linearity:** The discrete-time Fourier transform is a linear transformation; that is,

$$\mathcal{F}[\alpha x_1(n) + \beta x_2(n)] = \alpha \mathcal{F}[x_1(n)] + \beta \mathcal{F}[x_2(n)] \qquad (3.5)$$

for every α, β, $x_1(n)$, and $x_2(n)$.

2. **Time shifting:** A shift in the time domain corresponds to the phase shifting.

$$\mathcal{F}[x(n-k)] = X(e^{j\omega})e^{-j\omega k} \qquad (3.6)$$

3. **Frequency shifting:** Multiplication by a complex exponential corresponds to a shift in the frequency domain.

$$\mathcal{F}\left[x(n)e^{j\omega_0 n}\right] = X\left(e^{j(\omega-\omega_0)}\right) \qquad (3.7)$$

4. **Conjugation:** Conjugation in the time domain corresponds to the folding and conjugation in the frequency domain.

$$\mathcal{F}[x^*(n)] = X^*(e^{-j\omega}) \qquad (3.8)$$

5. **Folding:** Folding in the time domain corresponds to the folding in the frequency domain.

$$\mathcal{F}[x(-n)] = X(e^{-j\omega}) \qquad (3.9)$$

6. **Symmetries in real sequences:** We have already studied the conjugate symmetry of real sequences. These real sequences can be decomposed into their even and odd parts as we discussed in Chapter 2.

$$x(n) = x_e(n) + x_o(n)$$

Then

$$\begin{aligned} \mathcal{F}[x_e(n)] &= \operatorname{Re}\left[X(e^{j\omega})\right] \\ \mathcal{F}[x_o(n)] &= j\operatorname{Im}\left[X(e^{j\omega})\right] \end{aligned} \qquad (3.10)$$

Implication: If the sequence $x(n)$ is real and even, then $X(e^{j\omega})$ is also real and even. Hence only one plot over $[0, \pi]$ is necessary for its complete representation.

A similar property for complex-valued sequences is explored in Problem 3.7.

7. **Convolution:** This is one of the most useful properties that makes system analysis convenient in the frequency domain.

$$\mathcal{F}[x_1(n) * x_2(n)] = \mathcal{F}[x_1(n)] \mathcal{F}[x_2(n)] = X_1(e^{j\omega})X_2(e^{j\omega}) \qquad (3.11)$$

8. **Multiplication:** This is a dual of the convolution property.

$$\mathcal{F}\left[x_1(n)\cdot x_2(n)\right] = \mathcal{F}\left[x_1(n)\right] \circledast \mathcal{F}\left[x_2(n)\right] \triangleq \frac{1}{2\pi}\int X_1(e^{j\theta})X_2(e^{j(\omega-\theta)})d\theta$$

(3.12)

The convolution-like operation above is called a *periodic convolution* and hence denoted by $\circledast$. It is discussed (in its discrete form) in Chapter 5.

9. **Energy:** The energy of the sequence $x(n)$ can be written as

$$\mathcal{E}_x = \sum_{-\infty}^{\infty}|x(n)|^2 = \frac{1}{2\pi}\int_{-\pi}^{\pi}\left|X(e^{j\omega})\right|^2 d\omega$$

(3.13)

$$= \int_{0}^{\pi}\frac{\left|X(e^{j\omega})\right|^2}{\pi}d\omega \quad \text{(for real sequences using even symmetry)}$$

This is also known as Parseval's Theorem. From (3.13) the *energy density spectrum* of $x(n)$ is defined as

$$\Phi_x(\omega) \triangleq \frac{\left|X(e^{j\omega})\right|^2}{\pi}$$

(3.14)

Then the energy of $x(n)$ in the $[\omega_1, \omega_2]$ band is given by

$$\int_{\omega_1}^{\omega_2}\Phi_x(\omega)d\omega, \quad 0 \le \omega_1 < \omega_2 \le \pi$$

In the next several examples we will verify some of these properties using finite-duration sequences. We will follow our numerical procedure to compute discrete-time Fourier transforms in each case. Although this does not analytically prove the validity of each property, it provides us with an experimental tool in practice.

☐ **EXAMPLE 3.7** In this example we will verify the linearity property (3.5) using real-valued finite-duration sequences. Let $x_1(n)$ and $x_2(n)$ be two random sequences uniformly distributed between $[0, 1]$ over $0 \le n \le 10$. Then we can use our numerical discrete-time Fourier transform procedure as follows.

```
>> x1 = rand(1,11); x2 = rand(1,11); n = 0:10;
>> alpha = 2; beta = 3;
>> k = 0:500; w = (pi/500)*k;
>> X1 = x1 * (exp(-j*pi/500)).^(n'*k);   % DTFT of x1
>> X2 = x2 * (exp(-j*pi/500)).^(n'*k);   % DTFT of x2
>> x = alpha*x1 + beta*x2;               % Linear combination of x1 & x2
```

```
>> X = x * (exp(-j*pi/500)).^(n'*k);      % DTFT of x
>> % verification
>> X_check = alpha*X1 + beta*X2;          % Linear Combination of X1 & X2
>> error = max(abs(X-X_check))            % Difference
error =
   7.1054e-015
```

Since the maximum absolute error between the two Fourier transform arrays is less than 10^{-14}, the two arrays are identical within the limited numerical precision of MATLAB. □

□ **EXAMPLE 3.8** Let $x(n)$ be a random sequence uniformly distributed between $[0, 1]$ over $0 \leq n \leq 10$ and let $y(n) = x(n - 2)$. Then we can verify the sample shift property (3.6) as follows.

```
>> x = rand(1,11); n = 0:10;
>> k = 0:500; w = (pi/500)*k;
>> X = x * (exp(-j*pi/500)).^(n'*k);      % DTFT of x
>> % signal shifted by two samples
>> y = x; m = n+2;
>> Y = y * (exp(-j*pi/500)).^(m'*k);      % DTFT of y
>> % verification
>> Y_check = (exp(-j*2).^w).*X;           % multiplication by exp(-j2w)
>> error = max(abs(Y-Y_check))            % Difference
error =
   5.7737e-015
```
  □

□ **EXAMPLE 3.9** To verify the frequency shift property (3.7), we will use the graphical approach. Let

$$x(n) = \cos(\pi n/2), \quad 0 \leq n \leq 100 \quad \text{and} \quad y(n) = e^{j\pi n/4}x(n)$$

Then using MATLAB,

```
>> n = 0:100; x = cos(pi*n/2);
>> k = -100:100; w = (pi/100)*k;          % frequency between -pi and +pi
>> X = x * (exp(-j*pi/100)).^(n'*k);      % DTFT of x
%
>> y = exp(j*pi*n/4).*x;                  % signal multiplied by exp(j*pi*n/4)
>> Y = y * (exp(-j*pi/100)).^(n'*k);      % DTFT of y
% Graphical verification
>> subplot(1,1,1)
>> subplot(2,2,1); plot(w/pi,abs(X)); grid; axis([-1,1,0,60])
>> xlabel('frequency in pi units'); ylabel('|X|')
>> title('Magnitude of X')
>> subplot(2,2,2); plot(w/pi,angle(X)/pi); grid; axis([-1,1,-1,1])
>> xlabel('frequency in pi units'); ylabel('radiands/pi')
>> title('Angle of X')
>> subplot(2,2,3); plot(w/pi,abs(Y)); grid; axis([-1,1,0,60])
```

```
>> xlabel('frequency in pi units'); ylabel('|Y|')
>> title('Magnitude of Y')
>> subplot(2,2,4); plot(w/pi,angle(Y)/pi); grid; axis([-1,1,-1,1])
>> xlabel('frequency in pi units'); ylabel('radians/pi')
>> title('Angle of Y')
```

and from plots in Figure 3.5 we observe that $X(e^{j\omega})$ is indeed shifted by $\pi/4$ in both magnitude and angle. □

□ **EXAMPLE 3.10** To verify the conjugation property (3.8), let $x(n)$ be a complex-valued random sequence over $-5 \le n \le 10$ with real and imaginary parts uniformly distributed between $[0,1]$. The MATLAB verification is as follows.

```
>> n = -5:10; x = rand(1,length(n)) + j*rand(1,length(n));
>> k = -100:100; w = (pi/100)*k;          % frequency between -pi and +pi
>> X = x * (exp(-j*pi/100)).^(n'*k);      % DTFT of x
% conjugation property
>> y = conj(x);                           % signal conjugation
>> Y = y * (exp(-j*pi/100)).^(n'*k);      % DTFT of y
% verification
>> Y_check = conj(fliplr(X));             % conj(X(-w))
```

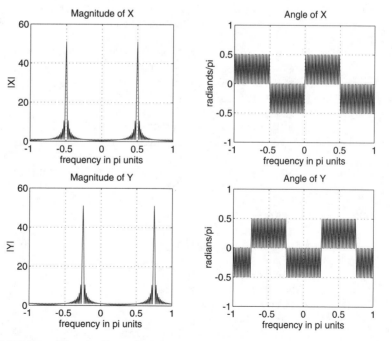

FIGURE 3.5 *Plots in Example 3.9*

```
>> error = max(abs(Y-Y_check))              % Difference
error =
      0                                                                    □
```

□   **EXAMPLE 3.11**   To verify the folding property (3.9), let $x(n)$ be a random sequence over $-5 \leq n \leq 10$ uniformly distributed between $[0, 1]$. The MATLAB verification is as follows.

```
>> n = -5:10; x = rand(1,length(n));
>> k = -100:100; w = (pi/100)*k;         % frequency between -pi and +pi
>> X = x * (exp(-j*pi/100)).^(n'*k);     % DTFT of x
% folding property
>> y = fliplr(x); m = -fliplr(n);        % signal folding
>> Y = y * (exp(-j*pi/100)).^(m'*k);     % DTFT of y
% verification
>> Y_check = fliplr(X);                   % X(-w)
>> error = max(abs(Y-Y_check))            % Difference
error =
      0                                                                    □
```

□   **EXAMPLE 3.12**   In this problem we verify the symmetry property (3.10) of real signals. Let

$$x(n) = \sin(\pi n/2), \quad -5 \leq n \leq 10$$

Then using the `evenodd` function developed in Chapter 2, we can compute the even and odd parts of $x(n)$ and then evaluate their discrete-time Fourier transforms. We will provide the numerical as well as graphical verification.

```
>> n = -5:10; x = sin(pi*n/2);
>> k = -100:100; w = (pi/100)*k;         % frequency between -pi and +pi
>> X = x * (exp(-j*pi/100)).^(n'*k);     % DTFT of x
% signal decomposition
>> [xe,xo,m] = evenodd(x,n);             % even and odd parts
>> XE = xe * (exp(-j*pi/100)).^(m'*k);   % DTFT of xe
>> XO = xo * (exp(-j*pi/100)).^(m'*k);   % DTFT of xo
% verification
>> XR = real(X);                          % real part of X
>> error1 = max(abs(XE-XR))               % Difference
error1 =
   1.8974e-019
>> XI = imag(X);                          % imag part of X
>> error2 = max(abs(XO-j*XI))             % Difference
error2 =
   1.8033e-019
% graphical verification
>> subplot(1,1,1)
>> subplot(2,2,1); plot(w/pi,XR); grid; axis([-1,1,-2,2])
>> xlabel('frequency in pi units'); ylabel('Re(X)');
>> title('Real part of X')
```

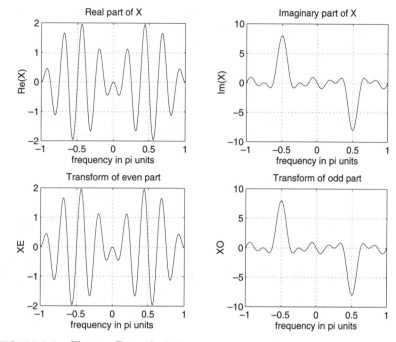

FIGURE 3.6 *Plots in Example 3.12*

```
>> subplot(2,2,2); plot(w/pi,XI); grid; axis([-1,1,-10,10])
>> xlabel('frequency in pi units'); ylabel('Im(X)');
>> title('Imaginary part of X')
>> subplot(2,2,3); plot(w/pi,real(XE)); grid; axis([-1,1,-2,2])
>> xlabel('frequency in pi units'); ylabel('XE');
>> title('Transform of even part')
>> subplot(2,2,4); plot(w/pi,imag(XO)); grid; axis([-1,1,-10,10])
>> xlabel('frequency in pi units'); ylabel('XO');
>> title('Transform of odd part')
```

From the plots in Figure 3.6 we observe that the real part of $X(e^{j\omega})$ (or the imaginary part of $X(e^{j\omega})$) is equal to the discrete-time Fourier transform of $x_e(n)$ (or $x_o(n)$). □

THE FREQUENCY DOMAIN REPRESENTATION OF LTI SYSTEMS

We earlier stated that the Fourier transform representation is the most useful signal representation for LTI systems. It is due to the following result.

RESPONSE TO A COMPLEX EXPONENTIAL $e^{j\omega_0 n}$

Let $x(n) = e^{j\omega_0 n}$ be the input to an LTI system represented by the impulse response $h(n)$.

$$e^{j\omega_0 n} \longrightarrow \boxed{h(n)} \longrightarrow h(n) * e^{j\omega_0 n}$$

Then

$$y(n) = h(n) * e^{j\omega_0 n} = \sum_{-\infty}^{\infty} h(k)e^{j\omega_0(n-k)}$$

$$= \left[\sum_{-\infty}^{\infty} h(k)e^{-j\omega_0 k}\right] e^{j\omega_0 n} \qquad (3.15)$$

$$= \left[\mathcal{F}[h(n)]\big|_{\omega=\omega_0}\right] e^{j\omega_0 n}$$

■ **DEFINITION 1** *Frequency Response*

The discrete-time Fourier transform of an impulse response is called the Frequency Response *(or* Transfer Function*) of an LTI system and is denoted by*

$$H(e^{j\omega}) \triangleq \sum_{-\infty}^{\infty} h(n)e^{-j\omega n} \qquad (3.16)$$

Then from (3.15) we can represent the system by

$$x(n) = e^{j\omega_0 n} \longrightarrow \boxed{H(e^{j\omega})} \longrightarrow y(n) = H(e^{j\omega_0}) \times e^{j\omega_0 n} \qquad (3.17)$$

Hence the output sequence is the input exponential sequence *modified* by the response of the system at frequency ω_0. This justifies the definition of $H(e^{j\omega})$ as a frequency response because it is what the complex exponential is multiplied by to obtain the output $y(n)$. This powerful result can be extended to a linear combination of complex exponentials using the linearity of LTI systems.

$$\sum_k A_k e^{j\omega_k n} \longrightarrow \boxed{h(n)} \longrightarrow \sum_k A_k H(e^{j\omega_k}) e^{j\omega_k n}$$

In general, the frequency response $H(e^{j\omega})$ is a complex function of ω. The magnitude $|H(e^{j\omega})|$ of $H(e^{j\omega})$ is called the *magnitude (or gain) response* function, and the angle $\angle H(e^{j\omega})$ is called the *phase response* function as we shall see below.

RESPONSE TO SINUSOIDAL SEQUENCES

Let $x(n) = A\cos(\omega_0 n + \theta_0)$ be an input to an LTI system $h(n)$. Then from (3.17) we can show that the response $y(n)$ is another sinusoid of the same frequency ω_0, with amplitude *gained* by $|H(e^{j\omega_0})|$ and phase *shifted* by $\angle H(e^{j\omega_0})$, that is,

$$y(n) = A\left|H(e^{j\omega_0})\right|\cos\left(\omega_0 n + \theta_0 + \angle H(e^{j\omega_0})\right) \qquad \text{(3.18)}$$

This response is called the *steady-state response* denoted by $y_{ss}(n)$. It can be extended to a linear combination of sinusoidal sequences.

$$\sum_k A_k \cos(\omega_k n + \theta_k) \longrightarrow \boxed{H(e^{j\omega})} \longrightarrow \sum_k A_k\left|H(e^{j\omega_k})\right|\cos\left(\omega_k n + \theta_k + \angle H(e^{j\omega_k})\right)$$

RESPONSE TO ARBITRARY SEQUENCES

Finally, (3.17) can be generalized to arbitrary *absolutely summable* sequences. Let $X(e^{j\omega}) = \mathcal{F}[x(n)]$ and $Y(e^{j\omega}) = \mathcal{F}[y(n)]$; then using the convolution property (3.11), we have

$$Y(e^{j\omega}) = H(e^{j\omega})\,X(e^{j\omega}) \qquad \text{(3.19)}$$

Therefore an LTI system can be represented in the frequency domain by

$$X(e^{j\omega}) \longrightarrow \boxed{H(e^{j\omega})} \longrightarrow Y(e^{j\omega}) = H(e^{j\omega})\,X(e^{j\omega})$$

The output $y(n)$ is then computed from $Y(e^{j\omega})$ using the inverse discrete-time Fourier transform (3.2). This requires an integral operation, which is not a convenient operation in MATLAB. As we shall see in Chapter 4, there is an alternate approach to the computation of output to arbitrary inputs using the z-transform and partial fraction expansion. In this chapter we will concentrate on computing the steady-state response.

☐ **EXAMPLE 3.13**   Determine the frequency response $H(e^{j\omega})$ of a system characterized by $h(n) = (0.9)^n u(n)$. Plot the magnitude and the phase responses.

Solution   Using (3.16),

$$H(e^{j\omega}) = \sum_{-\infty}^{\infty} h(n)e^{-j\omega n} = \sum_{0}^{\infty}(0.9)^n e^{-j\omega n}$$

$$= \sum_{0}^{\infty}(0.9e^{-j\omega})^n = \frac{1}{1 - 0.9e^{-j\omega}}$$

Hence

$$\left|H(e^{j\omega})\right| = \sqrt{\frac{1}{(1 - 0.9\cos\omega)^2 + (0.9\sin\omega)^2}} = \frac{1}{\sqrt{1.81 - 1.8\cos\omega}}$$

and

$$\angle H(e^{j\omega}) = -\arctan\left[\frac{0.9\sin\omega}{1 - 0.9\cos\omega}\right]$$

To plot these responses, we can either implement the $\left|H(e^{j\omega})\right|$ and $\angle H(e^{j\omega})$ functions or the frequency response $H(e^{j\omega})$ and then compute its magnitude and phase. The latter approach is more useful from a practical viewpoint (as shown in (3.18)).

```
>> w = [0:1:500]*pi/500;  % [0, pi] axis divided into 501 points.
>> H = exp(j*w) ./ (exp(j*w) - 0.9*ones(1,501));
>> magH = abs(H); angH = angle(H);
>> subplot(2,1,1); plot(w/pi,magH); grid;
>> xlabel('frequency in pi units'); ylabel('|H|');
>> title('Magnitude Response');
>> subplot(2,1,2); plot(w/pi,angH/pi); grid
>> xlabel('frequency in pi units'); ylabel('Phase in pi Radians');
>> title('Phase Response');
```

The plots are shown in Figure 3.7.   □

□   **EXAMPLE 3.14**   Let an input to the system in Example 3.13 be $0.1u(n)$. Determine the steady-state response $y_{ss}(n)$.

Solution   Since the input is not absolutely summable, the discrete-time Fourier transform is not particularly useful in computing the complete response. However, it can be used to compute the steady-state response. In the steady state (i.e., $n \to \infty$)

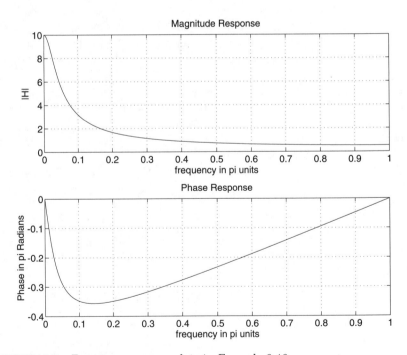

FIGURE 3.7   *Frequency response plots in Example 3.13*

the input is a constant sequence (or a sinusoid with $\omega_0 = \theta_0 = 0$). Then the output is

$$y_{ss}(n) = 0.1 \times H(e^{j0}) = 0.1 \times 10 = 1$$

where the gain of the system at $\omega = 0$ (also called the DC gain) is $H(e^{j0}) = 10$, which is obtained from Figure 3.7. □

FREQUENCY RESPONSE FUNCTION FROM DIFFERENCE EQUATIONS

When an LTI system is represented by the difference equation

$$y(n) + \sum_{\ell=1}^{N} a_\ell y(n - \ell) = \sum_{m=0}^{M} b_m x(n - m) \tag{3.20}$$

then to evaluate its frequency response from (3.16), we would need the impulse response $h(n)$. However, using (3.17), we can easily obtain $H(e^{j\omega})$. We know that when $x(n) = e^{j\omega n}$, then $y(n)$ must be $H(e^{j\omega})e^{j\omega n}$. Substituting in (3.20), we have

$$H(e^{j\omega})e^{j\omega n} + \sum_{\ell=1}^{N} a_\ell H(e^{j\omega})e^{j\omega(n-\ell)} = \sum_{m=0}^{M} b_m \, e^{j\omega(n-m)}$$

or

$$H(e^{j\omega}) = \frac{\displaystyle\sum_{m=0}^{M} b_m \, e^{-j\omega m}}{1 + \displaystyle\sum_{\ell=1}^{N} a_\ell \, e^{-j\omega \ell}} \tag{3.21}$$

after canceling the common factor $e^{j\omega n}$ term and rearranging. This equation can easily be implemented in MATLAB, given the difference equation parameters.

□ **EXAMPLE 3.15**   An LTI system is specified by the difference equation

$$y(n) = 0.8y(n - 1) + x(n)$$

   a. Determine $H(e^{j\omega})$.
   b. Calculate and plot the steady-state response $y_{ss}(n)$ to

$$x(n) = \cos(0.05\pi n)u(n)$$

Solution   Rewrite the difference equation as $y(n) - 0.8y(n - 1) = x(n)$.

   a. Using (3.21), we obtain

$$H(e^{j\omega}) = \frac{1}{1 - 0.8e^{-j\omega}} \tag{3.22}$$

b. In the steady state the input is $x(n) = \cos(0.05\pi n)$ with frequency $\omega_0 = 0.05\pi$ and $\theta_0 = 0°$. The response of the system is

$$H(e^{j0.05\pi}) = \frac{1}{1 - 0.8e^{-j0.05\pi}} = 4.0928e^{-j0.5377}$$

Therefore

$$y_{ss}(n) = 4.0928\cos(0.05\pi n - 0.5377) = 4.0928\cos\left[0.05\pi(n - 3.42)\right]$$

This means that at the output the sinusoid is scaled by 4.0928 and shifted by 3.42 samples. This can be verified using MATLAB.

```
>> subplot(1,1,1)
>> b = 1; a = [1,-0.8];
>> n=[0:100];x = cos(0.05*pi*n);
>> y = filter(b,a,x);
>> subplot(2,1,1); stem(n,x);
>> xlabel('n'); ylabel('x(n)'); title('Input sequence')
>> subplot(2,1,2); stem(n,y);
>> xlabel('n'); ylabel('y(n)'); title('Output sequence')
```

From the plots in Figure 3.8 we note that the amplitude of $y_{ss}(n)$ is approximately 4. To determine the shift in the output sinusoid, we can compare zero

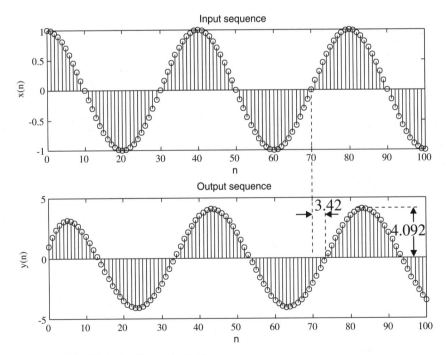

FIGURE 3.8 *Plots in Example 3.15*

crossings of the input and the output. This is shown in Figure 3.8, from which the shift is approximately 3.5 samples. □

In Example 3.15 the system was characterized by a first-order difference equation. It is fairly straightforward to implement (3.22) in MATLAB as we did in Example 3.13. In practice the difference equations are of large order and hence we need a compact procedure to implement the general expression (3.21). This can be done using a simple matrix-vector multiplication. If we evaluate $H(e^{j\omega})$ at $k = 0, 1, \ldots, K$ equispaced frequencies over $[0, \pi]$, then

$$H(e^{j\omega_k}) = \frac{\displaystyle\sum_{m=0}^{M} b_m \, e^{-j\omega_k m}}{1 + \displaystyle\sum_{\ell=1}^{N} a_\ell \, e^{-j\omega_k \ell}}, \quad k = 0, 1, \ldots, K \qquad (3.23)$$

If we let $\{b_m\}$, $\{a_\ell\}$ (with $a_0 = 1$), $\{m = 0, \ldots, M\}$, $\{\ell = 0, \ldots, N\}$, and $\{\omega_k\}$ be arrays (or row vectors), then the numerator and the denominator of (3.23) become

$$\underline{b} \exp\left(-j\underline{m}^T \underline{\omega}\right); \quad \underline{a} \exp\left(-j\underline{\ell}^T \underline{\omega}\right)$$

respectively. Now the array $H(e^{j\omega_k})$ in (3.23) can be computed using a ./ operation. This procedure can be implemented in a MATLAB function to determine the frequency response function, given $\{b_m\}$ and $\{a_\ell\}$ arrays. We will explore this in Example 3.16 and in Problem 3.15.

□ **EXAMPLE 3.16**   A 3rd-order lowpass filter is described by the difference equation

$$y(n) = 0.0181x(n) + 0.0543x(n-1) + 0.0543x(n-2) + 0.0181x(n-3)$$

$$+ 1.76y(n-1) - 1.1829y(n-2) + 0.2781y(n-3)$$

Plot the magnitude and the phase response of this filter and verify that it is a lowpass filter.

Solution   We will implement the above procedure in MATLAB and then plot the filter responses.

```
>> b = [0.0181,  0.0543, 0.0543,  0.0181]; % filter coefficient array b
>> a = [1.0000, -1.7600, 1.1829, -0.2781]; % filter coefficient array a
>> m = 0:length(b)-1; l = 0:length(a)-1;   % index arrays m and l
>> K = 500; k = 0:1:K;                      % index array k for frequencies
>> w = pi*k/K;                              % [0, pi] axis divided into 501 points.
>> num = b * exp(-j*m'*w);                  % Numerator calculations
```

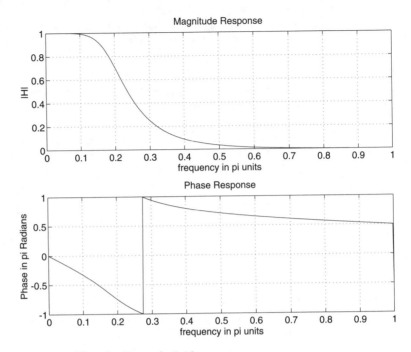

FIGURE 3.9 *Plots for Example 3.16*

```
>> den = a * exp(-j*l'*w);              % Denominator calculations
>> H = num ./ den;                      % Frequency response
>> magH = abs(H); angH = angle(H);      % mag and phase responses
>> subplot(1,1,1);
>> subplot(2,1,1); plot(w/pi,magH); grid; axis([0,1,0,1])
>> xlabel('frequency in pi units'); ylabel('|H|');
>> title('Magnitude Response');
>> subplot(2,1,2); plot(w/pi,angH/pi); grid
>> xlabel('frequency in pi units'); ylabel('Phase in pi Radians');
>> title('Phase Response');
```

From the plots in Figure 3.9 we see that the filter is indeed a lowpass filter. □

SAMPLING AND RECONSTRUCTION OF ANALOG SIGNALS

In many applications—for example, in digital communications—real-world analog signals are converted into discrete signals using sampling and quantization operations (collectively called analog-to-digital conversion or ADC). These discrete signals are processed by digital signal processors, and the processed signals are converted into analog signals using a reconstruction operation (called digital-to-analog conversion or DAC). Using Fourier analysis, we can describe the sampling operation from the frequency-domain viewpoint, analyze its effects, and then address the reconstruction operation. We will also assume that the number of quantization levels is sufficiently large that the effect of quantization on discrete signals is negligible.

SAMPLING

Let $x_a(t)$ be an analog (absolutely integrable) signal. Its continuous-time Fourier transform (CTFT) is given by

$$X_a(j\Omega) \triangleq \int_{-\infty}^{\infty} x_a(t)e^{-j\Omega t}dt \qquad (3.24)$$

where Ω is an analog frequency in radians/sec. The inverse continuous-time Fourier transform is given by

$$x_a(t) = \frac{1}{2\pi} \int_{-\infty}^{\infty} X_a(j\Omega)e^{j\Omega t}d\Omega \qquad (3.25)$$

We now sample $x_a(t)$ at *sampling interval* T_s seconds apart to obtain the discrete-time signal $x(n)$.

$$x(n) \triangleq x_a(\ nT_s)$$

Let $X(e^{j\omega})$ be the discrete-time Fourier transform of $x(n)$. Then it can be shown [19] that $X(e^{j\omega})$ is a countable sum of amplitude-scaled, frequency-scaled, and translated versions of the Fourier transform $X_a(j\Omega)$.

$$X(e^{j\omega}) = \frac{1}{T_s} \sum_{\ell=-\infty}^{\infty} X_a\left[j\left(\frac{\omega}{T_s} - \frac{2\pi}{T_s}\ell\right)\right] \qquad (3.26)$$

The above relation is known as the *aliasing formula*. The analog and digital frequencies are related through T_s

$$\omega = \Omega T_s \qquad (3.27)$$

while the sampling frequency F_s is given by

$$F_s = \frac{1}{T_s}, \quad \text{sam/sec} \qquad (3.28)$$

The graphical illustration of (3.26) is shown in Figure 3.10, from which we observe that, in general, the discrete signal is an *aliased version* of the corresponding analog signal because higher frequencies are aliased into

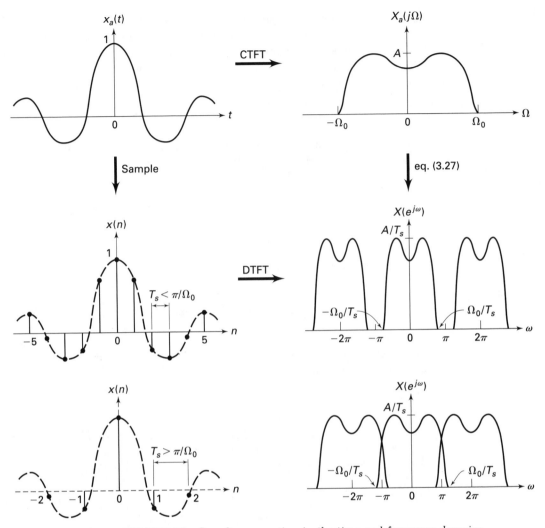

FIGURE 3.10 *Sampling operation in the time and frequency domains*

lower frequencies if there is an overlap. However, it is possible to recover the Fourier transform $X_a(j\Omega)$ from $X(e^{j\omega})$ (or equivalently, the analog signal $x_a(t)$ from its samples $x(n)$) if the infinite "replicas" of $X_a(j\Omega)$ do not overlap with each other to form $X(e^{j\omega})$. This is true for band-limited analog signals.

■ **DEFINITION 2** *Band-limited Signal*

A signal is band-limited if there exists a finite radian frequency Ω_0 such that $X_a(j\Omega)$ is zero for $|\Omega| > \Omega_0$. The frequency $F_0 = \Omega_0/2\pi$ is called the signal bandwidth in Hz.

Referring to Figure 3.10, if $\pi > \Omega_0 T_s$—or equivalently, $F_s/2 > F_0$—then

$$X(e^{j\omega}) = \frac{1}{T_s} X\left(j\frac{\omega}{T_s}\right); \quad -\frac{\pi}{T_s} < \frac{\omega}{T_s} \leq \frac{\pi}{T_s} \tag{3.29}$$

which leads to the sampling theorem for band limited signals.

■   THEOREM 3   *Sampling Principle*

A band-limited signal $x_a(t)$ with bandwidth F_0 can be reconstructed from its sample values $x(n) = x_a(nT_s)$ if the sampling frequency $F_s = 1/T_s$ is greater than twice the bandwidth F_0 of $x_a(t)$.

$$F_s > 2F_0$$

Otherwise aliasing would result in $x(n)$. The sampling rate of $2F_0$ for an analog band-limited signal is called the Nyquist rate.

It should be noted that after $x_a(t)$ is sampled, the highest analog frequency that $x(n)$ represents is $F_s/2$ Hz (or $\omega = \pi$). This agrees with the implication stated in Property 2 of the discrete-time Fourier transform in the first section of this chapter.

MATLAB
IMPLEMEN-
TATION

In a strict sense it is not possible to analyze analog signals using MATLAB unless we use the Symbolic toolbox. However, if we sample $x_a(t)$ on a fine grid that has a sufficiently small time increment to yield a smooth plot and a large enough maximum time to show all the modes, then we can approximate its analysis. Let Δt be the grid interval such that $\Delta t \ll T_s$. Then

$$x_G(m) \triangleq x_a(m\Delta t) \tag{3.30}$$

can be used as an array to simulate an analog signal. The sampling interval T_s should not be confused with the grid interval Δt, which is used strictly to represent an analog signal in MATLAB. Similarly, the Fourier transform relation (3.24) should also be approximated in light of (3.30) as follows

$$X_a(j\Omega) \approx \sum_m x_G(m)e^{-j\Omega m\Delta t}\Delta t = \Delta t \sum_m x_G(m)e^{-j\Omega m\Delta t} \tag{3.31}$$

Now if $x_a(t)$ (and hence $x_G(m)$) is of finite duration, then (3.31) is similar to the discrete-time Fourier transform relation (3.3) and hence can be implemented in MATLAB in a similar fashion to analyze the sampling phenomenon.

Sampling and Reconstruction of Analog Signals

□ **EXAMPLE 3.17**   Let $x_a(t) = e^{-1000|t|}$. Determine and plot its Fourier transform.

Solution   From (3.24)

$$X_a(j\Omega) = \int_{-\infty}^{\infty} x_a(t) e^{-j\Omega t} dt = \int_{-\infty}^{0} e^{1000t} e^{-j\Omega t} dt + \int_{0}^{\infty} e^{-1000t} e^{-j\Omega t} dt$$

$$= \frac{0.002}{1 + \left(\frac{\Omega}{1000}\right)^2} \qquad \textbf{(3.32)}$$

which is a real-valued function since $x_a(t)$ is a real and even signal. To evaluate $X_a(j\Omega)$ numerically, we have to first approximate $x_a(t)$ by a finite-duration grid sequence $x_G(m)$. Using the approximation $e^{-5} \approx 0$, we note that $x_a(t)$ can be approximated by a finite-duration signal over $-0.005 \leq t \leq 0.005$ (or equivalently, over $[-5, 5]$ msec). Similarly from (3.32), $X_a(j\Omega) \approx 0$ for $\Omega \geq 2\pi (2000)$. Hence choosing

$$\Delta t = 5 \times 10^{-5} \ll \frac{1}{2(2000)} = 25 \times 10^{-5}$$

we can obtain $x_G(m)$ and then implement (3.31) in MATLAB.

```
% Analog Signal
>> Dt = 0.00005; t = -0.005:Dt:0.005; xa = exp(-1000*abs(t));
% Continuous-time Fourier Transform
>>Wmax = 2*pi*2000; K = 500; k = 0:1:K; W = k*Wmax/K;
>>Xa = xa * exp(-j*t'*W) * Dt; Xa = real(Xa);
>>W = [-fliplr(W), W(2:501)]; % Omega from -Wmax to Wmax
>>Xa = [fliplr(Xa), Xa(2:501)]; % Xa over -Wmax to Wmax interval
>>subplot(1,1,1)
>>subplot(2,1,1);plot(t*1000,xa);
>>xlabel('t in msec.'); ylabel('xa(t)')
>>title('Analog Signal')
>>subplot(2,1,2);plot(W/(2*pi*1000),Xa*1000);
>>xlabel('Frequency in KHz'); ylabel('Xa(jW)*1000')
>>title('Continuous-time Fourier Transform')
```

Figure 3.11 shows the plots of $x_a(t)$ and $X_a(j\Omega)$. Note that to reduce the number of computations, we computed $X_a(j\Omega)$ over $[0, 4000\pi]$ radians/sec (or equivalently, over $[0, 2]$ KHz) and then duplicated it over $[-4000\pi, 0]$ for plotting purposes. The displayed plot of $X_a(j\Omega)$ agrees with (3.32).   □

□ **EXAMPLE 3.18**   To study the effect of sampling on the frequency-domain quantities, we will sample $x_a(t)$ in Example 3.17 at two different sampling frequencies.

   a. Sample $x_a(t)$ at $F_s = 5000$ sam/sec to obtain $x_1(n)$. Determine and plot $X_1(e^{j\omega})$.

   b. Sample $x_a(t)$ at $F_s = 1000$ sam/sec to obtain $x_2(n)$. Determine and plot $X_2(e^{j\omega})$.

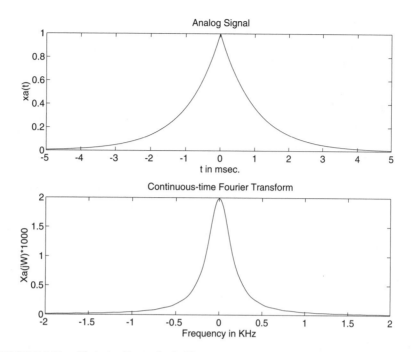

FIGURE 3.11 *Plots in Example 3.17*

Solution

a. Since the bandwidth of $x_a(t)$ is 2KHz, the Nyquist rate is 4000 sam/sec, which is less than the given F_s. Therefore aliasing will be (almost) nonexistent.

```
% Analog Signal
>> Dt = 0.00005; t = -0.005:Dt:0.005; xa = exp(-1000*abs(t));
% Discrete-time Signal
>> Ts = 0.0002; n = -25:1:25; x = exp(-1000*abs(n*Ts));
% Discrete-time Fourier transform
>> K = 500; k = 0:1:K; w = pi*k/K;
>> X = x * exp(-j*n'*w); X = real(X);
>> w = [-fliplr(w), w(2:K+1)];
>> X = [fliplr(X), X(2:K+1)];
>> subplot(1,1,1)
>> subplot(2,1,1);plot(t*1000,xa);
>> xlabel('t in msec.'); ylabel('x1(n)')
>> title('Discrete Signal'); hold on
>> stem(n*Ts*1000,x); gtext('Ts=0.2 msec'); hold off
>> subplot(2,1,2);plot(w/pi,X);
>> xlabel('Frequency in pi units'); ylabel('X1(w)')
>> title('Discrete-time Fourier Transform')
```

In the top plot in Figure 3.12 we have superimposed the discrete signal $x_1(n)$ over $x_a(t)$ to emphasize the sampling. The plot of $X_1(e^{j\omega})$ shows that it is a scaled version (scaled by $F_s = 5000$) of $X_a(j\Omega)$. Clearly there is no aliasing.

Sampling and Reconstruction of Analog Signals

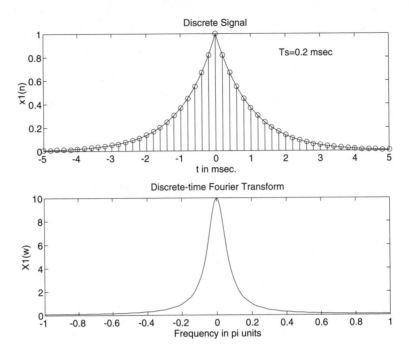

FIGURE 3.12 *Plots in Example 3.18a*

b. Here $F_s = 1000 < 4000$. Hence there will be a considerable amount of aliasing. This is evident from Figure 3.13, in which the shape of $X_2(e^{j\omega})$ is different from that of $X_a(j\Omega)$ and can be seen to be a result of adding overlapping replicas of $X_a(j\Omega)$. □

RECONSTRUC-
TION

From the sampling theorem and the above examples it is clear that if we sample band-limited $x_a(t)$ above its Nyquist rate, then we can reconstruct $x_a(t)$ from its samples $x(n)$. This reconstruction can be thought of as a two-step process:

- First the samples are converted into a weighted impulse train.

$$\sum_{n=-\infty}^{\infty} x(n)\delta(t-nT_s) = \cdots + x(-1)\delta(n+T_s) + x(0)\delta(t) + x(1)\delta(n-T_s) + \cdots$$

- Then the impulse train is filtered through an ideal analog lowpass filter band-limited to the $[-F_s/2, F_s/2]$ band.

$$x(n) \longrightarrow \boxed{\begin{array}{c}\text{Impulse train}\\\text{conversion}\end{array}} \longrightarrow \boxed{\begin{array}{c}\text{Ideal lowpass}\\\text{filter}\end{array}} \longrightarrow x_a(t)$$

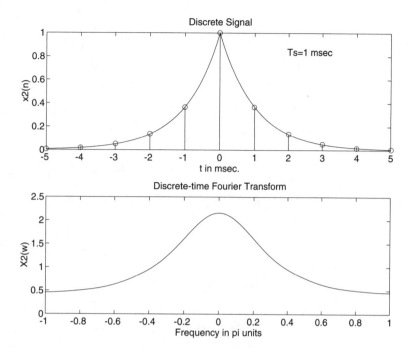

FIGURE 3.13 *Plots in Example 3.18b*

This two-step procedure can be described mathematically using an interpolating formula [19]

$$x_a(t) = \sum_{n=-\infty}^{\infty} x(n) \operatorname{sinc}\left[F_s(t - nT_s)\right] \tag{3.33}$$

where $\operatorname{sinc}(x) = \frac{\sin \pi x}{\pi x}$ is an interpolating function. The physical interpretation of the above reconstruction (3.33) is given in Figure 3.14, from which we observe that this *ideal* interpolation is not practically feasible because the entire system is noncausal and hence not realizable.

Practical D/A converters   In practice we need a different approach than (3.33). The two-step procedure is still feasible, but now we replace the ideal lowpass filter by a practical analog lowpass filter. Another interpretation of (3.33) is that it is an infinite-order interpolation. We want finite-order (and in fact low-order) interpolations. There are several approaches to do this.

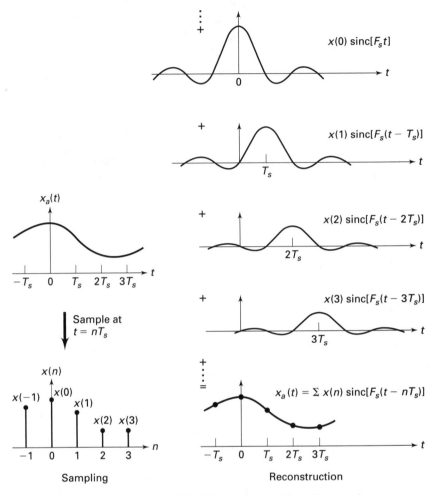

FIGURE 3.14 *Reconstruction of band-limited signal from its samples*

• **Zero-order-hold (ZOH) interpolation:** In this interpolation a given sample value is held for the sample interval until the next sample is received.

$$\hat{x}_a(t) = x(n), \quad nT_s \le n < (n+1)T_s$$

which can be obtained by filtering the impulse train through an interpolating filter of the form

$$h_0(t) = \begin{cases} 1, & 0 \le t \le T_s \\ 0, & \text{otherwise} \end{cases}$$

which is a rectangular pulse. The resulting signal is a piecewise-constant (staircase) waveform which requires an appropriately designed analog post-filter for accurate waveform reconstruction.

$$x(n) \longrightarrow \boxed{\text{ZOH}} \longrightarrow \hat{x}_a(t) \longrightarrow \boxed{\text{Post-Filter}} \longrightarrow x_a(t)$$

• **First-order-hold (FOH) interpolation:** In this case the adjacent samples are joined by straight lines. This can be obtained by filtering the impulse train through

$$h_1(t) = \begin{cases} 1 + \dfrac{t}{T_s}, & 0 \le t \le T_s \\[2mm] 1 - \dfrac{t}{T_s}, & T_s \le t \le 2T_s \\[2mm] 0, & \text{otherwise} \end{cases}$$

Once again an appropriately designed analog postfilter is required for accurate reconstruction. These interpolations can be extended to higher orders. One particularly useful interpolation employed by MATLAB is the following.

• **Cubic spline interpolation:** This approach uses spline interpolants for a smoother, but not necessarily more accurate, estimate of the analog signals between samples. Hence this interpolation does not require an analog postfilter. The smoother reconstruction is obtained by using a set of piecewise continuous third-order polynomials called *cubic splines*, given by [5]

$$x_a(t) = \alpha_0(n) + \alpha_1(n)(t - nT_s) + \alpha_2(n)(t - nT_s)^2$$
$$+ \alpha_3(n)(t - nT_s)^3, \quad nT_s \le n < (n+1)T_s \qquad \textbf{(3.34)}$$

where $\{\alpha_i(n), 0 \le i \le 3\}$ are the polynomial coefficients, which are determined by using least-squares analysis on the sample values. (Strictly speaking, this is not a causal operation but is a convenient one in MATLAB.)

MATLAB
IMPLEMEN-
TATION

For interpolation between samples MATLAB provides several approaches. The function sinc(x), which generates the $(\sin \pi x)/\pi x$ function, can be used to implement (3.33), given a finite number of samples. If $\{x(n), n_1 \le n \le n_2\}$ is given, and if we want to interpolate $x_a(t)$ on a very fine grid with the grid interval Δt, then from (3.33)

$$x_a(m\Delta t) \approx \sum_{n=n_1}^{n_2} x(n) \operatorname{sinc}\left[F_s(m\Delta t - nT_s)\right], \quad t_1 \le m\Delta t \le t_2 \qquad \textbf{(3.35)}$$

which can be implemented as a matrix-vector multiplication operation as shown below.

```
>> n = n1:n2; t = t1:t2; Fs = 1/Ts; nTs = n*Ts;  % Ts is the sampling interval
>> xa = x * sinc(Fs*(ones(length(n),1)*t-nTs'*ones(1,length(t))));
```

Note that it is not possible to obtain an *exact* analog $x_a(t)$ in light of the fact that we have assumed a finite number of samples. We now demonstrate the use of the sinc function in the following two examples and also study the aliasing problem in the time domain.

☐ **EXAMPLE 3.19** From the samples $x_1(n)$ in Example 3.18a, reconstruct $x_a(t)$ and comment on the results.

Solution Note that $x_1(n)$ was obtained by sampling $x_a(t)$ at $T_s = 1/F_s = 0.0002$ sec. We will use the grid spacing of 0.00005 sec over $-0.005 \le t \le 0.005$, which gives $x(n)$ over $-25 \le n \le 25$.

```
% Discrete-time Signal x1(n)
>> Ts = 0.0002; n = -25:1:25; nTs = n*Ts;
>> x = exp(-1000*abs(nTs));
% Analog Signal reconstruction
>> Dt = 0.00005; t = -0.005:Dt:0.005;
>> xa = x * sinc(Fs*(ones(length(n),1)*t-nTs'*ones(1,length(t))));
% check
>> error = max(abs(xa - exp(-1000*abs(t))))
error =
     0.0363
```

The maximum error between the reconstructed and the actual analog signal is 0.0363, which is due to the fact that $x_a(t)$ is not strictly band-limited (and also we have a finite number of samples). From Figure 3.15 we note that visually the reconstruction is excellent. ☐

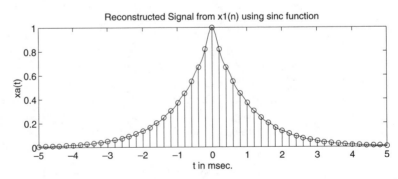

FIGURE 3.15 *Reconstructed signal in Example 3.19*

□ **EXAMPLE 3.20**   From the samples $x_2(n)$ in Example 3.18b reconstruct $x_a(t)$ and comment on the results.

Solution   In this case $x_2(n)$ was obtained by sampling $x_a(t)$ at $T_s = 1/F_s = 0.001$ sec. We will again use the grid spacing of 0.00005 sec over $-0.005 \leq t \leq 0.005$, which gives $x(n)$ over $-5 \leq n \leq 5$.

```
% Discrete-time Signal x2(n)
>> Ts = 0.001; n = -5:1:5; nTs = n*Ts;
>> x = exp(-1000*abs(nTs));
% Analog Signal reconstruction
>> Dt = 0.00005; t = -0.005:Dt:0.005;
>> xa = x * sinc(Fs*(ones(length(n),1)*t-nTs'*ones(1,length(t))));
% check
>> error = max(abs(xa - exp(-1000*abs(t))))
error =
      0.1852
```

The maximum error between the reconstructed and the actual analog signal is 0.1852, which is significant and cannot be attributed to the nonband-limitedness of $x_a(t)$ alone. From Figure 3.16 observe that the reconstructed signal differs from the actual one in many places over the interpolated regions. This is the visual demonstration of aliasing in the time domain.   □

The second MATLAB approach for signal reconstruction is a plotting approach. The **stairs** function plots a staircase (ZOH) rendition of the analog signal, given its samples, while the **plot** function depicts a linear (FOH) interpolation between samples.

□ **EXAMPLE 3.21**   Plot the reconstructed signal from the samples $x_1(n)$ in Example 3.18 using the ZOH and the FOH interpolations. Comment on the plots.

Solution   Note that in this reconstruction we do not compute $x_a(t)$ but merely plot it using its samples.

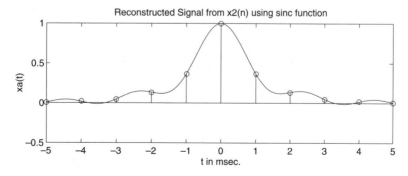

FIGURE 3.16   *Reconstructed signal in Example 3.20*

Sampling and Reconstruction of Analog Signals   71

```
% Discrete-time Signal x1(n) : Ts = 0.0002
>> Ts = 0.0002; n = -25:1:25; nTs = n*Ts;
>> x = exp(-1000*abs(nTs));
% Plots
>> subplot(2,1,1); stairs(nTs*1000,x);
>> xlabel('t in msec.'); ylabel('xa(t)')
>> title('Reconstructed Signal from x1(n) using zero-order-hold'); hold on
>> stem(n*Ts*1000,x); hold off
%
% Discrete-time Signal x2(n) : Ts = 0.001
>> Ts = 0.001; n = -5:1:5; nTs = n*Ts;
>> x = exp(-1000*abs(nTs));
% Plots
>> subplot(2,1,2); stairs(nTs*1000,x);
>> xlabel('t in msec.'); ylabel('xa(t)')
>> title('Reconstructed Signal from x2(n) using zero-order-hold'); hold on
>> stem(n*Ts*1000,x); hold off
```

The plots are shown in Figure 3.17, from which we observe that the ZOH reconstruction is a crude one and that the further processing of analog signal is necessary. The FOH reconstruction appears to be a good one, but a careful observation near $t = 0$ reveals that the peak of the signal is not correctly repro-

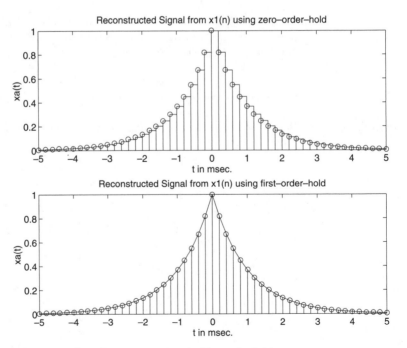

FIGURE 3.17 *Signal reconstruction in Example 3.21*

duced. In general, if the sampling frequency is much higher than the Nyquist rate, then the FOH interpolation provides an acceptable reconstruction. □

The third approach of reconstruction in MATLAB involves the use of cubic spline functions. The `spline` function implements interpolation between sample points. It is invoked by `xa = spline(nTs,x,t)`, in which `x` and `nTs` are arrays containing samples $x(n)$ at nT_s instances, respectively, and `t` array contains a fine grid at which $x_a(t)$ values are desired. Note once again that it is not possible to obtain an *exact* analog $x_a(t)$.

□ **EXAMPLE 3.22** From the samples $x_1(n)$ and $x_2(n)$ in Example 3.18, reconstruct $x_a(t)$ using the `spline` function. Comment on the results.

Solution This example is similar to Examples 3.19 and 3.20. Hence sampling parameters are the same as before.

```
% a) Discrete-time Signal x1(n): Ts = 0.0002
>> Ts = 0.0002; n = -25:1:25; nTs = n*Ts;
>> x = exp(-1000*abs(nTs));
% Analog Signal reconstruction
>> Dt = 0.00005; t = -0.005:Dt:0.005;
>> xa = spline(nTs,x,t);
% check
>> error = max(abs(xa - exp(-1000*abs(t))))
error = 0.0317
```

The maximum error between the reconstructed and the actual analog signal is 0.0317, which is due to the nonideal interpolation and the fact that $x_a(t)$ is nonband-limited. Comparing this error with that from the sinc (or ideal) interpolation, we note that this error is lower. The ideal interpolation generally suffers more from time-limitedness (or from a finite number of samples). From the top plot in Figure 3.18 we observe that visually the reconstruction is excellent.

```
% Discrete-time Signal x2(n): Ts = 0.001
>> Ts = 0.001; n = -5:1:5; nTs = n*Ts;
>> x = exp(-1000*abs(nTs));
% Analog Signal reconstruction
>> Dt = 0.00005; t = -0.005:Dt:0.005;
>> xa = spline(nTs,x,t);
% check
>> error = max(abs(xa - exp(-1000*abs(t))))
error = 0.1679
```

The maximum error in this case is 0.1679, which is significant and cannot be attributed to the nonideal interpolation or nonband-limitedness of $x_a(t)$. From the bottom plot in Figure 3.18 observe that the reconstructed signal again differs from the actual one in many places over the interpolated regions. □

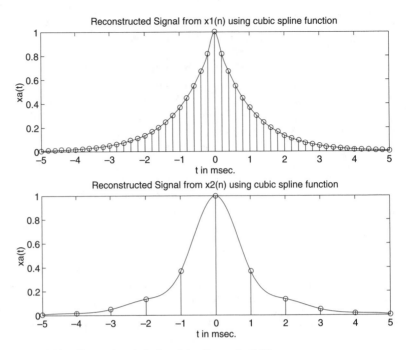

FIGURE 3.18   *Reconstructed signal in Example 3.22*

From these examples it is clear that for practical purposes the `spline` interpolation provides the best results.

PROBLEMS

P3.1   Write a MATLAB function to compute the DTFT of a finite-duration sequence. The format of the function should be

```
function [X] = dtft(x,n,w)
% Computes Discrete-time Fourier Transform
% [X] = dtft(x,n,w)
%
% X = DTFT values computed at w frequencies
% x = finite duration sequence over n
% n = sample position vector
% w = frequency location vector
```

Use this function to compute the DTFT in the following problems (wherever required).

P3.2   For each of the following sequences, determine the DTFT $X\left(e^{j\omega}\right)$. Plot the magnitude and angle of $X\left(e^{j\omega}\right)$.

a. $x(n) = 2(0.8)^n [u(n) - u(n - 20)]$

b. $x(n) = n(0.9)^n [u(n) - u(50)]$

c. $x(n) = \{4, 3, 2, 1, 2, 3, 4\}$. Comment on the angle plot.
$\qquad\uparrow$

d. $x(n) = \{4, 3, 2, 1, 1, 2, 3, 4\}$. Comment on the angle plot.
$\qquad\uparrow$

e. $x(n) = \{4, 3, 2, 1, 0, -1, -2, -3, -4\}$. Comment on the angle plot.
$\qquad\uparrow$

f. $x(n) = \{4, 3, 2, 1, -1, -2, -3, -4\}$. Comment on the angle plot.
$\qquad\uparrow$

P3.3 Determine analytically the DTFT of each of the following sequences. Plot the magnitude and angle of $X(e^{j\omega})$ using MATLAB.

a. $x(n) = 3(0.9)^n u(n)$

b. $x(n) = 2(0.8)^{n+2} u(n - 2)$

c. $x(n) = n(0.5)^n u(n)$

d. $x(n) = (n + 2)(-0.7)^{n-1} u(n - 2)$

e. $x(n) = 5(-0.9)^n \cos(0.1\pi n) u(n)$

P3.4 A symmetric rectangular pulse is given by

$$\mathcal{R}_N(n) = \begin{cases} 1, & -N \leq n \leq N \\ 0, & \text{otherwise} \end{cases}$$

Determine the DTFT for $N = 5, 15, 25, 100$. Scale the DTFT so that $X(e^{j0}) = 1$. Plot the normalized DTFT over $[-\pi, \pi]$. Study these plots and comment on their behavior as a function of N.

P3.5 Repeat Problem 3.4 for a symmetric triangular pulse that is given by

$$\mathcal{T}_N(n) = \left[1 - \frac{|n|}{N}\right] \mathcal{R}_N(n)$$

P3.6 Repeat Problem 3.4 for a symmetric raised cosine pulse that is given by

$$\mathcal{C}_N(n) = \left[0.5 + 0.5 \cos\left(\frac{\pi n}{N}\right)\right] \mathcal{R}_N(n)$$

P3.7 A complex-valued sequence $x(n)$ can be decomposed into a conjugate symmetric part $x_e(n)$ and a conjugate-antisymmetric part $x_o(n)$ as discussed in Chapter 2. Show that

$$\mathcal{F}[x_e(n)] = X_R(e^{j\omega}) \qquad \text{and} \qquad \mathcal{F}[x_o(n)] = jX_I(e^{j\omega})$$

where $X_R(e^{j\omega})$ and $X_I(e^{j\omega})$ are the real and imaginary parts of the DTFT $X(e^{j\omega})$, respectively. Verify this property on

$$x(n) = e^{j0.1\pi n} [u(n) - u(n - 20)]$$

using the MATLAB functions developed in Chapter 2.

P3.8 A complex-valued DTFT $X(e^{j\omega})$ can also be decomposed into its conjugate-symmetric part $X_e(e^{j\omega})$ and conjugate-antisymmetric part $X_o(e^{j\omega})$; that is,

$$X(e^{j\omega}) = X_e(e^{j\omega}) + X_o(e^{j\omega})$$

where

$$X_e\left(e^{j\omega}\right) = \tfrac{1}{2}\left[X\left(e^{j\omega}\right) + X^*\left(e^{-j\omega}\right)\right] \qquad \text{and} \qquad X_0\left(e^{j\omega}\right) = \tfrac{1}{2}\left[X\left(e^{j\omega}\right) - X^*\left(e^{-j\omega}\right)\right]$$

Show that

$$\mathcal{F}^{-1}\left[X_e\left(e^{j\omega}\right)\right] = x_R\left(n\right) \qquad \text{and} \qquad \mathcal{F}^{-1}\left[X_0\left(e^{j\omega}\right)\right] = x_I\left(n\right)$$

where $x_R\left(n\right)$ and $x_I\left(n\right)$ are the real and imaginary parts of $x\left(n\right)$. Verify this property on

$$x\left(n\right) = e^{j0.1\pi n}\left[u\left(n\right) - u\left(n - 20\right)\right]$$

using the MATLAB functions developed in Chapter 2.

P3.9 Using the frequency-shifting property, show that the DTFT of a sinusoidal pulse

$$x\left(n\right) = \left(\cos\omega_o n\right)\mathcal{R}_N\left(n\right)$$

is given by

$$X\left(e^{j\omega}\right) = \frac{1}{2}\left[\frac{\sin\left\{\left(\omega - \omega_0\right)N/2\right\}}{\sin\left\{\left(\omega - \omega_0\right)/2\right\}}\right] + \frac{1}{2}\left[\frac{\sin\left\{\left(\omega + \omega_0\right)N/2\right\}}{\sin\left\{\left(\omega + \omega_0\right)/2\right\}}\right]$$

where $\mathcal{R}_N\left(n\right)$ is the rectangular pulse given in Problem 3.4. Compute and plot $X\left(e^{j\omega}\right)$ for $\omega_o = \pi/2$ and $N = 5, 15, 25, 100$. Use the plotting interval $[-\pi, \pi]$. Comment on your results.

P3.10 Let $x\left(n\right) = T_{10}\left(n\right)$ be a triangular pulse given in Problem 3.5. Using properties of the DTFT, determine and plot the DTFT of the following sequences.

a. $x\left(n\right) = T_{10}\left(-n\right)$

b. $x\left(n\right) = T_{10}\left(n\right) - T_{10}\left(n - 10\right)$

c. $x\left(n\right) = T_{10}\left(n\right) * T_{10}\left(-n\right)$

d. $x\left(n\right) = T_{10}\left(n\right)e^{j\pi n}$

e. $x\left(n\right) = T_{10}\left(n\right) \cdot T_{10}\left(n\right)$

P3.11 For each of the linear time-invariant systems described by the impulse response, determine the frequency response function $H\left(e^{j\omega}\right)$ and plot the magnitude response $\left|H\left(e^{j\omega}\right)\right|$ and the phase response $\angle H\left(e^{j\omega}\right)$.

a. $h\left(n\right) = \left(0.9\right)^{|n|}$

b. $h\left(n\right) = \text{sinc}\left(0.2n\right)\left[u\left(n + 20\right) - u\left(n - 20\right)\right]$, where $\text{sinc}\,0 = 1$.

c. $h\left(n\right) = \text{sinc}\left(0.2n\right)\left[u\left(n\right) - u\left(n - 40\right)\right]$

d. $h\left(n\right) = \left[\left(0.5\right)^n + \left(0.4\right)^n\right]u\left(n\right)$

e. $h\left(n\right) = \left(0.5\right)^{|n|}\cos\left(0.1\pi n\right)$

P3.12 Let $x\left(n\right) = 3\cos\left(0.5\pi n + 60°\right) + 2\sin\left(0.3\pi n\right)$ be the input to each of the systems described in Problem 3.11. In each case determine the output $y\left(n\right)$.

P3.13 An ideal lowpass filter is described in the frequency domain by

$$H_d\left(e^{j\omega}\right) = \begin{cases} 1 \cdot e^{-j\alpha\omega}, & |\omega| \leq \omega_c \\ 0, & \omega_c < |\omega| \leq \pi \end{cases}$$

where ω_c is called the cutoff frequency and α is called the phase delay.

Chapter 3 ■ THE DISCRETE-TIME FOURIER ANALYSIS

a. Determine the ideal impulse response $h_d(n)$ using the IDTFT relation (3.2).

b. Determine and plot the truncated impulse response

$$h(n) = \begin{cases} h_d(n), & 0 \le n \le N-1 \\ 0, & \text{otherwise} \end{cases}$$

for $N = 41$, $\alpha = 20$, and $\omega_c = 0.5\pi$.

c. Determine and plot the frequency response function $H\left(e^{j\omega}\right)$ and compare it with the ideal lowpass filter response $H_d\left(e^{j\omega}\right)$. Comment on your observations.

P3.14 An ideal highpass filter is described in the frequency domain by

$$H_d\left(e^{j\omega}\right) = \begin{cases} 1 \cdot e^{-j\alpha\omega}, & \omega_c < |\omega| \le \pi \\ 0, & |\omega| \le \omega_c \end{cases}$$

where ω_c is called the cutoff frequency and α is called the phase delay.

a. Determine the ideal impulse response $h_d(n)$ using the IDTFT relation (3.2).

b. Determine and plot the truncated impulse response

$$h(n) = \begin{cases} h_d(n), & 0 \le n \le N-1 \\ 0, & \text{otherwise} \end{cases}$$

for $N = 31$, $\alpha = 15$, and $\omega_c = 0.5\pi$.

c. Determine and plot the frequency response function $H\left(e^{j\omega}\right)$ and compare it with the ideal highpass filter response $H_d\left(e^{j\omega}\right)$. Comment on your observations.

P3.15 For a linear time-invariant system described by the difference equation

$$y(n) = \sum_{m=0}^{M} b_m x(n-m) - \sum_{\ell=1}^{N} a_\ell y(n-\ell)$$

the frequency response function is given by

$$H\left(e^{j\omega}\right) = \frac{\sum_{m=0}^{M} b_m e^{-j\omega m}}{1 + \sum_{\ell=1}^{N} a_\ell e^{-j\omega\ell}}$$

Write a MATLAB function freqresp to implement the above relation. The format of this function should be

```
function [H] = freqresp(b,a,w)
% Frequency response function from difference equation
%   [H] = freqresp(b,a,w)
% H = frequency response array evaluated at w frequencies
% b = numerator coefficient array
% a = denominator coefficient array (a(1)=1)
% w = frequency location array
```

P3.16 Determine $H\left(e^{j\omega}\right)$ and plot its magnitude and phase for each of the following systems.

a. $y(n) = \sum_{m=0}^{6} x(n-m)$

b. $y(n) = x(n) + 2x(n-1) + x(n-2) - 0.5y(n-1) - 0.25y(n-2)$

---

c. $y(n) = 2x(n) + x(n-1) - 0.25y(n-1) + 0.25y(n-2)$

d. $y(n) = x(n) + x(n-2) - 0.81y(n-2)$

e. $y(n) = x(n) - \sum_{\ell=1}^{5}(0.5)^{\ell} y(n-\ell)$

P3.17 A linear time-invariant system is described by the difference equation

$$y(n) = \sum_{m=0}^{3} x(n-2m) - \sum_{\ell=1}^{3}(0.81)^{\ell} y(n-2\ell)$$

Determine the steady-state response of the system to the following inputs:

a. $x(n) = 5 + 10(-1)^n$

b. $x(n) = 1 + \cos(0.5\pi n + \pi/2)$

c. $x(n) = 2\sin(\pi n/4) + 3\cos(3\pi n/4)$

d. $x(n) = \sum_{k=0}^{5}(k+1)\cos(\pi k n/4)$

e. $x(n) = \cos(\pi n)$

In each case generate $x(n)$, $0 \le n \le 200$ and process it through the `filter` function to obtain $y(n)$. Compare your $y(n)$ with the steady-state responses in each case.

P3.18 An analog signal $x_a(t) = \sin(1000\pi t)$ is sampled using the following sampling intervals. In each case plot the spectrum of the resulting discrete-time signal.

a. $T_s = 0.1$ ms

b. $T_s = 1$ ms

c. $T_s = 0.01$ sec

P3.19 We have the following analog filter, which is realized using a discrete filter.

$$x_a(t) \longrightarrow \boxed{\text{A/D}} \xrightarrow{x(n)} \boxed{h(n)} \xrightarrow{y(n)} \boxed{\text{D/A}} \longrightarrow y_a(t)$$

The sampling rate in the A/D and D/A is 100 sam/sec, and the impulse response is $h(n) = (0.5)^n u(n)$.

a. What is the digital frequency in $x(n)$ if $x_a(t) = 3\cos(20\pi t)$?

b. Find the steady-state output $y_a(t)$ if $x_a(t) = 3\cos(20\pi t)$.

c. Find the steady-state output $y_a(t)$ if $x_a(t) = 3u(t)$.

d. Find two other analog signals $x_a(t)$, with different analog frequencies, that will give the same steady-state output $y_a(t)$ when $x_a(t) = 3\cos(20\pi t)$ is applied.

e. To prevent aliasing, a prefilter would be required to process $x_a(t)$ before it passes to the A/D converter. What type of filter should be used, and what should be the largest cutoff frequency that would work for the given configuration?

P3.20 Consider an analog signal $x_a(t) = \sin(20\pi t)$, $0 \le t \le 1$. It is sampled at $T_s = 0.01$, 0.05, and 0.1 sec intervals to obtain $x(n)$.

a. For each T_s plot $x(n)$.

b. Reconstruct the analog signal $y_a(t)$ from the samples $x(n)$ using the sinc interpolation (use $\Delta t = 0.001$) and determine the frequency in $y_a(t)$ from your plot. (Ignore the end effects.)

c. Reconstruct the analog signal $y_a(t)$ from the samples $x(n)$ using the cubic spline interpolation and determine the frequency in $y_a(t)$ from your plot. (Ignore the end effects.)

d. Comment on your results.

P3.21 Consider the analog signal $x_a(t) = \sin(20\pi t + \pi/4)$, $0 \le t \le 1$. It is sampled at $T_s = 0.05$ sec intervals to obtain $x(n)$.

a. Plot $x_a(t)$ and superimpose $x(n)$ on it using the `plot(n,x,'o')` function.

b. Reconstruct the analog signal $y_a(t)$ from the samples $x(n)$ using the sinc interpolation (use $\Delta t = 0.001$) and superimpose $x(n)$ on it.

c. Reconstruct the analog signal $y_a(t)$ from the samples $x(n)$ using the cubic spline interpolation and superimpose $x(n)$ on it.

d. You should observe that the resultant reconstruction in each case has the correct frequency but a different amplitude. Explain this observation. Comment on the role of phase of $x_a(t)$ on sampling and reconstruction of signals.

4

THE Z-TRANSFORM

In Chapter 3 we studied the discrete-time Fourier transform approach for representing discrete signals using complex exponential sequences. This representation clearly has advantages for LTI systems because it describes systems in the frequency domain using the frequency response function $H(e^{j\omega})$. The computation of the sinusoidal steady-state response is greatly facilitated by the use of $H(e^{j\omega})$. Furthermore, response to any arbitrary absolutely summable sequence $x(n)$ can easily be computed in the frequency domain by multiplying the transform $X(e^{j\omega})$ and the frequency response $H(e^{j\omega})$. However, there are *two* shortcomings to the Fourier transform approach. First, there are many useful signals in practice— such as $u(n)$ and $nu(n)$—for which the discrete-time Fourier transform does not exist. Second, the transient response of a system due to initial conditions or due to changing inputs cannot be computed using the discrete-time Fourier transform approach.

Therefore we now consider an extension of the discrete-time Fourier transform to address the above two problems. This extension is called the z-transform. Its bilateral (or two-sided) version provides another domain in which a larger class of sequences and systems can be analyzed, while its unilateral (or one-sided) version can be used to obtain system responses with initial conditions or changing inputs.

THE BILATERAL z-TRANSFORM

The z-transform of a sequence $x(n)$ is given by

$$X(z) \triangleq \mathcal{Z}\left[x(n)\right] = \sum_{n=-\infty}^{\infty} x(n)z^{-n} \qquad (4.1)$$

where z is a complex variable. The set of z values for which $X(z)$ exists is called the *region of convergence (ROC)* and is given by

$$R_{x-} < |z| < R_{x+} \qquad (4.2)$$

for some positive numbers R_{x-} and R_{x+}.

The inverse z-transform of a complex function $X(z)$ is given by

$$x(n) \triangleq \mathcal{Z}^{-1}[X(z)] = \frac{1}{2\pi j} \oint_C X(z) z^{n-1} dz \qquad (4.3)$$

where C is a **counterclockwise contour** encircling the origin and lying in the ROC.

Comments:

1. The complex variable z is called the *complex frequency* given by $z = |z| e^{j\omega}$, where $|z|$ is the attenuation and ω is the real frequency.

2. Since the ROC (4.2) is defined in terms of the magnitude $|z|$, the shape of the ROC is an open ring as shown in Figure 4.1. Note that R_{x-} may be equal to zero and/or R_{x+} could possibly be ∞.

3. If $R_{x+} < R_{x-}$, then the ROC is a *null space* and the z-transform *does not exist*.

4. The function $|z| = 1$ (or $z = e^{j\omega}$) is a circle of unit radius in the z-plane and is called the *unit circle*. If the ROC contains the unit circle, then we can evaluate $X(z)$ on the unit circle.

$$X(z)|_{z=e^{j\omega}} = X(e^{j\omega}) = \sum_{n=-\infty}^{\infty} x(n) e^{-j\omega} = \mathcal{F}[x(n)]$$

Therefore the discrete-time Fourier transform $X(e^{j\omega})$ may be viewed as a special case of the z-transform $X(z)$.

☐ **EXAMPLE 4.1** Let $x_1(n) = a^n u(n)$, $\quad 0 < |a| < \infty$. (This sequence is called a *positive-time* sequence). Then

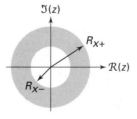

FIGURE 4.1 *A general region of convergence*

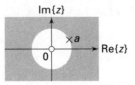

FIGURE 4.2 *The ROC in Example 4.1*

$$X_1(z) = \sum_0^\infty a^n z^{-n} = \sum_0^\infty \left(\frac{a}{z}\right)^n = \frac{1}{1 - az^{-1}}; \text{ if } \left|\frac{a}{z}\right| < 1$$

$$= \frac{z}{z - a}, \quad |z| > |a| \Rightarrow \text{ROC}_1: \underbrace{|a|}_{R_{x-}} < |z| < \underbrace{\infty}_{R_{x+}}$$

Note: $X_1(z)$ in this example is a rational function; that is,

$$X_1(z) \triangleq \frac{B(z)}{A(z)} = \frac{z}{z - a}$$

where $B(z) = z$ is the *numerator polynomial* and $A(z) = z-a$ is the *denominator polynomial*. The roots of $B(z)$ are called the *zeros* of $X(z)$, while the roots of $A(z)$ are called the *poles* of $X(z)$. In this example $X_1(z)$ has a zero at the origin $z = 0$ and a pole at $z = a$. Hence $x_1(n)$ can also be represented by a *pole-zero diagram* in the z-plane in which zeros are denoted by 'o' and poles by '×' as shown in Figure 4.2. □

□   **EXAMPLE 4.2**   Let $x_2(n) = -b^n u(-n-1), 0 < |b| < \infty$. (This sequence is called a *negative-time sequence*.) Then

$$X_2(z) = -\sum_{-\infty}^{-1} b^n z^{-n} = -\sum_{-\infty}^{-1} \left(\frac{b}{z}\right)^n$$

$$= -\sum_1^\infty \left(\frac{z}{b}\right)^n = 1 - \sum_0^\infty \left(\frac{z}{b}\right)^n$$

$$= 1 - \frac{1}{1 - z/b} = \frac{z}{z - b}, \quad \text{ROC}_2: \underbrace{0}_{R_{x-}} < |z| < \underbrace{|b|}_{R_{x+}}$$

The ROC$_2$ and the pole-zero plot for this $x_2(n)$ are shown in Figure 4.3.

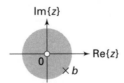

FIGURE 4.3 *The ROC in Example 4.2*

Chapter 4 ■ THE z-TRANSFORM

Note: If $b = a$ in this example, then $X_2(z) = X_1(z)$ except for their respective ROCs; that is, $\text{ROC}_1 \neq \text{ROC}_2$. This implies that the ROC is a distinguishing feature that guarantees the uniqueness of the z-transform. Hence it plays a very important role in system analysis. □

□ **EXAMPLE 4.3**   Let $x_3(n) = x_1(n) + x_2(n) = a^n u(n) - b^n u(-n-1)$ (This sequence is called a *two-sided sequence.*) Then using the above two examples,

$$X_3(z) = \sum_{n=0}^{\infty} a^n z^{-n} - \sum_{-\infty}^{-1} b^n z^{-n}$$

$$= \left\{ \frac{z}{z-a}, \text{ROC}_1: |z| > |a| \right\} + \left\{ \frac{z}{z-b}, \text{ROC}_1: |z| < |b| \right\}$$

$$= \frac{z}{z-a} + \frac{z}{z-b}; \quad \text{ROC}_3: \text{ROC}_1 \cap \text{ROC}_2$$

If $|b| < |a|$, the ROC_3 is a null space and $X_3(z)$ does not exist. If $|a| < |b|$, then the ROC_3 is $|a| < |z| < |b|$ and $X_3(z)$ exists in this region as shown in Figure 4.4. □

PROPERTIES
OF THE ROC

From the observation of the ROCs in the above three examples, we state the following properties.

1. The ROC is **always bounded by a circle** since the convergence condition is on the magnitude $|z|$.

2. The sequence $x_1(n) = a^n u(n)$ in Example 4.1 is a special case of a *right-sided sequence*, defined as a sequence $x(n)$ that is zero for some $n < n_0$. From Example 4.1 the ROC for right-sided sequences is **always outside of a circle of radius R_{x-}**. If $n_0 \geq 0$, then the right-sided sequence is also called a *causal* sequence.

3. The sequence $x_2(n) = -b^n u(-n-1)$ in Example 4.2 is a special case of a *left-sided* sequence, defined as a sequence $x(n)$ that is zero for some $n > n_0$. If $n_0 \leq 0$, the resulting sequence is called an *anticausal* sequence. From Example 4.2 the ROC for left-sided sequences is **always inside of a circle of radius R_{x+}**.

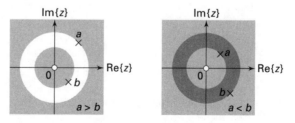

FIGURE 4.4   *The ROC in Example 4.3*

4. The sequence $x_3(n)$ in Example 4.3 is a two-sided sequence. The ROC for two-sided sequences is **always an open ring** $\boldsymbol{R_{x-} < |z| < R_{x+}}$ if it exists.

5. The sequences that are zero for $n < n_1$ and $n > n_2$ are called *finite-duration sequences*. The ROC for such sequences is **the entire z-plane**. If $n_1 < 0$, then $z = \infty$ is not in the ROC. If $n_2 > 0$, then $z = 0$ is not in the ROC.

6. The ROC cannot include a pole since $X(z)$ converges uniformly in there.

7. There is at least one pole on the boundary of a ROC of a rational $X(z)$.

8. The ROC is one contiguous region; that is, the ROC does not come in pieces.

In digital signal processing, signals are assumed to be causal since almost every digital data is acquired in real time. Therefore the only ROC of interest to us is the one given in 2 above.

IMPORTANT PROPERTIES OF THE z-TRANSFORM

The properties of the z-transform are generalizations of the properties of the discrete-time Fourier transform that we studied in Chapter 3. We state the following important properties of the z-transform without proof.

1. **Linearity:**

$$\mathcal{Z}\left[a_1 x_1(n) + a_2 x_2(n)\right] = a_1 X_1(z) + a_2 X_2(z); \quad \text{ROC: ROC}_{x_1} \cap \text{ROC}_{x_2}$$

$$\text{(4.4)}$$

2. **Sample shifting:**

$$\mathcal{Z}\left[x(n - n_0)\right] = z^{-n_0} X(z); \quad \text{ROC: ROC}_x \qquad \text{(4.5)}$$

3. **Frequency shifting:**

$$\mathcal{Z}\left[a^n x(n)\right] = X\left(\frac{z}{a}\right); \quad \text{ROC: ROC}_x \text{ scaled by } |a| \qquad \text{(4.6)}$$

4. **Folding:**

$$\mathcal{Z}\left[x(-n)\right] = X(1/z); \quad \text{ROC: Inverted ROC}_x \qquad \text{(4.7)}$$

5. **Complex conjugation:**

$$\mathcal{Z}\left[x^*(n)\right] = X^*(z^*); \quad \text{ROC: ROC}_x \qquad \text{(4.8)}$$

6. **Differentiation in the z-domain:**

$$\mathcal{Z}\left[nx(n)\right] = -z\frac{dX(z)}{dz}; \quad \text{ROC: ROC}_x \tag{4.9}$$

This property is also called "multiplication by a ramp" property.

7. **Multiplication:**

$$\mathcal{Z}\left[x_1(n)x_2(n)\right] = \frac{1}{2\pi j}\oint_C X_1(\nu)\,X_2(z/\nu)\,\nu^{-1}d\nu; \tag{4.10}$$

$$\text{ROC: ROC}_{x_1}\cap\text{Inverted ROC}_{x_2}$$

where C is a closed contour that encloses the origin and lies in the common ROC.

8. **Convolution:**

$$\mathcal{Z}\left[x_1(n)*x_2(n)\right] = X_1(z)X_2(z); \quad \text{ROC: ROC}_{x_1}\cap\text{ROC}_{x_2} \tag{4.11}$$

This last property transforms the time-domain convolution operation into a multiplication between two functions. It is a significant property in many ways. First, if $X_1(z)$ and $X_2(z)$ are two polynomials, then their product can be implemented using the **conv** function in MATLAB.

☐ **EXAMPLE 4.4**   Let $X_1(z) = 2 + 3z^{-1} + 4z^{-2}$ and $X_2(z) = 3 + 4z^{-1} + 5z^{-2} + 6z^{-3}$. Determine $X_3(z) = X_1(z)X_2(z)$.

Solution   From the definition of the z-transform we observe that

$$x_1(n) = \{2,3,4\} \qquad \text{and} \qquad x_2(n) = \{3,4,5,6\}$$
$$\quad\uparrow \qquad\qquad\qquad\qquad\qquad \uparrow$$

Then the convolution of the above two sequences will give the coefficients of the required polynomial product.

```
>> x1 = [2,3,4]; x2 = [3,4,5,6];
>> x3 = conv(x1,x2)
x3 =     6    17    34    43    38    24
```

Hence

$$X_3(z) = 6 + 17z^{-1} + 34z^{-2} + 43z^{-3} + 38z^{-4} + 24z^{-5}$$

Using the **conv_m** function developed in Chapter 2, we can also multiply two z-domain polynomials corresponding to noncausal sequences.   ☐

☐ **EXAMPLE 4.5**   Let $X_1(z) = z + 2 + 3z^{-1}$ and $X_2(z) = 2z^2 + 4z + 3 + 5z^{-1}$. Determine $X_3(z) = X_1(z)X_2(z)$.

Solution

Note that

$$x_1(n) = \{1, 2, 3\} \quad \text{and} \quad x_2(n) = \{2, 4, 3, 5\}$$

Using MATLAB,

```
>> x1 = [1,2,3]; n1 = [-1:1];
>> x2 = [2,4,3,5]; n2 = [-2:1];
>> [x3,n3] = conv_m(x1,n1,x2,n2)
x3 =
     2     8    17    23    19    15
n3 =
    -3    -2    -1     0     1     2
```

we have

$$X_3(z) = 2z^3 + 8z^2 + 17z + 23 + 19z^{-1} + 15z^{-2} \qquad \square$$

In passing we note that to divide one polynomial by another one, we would require an inverse operation called *deconvolution* [19, Chapter 6]. In MATLAB [p,r] = deconv(b,a) computes the result of dividing b by a in a polynomial part p and a remainder r. For example, if we divide the polynomial $X_3(z)$ in Example 4.4 by $X_1(z)$,

```
>> x3 = [6,17,34,43,38,24]; x1 = [2,3,4];
>> [x2,r] = deconv(x3,x1)
x2 =
     3     4     5     6
r =
     0     0     0     0     0     0
```

then we obtain the coefficients of the polynomial $X_2(z)$ as expected. To obtain the sample index, we will have to modify the **deconv** function as we did in the **conv_m** function. This is explored in Problem 4.8. This operation is useful in obtaining a *proper* rational part from an *improper* rational function.

The second important use of the convolution property is in system output computations as we shall see in a later section. This interpretation is particularly useful for verifying the z-transform expression $X(z)$ using MATLAB. Note that since MATLAB is a numerical processor (unless the Symbolic toolbox is used), it cannot be used for direct z-transform calculations. We will now elaborate on this. Let $x(n)$ be a sequence with a rational transform

$$X(z) = \frac{B(z)}{A(z)}$$

where $B(z)$ and $A(z)$ are polynomials in z^{-1}. If we use the coefficients of $B(z)$ and $A(z)$ as the **b** and **a** arrays in the `filter` routine and excite this filter by the impulse sequence $\delta(n)$, then from (4.11) and using $\mathcal{Z}[\delta(n)] = 1$, the output of the filter will be $x(n)$. (This is a numerical approach of computing the inverse z-transform; we will discuss the analytical approach in the next section.) We can compare this output with the given $x(n)$ to verify that $X(z)$ is indeed the transform of $x(n)$. This is illustrated in Example 4.6.

SOME COMMON z-TRANSFORM PAIRS

Using the definition of z-transform and its properties, one can determine z-transforms of common sequences. A list of some of these sequences is given in Table 4.1.

☐ **EXAMPLE 4.6** Using z-transform properties and the z-transform table, determine the z-transform of

$$x(n) = (n-2)(0.5)^{(n-2)} \cos\left[\frac{\pi}{3}(n-2)\right] u(n-2)$$

TABLE 4.1 *Some common z-transform pairs*

Sequence	Transform	ROC
$\delta(n)$	1	$\forall z$
$u(n)$	$\dfrac{1}{1-z^{-1}}$	$\lvert z \rvert > 1$
$-u(-n-1)$	$\dfrac{1}{1-z^{-1}}$	$\lvert z \rvert < 1$
$a^n u(n)$	$\dfrac{1}{1-az^{-1}}$	$\lvert z \rvert > \lvert a \rvert$
$-b^n u(-n-1)$	$\dfrac{1}{1-bz^{-1}}$	$\lvert z \rvert < \lvert b \rvert$
$[a^n \sin\omega_0 n]\, u(n)$	$\dfrac{(a\sin\omega_0)z^{-1}}{1-(2a\cos\omega_0)z^{-1}+a^2 z^{-2}}$	$\lvert z \rvert > \lvert a \rvert$
$[a^n \cos\omega_0 n]\, u(n)$	$\dfrac{1-(a\cos\omega_0)z^{-1}}{1-(2a\cos\omega_0)z^{-1}+a^2 z^{-2}}$	$\lvert z \rvert > \lvert a \rvert$
$na^n u(n)$	$\dfrac{az^{-1}}{(1-az^{-1})^2}$	$\lvert z \rvert > \lvert a \rvert$
$-nb^n u(-n-1)$	$\dfrac{bz^{-1}}{(1-bz^{-1})^2}$	$\lvert z \rvert < \lvert b \rvert$

Solution

Applying the sample-shift property,

$$X(z) = \mathcal{Z}[x(n)] = z^{-2}\mathcal{Z}\left[n(0.5)^n \cos\left(\frac{\pi n}{3}\right)u(n)\right]$$

with no change in the ROC. Applying the multiplication by a ramp property,

$$X(z) = z^{-2}\left\{-z\frac{d\mathcal{Z}\left[(0.5)^n \cos\left(\frac{\pi}{3}n\right)u(n)\right]}{dz}\right\}$$

with no change in the ROC. Now the z-transform of $(0.5)^n \cos\left(\frac{\pi}{3}n\right)u(n)$ from Table 4.1 is

$$\mathcal{Z}\left[(0.5)^n \cos\left(\frac{\pi n}{3}\right)u(n)\right] = \frac{1 - \left(0.5\cos\frac{\pi}{3}\right)z^{-1}}{1 - 2\left(0.5\cos\frac{\pi}{3}\right)z^{-1} + 0.25z^{-2}}; \quad |z| > 0.5$$

$$= \frac{1 - 0.25z^{-1}}{1 - 0.5z^{-1} + 0.25z^{-2}}; \quad |z| > 0.5$$

Hence

$$X(z) = -z^{-1}\frac{d}{dz}\left\{\frac{1 - 0.25z^{-1}}{1 - 0.5z^{-1} + 0.25z^{-2}}\right\}, \qquad\qquad |z| > 0.5$$

$$= -z^{-1}\left\{\frac{-0.25z^{-2} + 0.5z^{-3} - 0.0625z^{-4}}{1 - z^{-1} + 0.75z^{-2} - 0.25z^{-3} + 0.0625z^{-4}}\right\}, \qquad |z| > 0.5$$

$$= \frac{0.25z^{-3} - 0.5z^{-4} + 0.0625z^{-5}}{1 - z^{-1} + 0.75z^{-2} - 0.25z^{-3} + 0.0625z^{-4}}, \qquad |z| > 0.5$$

MATLAB verification: To check that the above $X(z)$ is indeed the correct expression, let us compute the first 8 samples of the sequence $x(n)$ corresponding to $X(z)$ as discussed before.

```
>> b = [0,0,0,0.25,-0.5,0.0625]; a = [1,-1,0.75,-0.25,0.0625];
>> [delta,n]=impseq(0,0,7)
delta =
     1     0     0     0     0     0     0     0
n =
     0     1     2     3     4     5     6     7
>> x = filter(b,a,delta) % check sequence
x =
  Columns 1 through 4
            0                  0                  0   0.25000000000000
  Columns 5 through 8
  -0.25000000000000  -0.37500000000000  -0.12500000000000   0.07812500000000
>> x = [(n-2).*(1/2).^(n-2).*cos(pi*(n-2)/3)].*stepseq(2,0,7) % original sequence
x =
  Columns 1 through 4
            0                  0                  0   0.25000000000000
```

```
Columns 5 through 8
-0.25000000000000   -0.37500000000000   -0.12500000000000    0.07812500000000
```

This approach can be used to verify the z-transform computations.　　□

INVERSION OF THE z-TRANSFORM

From definition (4.3) the inverse z-transform computation requires an evaluation of a complex contour integral that, in general, is a complicated procedure. The most practical approach is to use the partial fraction expansion method. It makes use of the z-transform Table 4.1 (or similar tables available in many textbooks.) The z-transform, however, must be a rational function. This requirement is generally satisfied in digital signal processing.

Central Idea:　When $X(z)$ is a rational function of z^{-1}, it can be expressed as a sum of simple (first-order) factors using the partial fraction expansion. The individual sequences corresponding to these factors can then be written down using the z-transform table.

The inverse z-transform procedure can be summarized as follows:

Method:　Given

$$X(z) = \frac{b_0 + b_1 z^{-1} + \cdots + b_M z^{-M}}{1 + a_1 z^{-1} + \cdots + a_N z^{-N}}, \ R_{x-} < |z| < R_{x+} \qquad \textbf{(4.12)}$$

- express it as

$$X(z) = \underbrace{\frac{\tilde{b}_0 + \tilde{b}_1 z^{-1} + \cdots + \tilde{b}_{N-1} z^{-(N-1)}}{1 + a_1 z^{-1} + \cdots + a_N z^{-N}}}_{\text{Proper rational part}} + \underbrace{\sum_{k=0}^{M-N} C_k z^{-k}}_{\text{polynomial part if } M \geq N}$$

where the first term on the right-hand side is the proper rational part and the second term is the polynomial (finite-length) part. This can be obtained by performing polynomial division if $M \geq N$ using the `deconv` function.

- perform a partial fraction expansion on the proper rational part of $X(z)$ to obtain

$$X(z) = \sum_{k=1}^{N} \frac{R_k}{1 - p_k z^{-1}} + \underbrace{\sum_{k=0}^{M-N} C_k z^{-k}}_{M \geq N} \qquad \textbf{(4.13)}$$

where p_k is the kth pole of $X(z)$ and R_k is the residue at p_k. It is assumed that the poles are distinct for which the residues are given by

$$R_k = \frac{\tilde{b}_0 + \tilde{b}_1 z^{-1} + \cdots + \tilde{b}_{N-1} z^{-(N-1)}}{1 + a_1 z^{-1} + \cdots + a_N z^{-N}} \left(1 - p_k z^{-1}\right) \Bigg|_{z=p_k}$$

For repeated poles the expansion (4.13) has a more general form. If a pole p_k has multiplicity r, then its expansion is given by

$$\sum_{\ell=1}^{r} \frac{R_{k,\ell} z^{-(\ell-1)}}{(1 - p_k z^{-1})^\ell} = \frac{R_{k,1}}{1 - p_k z^{-1}} + \frac{R_{k,2} z^{-1}}{(1 - p_k z^{-1})^2} + \cdots + \frac{R_{k,r} z^{-(r-1)}}{(1 - p_k z^{-1})^r}$$

(4.14)

where the residues $R_{k,\ell}$ are computed using a more general formula, which is available in [19].
  • write $x(n)$ as

$$x(n) = \sum_{k=1}^{N} R_k \mathcal{Z}^{-1} \left[\frac{1}{1 - p_k z^{-1}} \right] + \underbrace{\sum_{k=0}^{M-N} C_k \delta(n-k)}_{M \geq N}$$

  • finally, use the relation from Table 4.1

$$\mathcal{Z}^{-1} \left[\frac{z}{z - p_k} \right] = \begin{cases} p_k^n u(n) & |z_k| \leq R_{x-} \\ -p_k^n u(-n-1) & |z_k| \geq R_{x+} \end{cases}$$

(4.15)

to complete $x(n)$.

□   **EXAMPLE 4.7**   Find the inverse z-transform of $X(z) = \dfrac{z}{3z^2 - 4z + 1}$.

Solution   Write

$$X(z) = \frac{z}{3(z^2 - \frac{4}{3}z + \frac{1}{3})} = \frac{\frac{1}{3}z^{-1}}{1 - \frac{4}{3}z^{-1} + \frac{1}{3}z^{-2}}$$

$$= \frac{\frac{1}{3}z^{-1}}{(1 - z^{-1})(1 - \frac{1}{3}z^{-1})} = \frac{\frac{1}{2}}{1 - z^{-1}} - \frac{\frac{1}{2}}{1 - \frac{1}{3}z^{-1}}$$

or

$$X(z) = \frac{1}{2} \left(\frac{1}{1 - z^{-1}} \right) - \frac{1}{2} \left(\frac{1}{1 - \frac{1}{3}z^{-1}} \right)$$

Now, $X(z)$ has two poles: $z_1 = 1$ and $z_2 = \frac{1}{3}$; and since the ROC is not specified, there are *three* possible ROCs as shown in Figure 4.5.

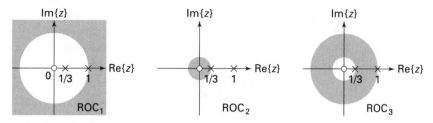

FIGURE 4.5 *The ROCs in Example 4.7*

a. ROC_1: $1 < |z| < \infty$. Here both poles are on the interior side of the ROC_1; that is, $|z_1| \leq R_{x-} = 1$ and $|z_2| \leq 1$. Hence from (4.15)

$$x_1(n) = \frac{1}{2}u(n) - \frac{1}{2}\left(\frac{1}{3}\right)^n u(n)$$

which is a right-sided sequence.

b. ROC_2: $0 < |z| < \frac{1}{3}$. Here both poles are on the exterior side of the ROC_2; that is, $|z_1| \geq R_{x+} = \frac{1}{3}$ and $|z_2| \geq \frac{1}{3}$. Hence from (4.15)

$$x_2(n) = \frac{1}{2}\left\{-u(-n-1)\right\} - \frac{1}{2}\left\{-\left(\frac{1}{3}\right)^n u(-n-1)\right\}$$

$$= \frac{1}{2}\left(\frac{1}{3}\right)^n u(-n-1) - \frac{1}{2}u(-n-1)$$

which is a left-sided sequence.

c. ROC_3: $\frac{1}{3} < |z| < 1$. Here pole z_1 is on the exterior side of the ROC_3—that is, $|z_1| \geq R_{x+} = 1$—while pole z_2 is on the interior side—that is, $|z_2| \leq \frac{1}{3}$. Hence from (4.15)

$$x_3(n) = -\frac{1}{2}u(-n-1) - \frac{1}{2}\left(\frac{1}{3}\right)^n u(n)$$

which is a two-sided sequence. □

MATLAB
IMPLEMEN-
TATION

A MATLAB function **residuez** is available to compute the residue part and the direct (or polynomial) terms of a rational function in z^{-1}. Let

$$X(z) = \frac{b_0 + b_1 z^{-1} + \cdots + b_M z^{-M}}{a_0 + a_1 z^{-1} + \cdots + a_N z^{-N}} = \frac{B(z)}{A(z)}$$

$$= \sum_{k=1}^{N} \frac{R_k}{1 - p_k z^{-1}} + \underbrace{\sum_{k=0}^{M-N} C_k z^{-k}}_{M \geq N}$$

be a rational function in which the numerator and the denominator polynomials are in *ascending* powers of z^{-1}. Then [R,p,C]=residuez(b,a)

---

finds the residues, poles, and direct terms of $X(z)$ in which two polynomials $B(z)$ and $A(z)$ are given in two vectors b and a, respectively. The returned column vector R contains the residues, column vector p contains the pole locations, and row vector C contains the direct terms. If p(k)=...=p(k+r-1) is a pole of multiplicity r, then the expansion includes the term of the form

$$\frac{R_k}{1 - p_k z^{-1}} + \frac{R_{k+1}}{(1 - p_k z^{-1})^2} + \cdots + \frac{R_{k+r-1}}{(1 - p_k z^{-1})^r} \qquad (4.16)$$

which is different from (4.14).

Similarly, [b,a]=residuez(R,p,C), with three input arguments and two output arguments, converts the partial fraction expansion back to polynomials with coefficients in row vectors b and a.

□   **EXAMPLE 4.8**   To check our residue functions, let us consider the rational function

$$X(z) = \frac{z}{3z^2 - 4z + 1}$$

given in Example 4.7.

Solution   First rearrange $X(z)$ so that it is a function in ascending powers of z^{-1}.

$$X(z) = \frac{z^{-1}}{3 - 4z^{-1} + z^{-2}} = \frac{0 + z^{-1}}{3 - 4z^{-1} + z^{-2}}$$

Now using MATLAB,

```
>> b = [0,1]; a = [3,-4,1];
>> [R,p,C] = residuez(b,a)
R =
     0.5000
    -0.5000
p =
     1.0000
     0.3333
C =
     []
```

we obtain

$$X(z) = \frac{\frac{1}{2}}{1 - z^{-1}} - \frac{\frac{1}{2}}{1 - \frac{1}{3}z^{-1}}$$

as before. Similarly, to convert back to the rational function form,

```
>> [b,a] = residuez(R,p,C)
b =
     0.0000
     0.3333
```

```
a =
    1.0000
   -1.3333
    0.3333
```

so that

$$X(z) = \frac{0 + \frac{1}{3}z^{-1}}{1 - \frac{4}{3}z^{-1} + \frac{1}{3}z^{-2}} = \frac{z^{-1}}{3 - 4z^{-1} + z^{-2}} = \frac{z}{3z^2 - 4z + 1}$$

as before. □

□ **EXAMPLE 4.9** Compute the inverse z-transform of

$$X(z) = \frac{1}{(1 - 0.9z^{-1})^2 (1 + 0.9z^{-1})}, \quad |z| > 0.9$$

Solution We can evaluate the denominator polynomial as well as the residues using MAT-LAB.

```
>> b = 1; a = poly([0.9,0.9,-0.9])
a =
    1.0000    -0.9000    -0.8100    0.7290
>> [R,p,C]=residuez(b,a)
R =
    0.2500
    0.5000
    0.2500
p =
    0.9000
    0.9000
   -0.9000
c =
    []
```

Note that the denominator polynomial is computed using MATLAB's polynomial function **poly**, which computes the polynomial coefficients, given its roots. We could have used the **conv** function, but the use of the **poly** function is more convenient for this purpose. From the residue calculations and using the order of residues given in (4.16), we have

$$X(z) = \frac{0.25}{1 - 0.9z^{-1}} + \frac{0.5}{(1 - 0.9z^{-1})^2} + \frac{0.25}{1 + 0.9z^{-1}}, \quad |z| > 0.9$$

$$= \frac{0.25}{1 - 0.9z^{-1}} + \frac{0.5}{0.9}z\frac{\left(0.9z^{-1}\right)}{(1 - 0.9z^{-1})^2} + \frac{0.25}{1 + 0.9z^{-1}}, \quad |z| > 0.9$$

Hence from Table 4.1 and using the z-transform property of time-shift,

$$x(n) = 0.25\,(0.9)^n\,u(n) + \frac{5}{9}\,(n+1)\,(0.9)^{n+1}\,u(n+1) + 0.25\,(-0.9)^n\,u(n)$$

which upon simplification becomes

$$x(n) = 0.75\,(0.9)^n\,u(n) + 0.5n\,(0.9)^n\,u(n) + 0.25\,(-0.9)^n\,u(n)$$

MATLAB verification:

```
>> [delta,n] = impseq(0,0,7);
>> x = filter(b,a,delta) % check sequence
x =
  Columns 1 through 4
    1.00000000000000    0.90000000000000    1.62000000000000    1.45800000000000
  Columns 5 through 8
    1.96830000000000    1.77147000000000    2.12576400000000    1.91318760000000
>> x = (0.75)*(0.9).^n + (0.5)*n.*(0.9).^n + (0.25)*(-0.9).^n % answer sequence
x =
  Columns 1 through 4
    1.00000000000000    0.90000000000000    1.62000000000000    1.45800000000000
  Columns 5 through 8
    1.96830000000000    1.77147000000000    2.12576400000000    1.91318760000000    □
```

□   **EXAMPLE 4.10**   Determine the inverse z-transform of

$$X(z) = \frac{1 + 0.4\sqrt{2}z^{-1}}{1 - 0.8\sqrt{2}z^{-1} + 0.64z^{-2}}$$

so that the resulting sequence is causal and contains no complex numbers.

Solution   We will have to find the poles of $X(z)$ in the polar form to determine the ROC of the causal sequence.

```
>> b = [1,0.4*sqrt(2)]; a=[1,-0.8*sqrt(2),0.64];
>> [R,p,C] = residuez(b,a)
R =
    0.5000 - 1.0000i
    0.5000 + 1.0000i
p =
    0.5657 + 0.5657i
    0.5657 - 0.5657i
C =
    []
>> Mp=abs(p')    % pole magnitudes
Mp =
    0.8000    0.8000
>> Ap=angle(p')/pi    % pole angles in pi units
Ap =
   -0.2500    0.2500
```

From the above calculations

$$X(z) = \frac{0.5 + j}{1 - |0.8| e^{-j\frac{\pi}{4}} z^{-1}} + \frac{0.5 - j}{1 - |0.8| e^{j\frac{\pi}{4}} z^{-1}}, \quad |z| > 0.8$$

and from Table 4.1 we have

$$x(n) = (0.5 + j) |0.8|^n e^{-j\frac{\pi}{4}n} u(n) + (0.5 - j) |0.8|^n e^{j\frac{\pi}{4}n} u(n)$$

$$= |0.8|^n \left[0.5 \left\{ e^{-j\frac{\pi}{4}n} + e^{j\frac{\pi}{4}n} \right\} + j \left\{ e^{-j\frac{\pi}{4}n} + e^{j\frac{\pi}{4}n} \right\} \right] u(n)$$

$$= |0.8|^n \left[\cos\left(\frac{\pi n}{4} \right) + 2 \sin\left(\frac{\pi n}{4} \right) \right] u(n)$$

MATLAB verification:

```
>> [delta, n] = impseq(0,0,6);
>> x = filter(b,a,delta) % check sequence
x =
  Columns 1 through 4
    1.00000000000000    1.69705627484771    1.28000000000000    0.36203867196751
  Columns 5 through 8
   -0.40960000000000   -0.69511425017762   -0.52428800000000   -0.14829104003789
>> x = ((0.8).^n).*(cos(pi*n/4)+2*sin(pi*n/4))
x =
  Columns 1 through 4
    1.00000000000000    1.69705627484771    1.28000000000000    0.36203867196751
  Columns 5 through 8
   -0.40960000000000   -0.69511425017762   -0.52428800000000   -0.14829104003789
```

□

SYSTEM REPRESENTATION IN THE z-DOMAIN

Similar to the frequency response function $H(e^{j\omega})$, we can define the z-domain function, $H(z)$, called the *system function*. However, unlike $H(e^{j\omega})$, $H(z)$ exists for systems that may not be BIBO stable.

■ DEFINITION 1   **The System Function**

The system function $H(z)$ is given by

$$H(z) \triangleq \mathcal{Z}[h(n)] = \sum_{-\infty}^{\infty} h(n) z^{-n}; \quad R_{h-} < |z| < R_{h+} \tag{4.17}$$

Using the convolution property (4.11) of the z-transform, the output transform $Y(z)$ is given by

$$Y(z) = H(z) X(z) \quad : \text{ROC}_y = \text{ROC}_h \cap \text{ROC}_x \tag{4.18}$$

provided ROC$_x$ overlaps with ROC$_h$. Therefore a linear and time-invariant system can be represented in the z-domain by

$$X(z) \longrightarrow \boxed{H(z)} \longrightarrow Y(z) = H(z)\,X(z)$$

SYSTEM FUNCTION FROM THE DIFFERENCE EQUATION REPRESENTATION

When LTI systems are described by a difference equation

$$y(n) + \sum_{k=1}^{N} a_k y(n-k) = \sum_{\ell=0}^{M} b_\ell x(n-\ell) \tag{4.19}$$

the system function $H(z)$ can easily be computed. Taking the z-transform of both sides, and using properties of the z-transform,

$$Y(z) + \sum_{k=1}^{N} a_k z^{-k} Y(z) = \sum_{\ell=0}^{M} b_\ell z^{-\ell} X(z)$$

or

$$H(z) \triangleq \frac{Y(z)}{X(z)} = \frac{\displaystyle\sum_{\ell=0}^{M} b_\ell z^{-\ell}}{1 + \displaystyle\sum_{k=1}^{N} a_k z^{-k}} = \frac{B(z)}{A(z)} \tag{4.20}$$

$$= \frac{b_0 z^{-M} \left(z^M + \cdots + \frac{b_M}{b_0} \right)}{z^{-N} \left(z^N + \cdots + a_N \right)}$$

After factorization, we obtain

$$H(z) = b_0 \, z^{N-M} \, \frac{\displaystyle\prod_{\ell=1}^{N} (z - z_\ell)}{\displaystyle\prod_{k=1}^{N} (z - p_k)} \tag{4.21}$$

where z_ℓ's are the system zeros and p_k's are the system poles. Thus $H(z)$ (and hence an LTI system) can also be represented in the z-domain using a pole-zero plot. This fact is useful in designing simple filters by proper placement of poles and zeros.

To determine zeros and poles of a rational $H(z)$, we can use the MATLAB function **roots** on both the numerator and the denominator polynomials. (Its inverse function **poly** determines polynomial coefficients from its roots as we discussed in the previous section.) It is also possible to use MATLAB to plot these roots for a visual display of a pole-zero plot. The function **zplane(b,a)** plots poles and zeros, given the numerator *row* vector **b** and the denominator *row* vector **a**. As before, the symbol "o" represents a zero and the symbol "**x**" represents a pole. The plot includes the unit circle for reference. Similarly, `zplane(z,p)` plots the zeros in *column* vector **z** and the poles in *column* vector **p**. Note very carefully the form of the input arguments for the proper use of this function.

TRANSFER FUNCTION REPRESENTATION

If the ROC of $H(z)$ includes a unit circle $(z = e^{j\omega})$, then we can evaluate $H(z)$ on the unit circle, resulting in a frequency response function or transfer function $H(e^{j\omega})$. Then from (4.21)

$$H(e^{j\omega}) = b_0 \, e^{j(N-M)\omega} \frac{\prod_1^M (e^{j\omega} - z_\ell)}{\prod_1^N (e^{j\omega} - p_k)} \tag{4.22}$$

The factor $(e^{j\omega} - z_\ell)$ can be interpreted as a *vector* in the complex z-plane from a zero z_ℓ to the unit circle at $z = e^{j\omega}$, while the factor $(e^{j\omega} - p_k)$ can be interpreted as a vector from a pole p_k to the unit circle at $z = e^{j\omega}$. This is shown in Figure 4.6. Hence the magnitude response function

$$\left| H(e^{j\omega}) \right| = |b_0| \frac{\left| e^{j\omega} - z_1 \right| \cdots \left| e^{j\omega} - z_M \right|}{\left| e^{j\omega} - p_1 \right| \cdots \left| e^{j\omega} - p_N \right|} \tag{4.23}$$

can be interpreted as a product of the lengths of vectors from zeros to the unit circle *divided* by the lengths of vectors from poles to the unit circle and *scaled* by $|b_0|$. Similarly, the phase response function

$$\angle H(e^{j\omega}) = \underbrace{[0 \text{ or } \pi]}_{\text{constant}} + \underbrace{[(N-M)\,\omega]}_{\text{linear}} + \underbrace{\sum_1^M \angle(e^{j\omega} - z_k) - \sum_1^N \angle(e^{j\omega} - p_k)}_{\text{nonlinear}}$$

$$\tag{4.24}$$

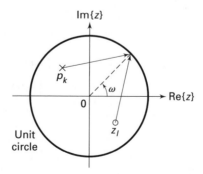

FIGURE 4.6 *Pole and zero vectors*

can be interpreted as a sum of a constant factor, a linear-phase factor, and a nonlinear-phase factor (angles from the "zero vectors" *minus* the sum of angles from the "pole vectors").

MATLAB
IMPLEMEN-
TATION

In Chapter 3 we plotted magnitude and phase responses in MATLAB by directly implementing their functional forms. MATLAB also provides a function called `freqz` for this computation, which uses the interpretation given above. In its simplest form this function is invoked by

```
[H,w] = freqz(b,a,N)
```

which returns the N-point frequency vector w and the N-point complex frequency response vector H of the system, given its numerator and denominator coefficients in vectors b and a. The frequency response is evaluated at N points equally spaced around the upper half of the unit circle. Note that the b and a vectors are the same vectors we use in the `filter` function or derived from the difference equation representation (4.19). The second form

```
[H,w] = freqz(b,a,N,'whole')
```

uses N points around the whole unit circle for computation. In yet another form

```
H = freqz(b,a,w)
```

it returns the frequency response at frequencies designated in vector w, normally between 0 and π.

☐ **EXAMPLE 4.11** Given a causal system

$$y(n) = 0.9y(n-1) + x(n)$$

a. Find $H(z)$ and sketch its pole-zero plot.
b. Plot $\left| H(e^{j\omega}) \right|$ and $\angle H(e^{j\omega})$.
c. Determine the impulse response $h(n)$.

Solution

The difference equation can be put in the form

$$y(n) - 0.9y(n-1) = x(n)$$

a. From (4.21)

$$H(z) = \frac{1}{1 - 0.9z^{-1}}; \quad |z| > 0.9$$

since the system is causal. There is one pole at 0.9 and one zero at the origin. We will use MATLAB to illustrate the use of the `zplane` function.

```
>> b = [1, 0]; a = [1, -0.9];
>> zplane(b,a)
```

Note that we specified b=[1,0] instead of b=1 because the zplane function assumes that scalars are zeros or poles. The resulting pole-zero plot is shown in Figure 4.7.

b. Using (4.23) and (4.24), we can determine the magnitude and phase of $H(e^{j\omega})$. Once again we will use MATLAB to illustrate the use of the freqz function. Using its first form, we will take 100 points along the upper half of the unit circle.

```
>> [H,w] = freqz(b,a,100);
>> magH = abs(H); phaH = angle(H);
>> subplot(2,1,1);plot(w/pi,magH);grid
>> xlabel('frequency in pi units'); ylabel('Magnitude');
>> title('Magnitude Response')
>> subplot(2,1,2);plot(w/pi,phaH/pi);grid
>> xlabel('frequency in pi units'); ylabel('Phase in pi units');
>> title('Phase Response')
```

The response plots are shown in Figure 4.8. If you study these plots carefully, you will observe that the plots are computed between $0 \le \omega \le 0.99\pi$ and fall short at $\omega = \pi$. This is due to the fact that in MATLAB the lower half of the

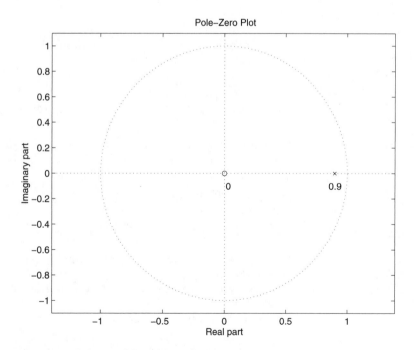

FIGURE 4.7 *Pole-zero plot of Example 4.11a*

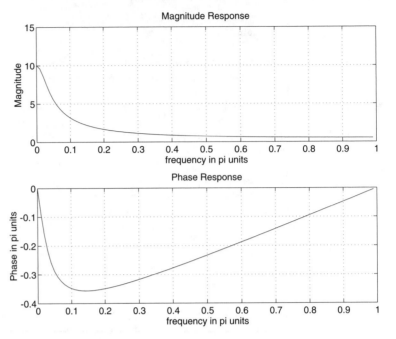

FIGURE 4.8 *Frequency response plots in Example 4.11*

unit circle begins at $\omega = \pi$. To overcome this problem, we will use the second form of the **freqz** function as follows.

```
>> [H,w] = freqz(b,a,200,'whole');
>> magH = abs(H(1:101)); phaH = angle(H(1:101));
```

Now the 101st element of the array H will correspond to $\omega = \pi$. A similar result can be obtained using the third form of the **freqz** function.

```
>> w = [0:1:100]*pi/100;
>> H = freqz(b,a,w);
>> magH = abs(H); phaH = angle(H);
```

In the future we will use any one of these forms, depending on our convenience. Also note that in the plots we divided the **w** and **phaH** arrays by **pi** so that the plot axes are in the units of π and easier to read. **This practice is strongly recommended.**

    c. From the z-transform Table 4.1

$$h(n) = \mathcal{Z}^{-1} \left[\frac{1}{1 - 0.9z^{-1}}, \; |z| > 0.9 \right] = (0.9)^n u(n) \qquad \qquad \Box$$

Given that

$$H(z) = \frac{z+1}{z^2 - 0.9z + 0.81}$$

is a causal system, find

  a. its transfer function representation,
  b. its difference equation representation, and
  c. its impulse response representation.

Solution

The poles of the system function are at $z = 0.9\angle \pm \pi/3$. Hence the ROC of the above causal system is $|z| > 0.9$. Therefore the unit circle is in the ROC, and the discrete-time Fourier transform $H(e^{j\omega})$ exists.

  a. Substituting $z = e^{j\omega}$ in $H(z)$,

$$H(e^{j\omega}) = \frac{e^{j\omega} + 1}{e^{j2\omega} - 0.9e^{j\omega} + 0.81} = \frac{e^{j\omega} + 1}{(e^{j\omega} - 0.9e^{j\pi/3})(e^{j\omega} - 0.9e^{-j\pi/3})}$$

  b. Using $H(z) = Y(z)/X(z)$,

$$\frac{Y(z)}{X(z)} = \frac{z+1}{z^2 - 0.9z + 0.81}\left(\frac{z^{-2}}{z^{-2}}\right) = \frac{z^{-1} + z^{-2}}{1 - 0.9z^{-1} + 0.81z^{-2}}$$

Cross multiplying,

$$Y(z) - 0.9z^{-1}Y(z) + 0.81z^{-2}Y(z) = z^{-1}X(z) + z^{-2}X(z)$$

Now taking the inverse z-transform,

$$y(n) - 0.9y(n-1) + 0.81y(n-2) = x(n-1) + x(n-2)$$

or

$$y(n) = 0.9y(n-1) - 0.81y(n-2) + x(n-1) + x(n-2)$$

  c. Using MATLAB,

```
>> b = [0,1,1]; a = [1,-0.9,0.81];
>> [R,p,C] = residuez(b,a)
R =
  -0.6173 + 0.9979i
  -0.6173 - 0.9979i
p =
   0.4500 - 0.7794i
   0.4500 + 0.7794i
C =
   1.2346
>> Mp = abs(p')
Mp =
   0.9000    0.9000
```

```
>> Ap = angle(p')/pi
Ap =
   -0.3333    0.3333
```

we have

$$H(z) = 1.2346 + \frac{-0.6173 + j0.9979}{1 - |0.9|\,e^{-j\pi/3}z^{-1}} + \frac{-0.6173 - j0.9979}{1 - |0.9|\,e^{j\pi/3}z^{-1}}, \quad |z| > 0.9$$

Hence from Table 4.1

$$h(n) = 1.2346\delta(n) + \left[(-0.6173 + j0.9979)\,|0.9|^n\,e^{-j\pi n/3}\right.$$

$$\left. + (-0.6173 - j0.9979)\,|0.9|^n\,e^{j\pi n/3}\right]u(n)$$

$$= 1.2346\delta(n) + |0.9|^n\left[-1.2346\cos(\pi n/3) + 1.9958\sin(\pi n/3)\right]u(n)$$

$$= |0.9|^n\left[-1.2346\cos(\pi n/3) + 1.9958\sin(\pi n/3)\right]u(n-1)$$

The last step results from the fact that $h(0) = 0$. □

RELATIONSHIPS BETWEEN SYSTEM REPRESENTATIONS

In this and the previous two chapters we developed several system representations. Figure 4.9 depicts the relationships between these representations in a graphical form.

STABILITY AND CAUSALITY

For LTI systems the BIBO stability is equivalent to $\sum_{-\infty}^{\infty}|h(k)| < \infty$. From the existence of the discrete-time Fourier transform this stability implies that $H(e^{j\omega})$ exists, which further implies that the unit circle $|z| = 1$ must be in the ROC of $H(z)$. This result is called the z-domain stability

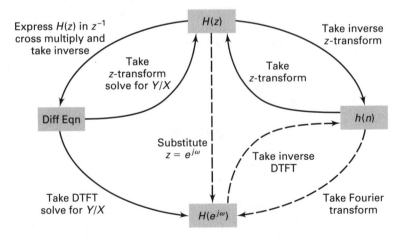

FIGURE 4.9 *System representations in pictorial form*

theorem; therefore the dashed paths in Figure 4.9 exist only if the system is stable.

■ **THEOREM 2** *z-Domain LTI Stability*
>An LTI system is stable if and only if the unit circle is in the ROC of $H(z)$.

For LTI causality we require that $h(n) = 0$, for $n < 0$ (i.e., a right-sided sequence). This implies that the ROC of $H(z)$ must be outside of some circle of radius R_{h-}. This is not a sufficient condition since any right-sided sequence has a similar ROC. However, when the system is stable, then its causality is easy to check.

■ **THEOREM 3** *z-Domain Causal LTI Stability*
>A causal LTI system is stable if and only if the system function $H(z)$ has all its poles inside the unit circle.

□ **EXAMPLE 4.13** A causal LTI system is described by the following difference equation:

$$y(n) = 0.81y(n-2) + x(n) - x(n-2)$$

Determine

    a. the system function $H(z)$,
    b. the unit impulse response $h(n)$,
    c. the unit step response $v(n)$, that is, the response to the unit step $u(n)$, and
    d. the frequency response function $H(e^{j\omega})$, and plot its magnitude and phase over $0 \le \omega \le \pi$.

Solution

Since the system is causal, the ROC will be outside of a circle with radius equal to the largest pole magnitude.

    a. Taking the z-transform of both sides of the difference equation and then solving for $Y(z)/X(z)$ or using (4.20), we obtain

$$H(z) = \frac{1 - z^{-2}}{1 - 0.81z^{-2}} = \frac{1 - z^{-2}}{(1 + 0.9z^{-1})(1 - 0.9z^{-1})}, \quad |z| > 0.9$$

    b. Using MATLAB for the partial fraction expansion,

```
>> b = [1,0,-1]; a = [1,0,-0.81];
>> [R,p,C] = residuez(b,a);
R =
  -0.1173
  -0.1173
p =
  -0.9000
```

$$C = \begin{matrix} 0.9000 \\ 1.2346 \end{matrix}$$

we have

$$H(z) = 1.2346 - 0.1173\frac{1}{1 + 0.9z^{-1}} - 0.1173\frac{1}{1 - 0.9z^{-1}}, \quad |z| > 0.9$$

or from Table 4.1

$$h(n) = 1.2346\delta(n) - 0.1173\left\{1 + (-1)^n\right\}(0.9)^n u(n)$$

c. From Table 4.1 $\mathcal{Z}\left[u(n)\right] = \dfrac{1}{1 - z^{-1}}$, $|z| > 1$. Hence

$$V(z) = H(z)U(z) = \left[\frac{\left(1 + z^{-1}\right)\left(1 - z^{-1}\right)}{\left(1 + 0.9z^{-1}\right)\left(1 - 0.9z^{-1}\right)}\right]\left[\frac{1}{1 - z^{-1}}\right], \quad |z| > 0.9 \cap |z| > 1$$

$$= \frac{1 + z^{-1}}{\left(1 + 0.9z^{-1}\right)\left(1 - 0.9z^{-1}\right)}, \quad |z| > 0.9$$

or

$$V(z) = 1.0556\frac{1}{1 - 0.9z^{-1}} - 0.0556\frac{1}{1 + 0.9z^{-1}}, \quad |z| > 0.9$$

Finally,

$$v(n) = \left[1.0556\,(0.9)^n - 0.0556\,(-0.9)^n\right]u(n)$$

Note that in the calculation of $V(z)$ there is a pole-zero cancellation at $z = 1$. This has two implications. First, the ROC of $V(z)$ is still $\{|z| > 0.9\}$ and not $\{|z| > 0.9 \cap |z| > 1 = |z| > 1\}$. Second, the step response $v(n)$ contains no steady-state term $u(n)$.

d. Substituting $z = e^{j\omega}$ in $H(z)$,

$$H(e^{j\omega}) = \frac{1 - e^{-j2\omega}}{1 - 0.81e^{-j2\omega}}$$

We will use MATLAB to compute and plot responses.

```
>> w = [0:1:500]*pi/500;
>> H = freqz(b,a,w);
>> magH = abs(H); phaH = angle(H);
>> subplot(2,1,1); plot(w/pi,magH); grid
>> xlabel('frequency in pi units');  ylabel('Magnitude')
>> title('Magnitude Response')
>> subplot(2,1,2); plot(w/pi,phaH/pi); grid
>> xlabel('frequency in pi units'); ylabel('Phase in pi units')
>> title('Phase Response')
```

The frequency response plots are shown in Figure 4.10. □

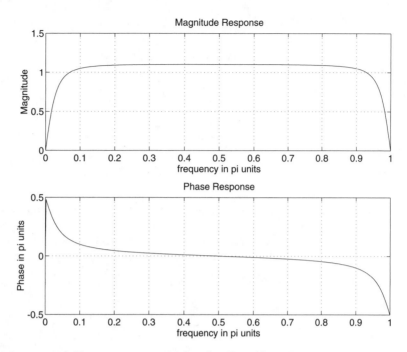

FIGURE 4.10 *Frequency response plots for Example 4.13*

SOLUTIONS OF THE DIFFERENCE EQUATIONS
▄

In Chapter 2 we mentioned two forms for the solution of linear constant coefficient difference equations. One form involved finding the particular and the homogeneous solutions, while the other form involved finding the zero-input (initial condition) and the zero-state responses. Using z-transforms, we now provide a method for obtaining these forms. In addition, we will also discuss the *transient* and the *steady-state* responses. In digital signal processing difference equations generally evolve in the positive n direction. Therefore our time frame for these solutions will be $n \geq 0$. For this purpose we define a version of the bilateral z-transform called the *one-sided z-transform*.

■ **DEFINITION 4** **The One-sided z Transform**
The one-sided z-transform of a sequence $x(n)$ is given by

$$\mathcal{Z}^+ [x(n)] \triangleq \mathcal{Z} [x(n)u(n)] \triangleq X^+ [z] = \sum_{n=0}^{\infty} x(n)z^{-n} \qquad \textbf{(4.25)}$$

Then the sample shifting property is given by

$$\mathcal{Z}^+ [x(n - k)] = \mathcal{Z} [x(n - k)u(n)]$$

$$= \sum_{n=0}^{\infty} x(n - k)z^{-n} = \sum_{m=-k}^{\infty} x(m)z^{-(m+k)}$$

$$= \sum_{m=-k}^{-1} x(m)z^{-(m+k)} + \left[\sum_{m=0}^{\infty} x(m)z^{-m} \right] z^{-k}$$

or

$$\mathcal{Z}^+ [x(n - k)] = x(-1)z^{1-k} + x(-2)z^{2-k} + \cdots + x(-k) + z^{-k}X^+(z) \quad \textbf{(4.26)}$$

This result can now be used to solve difference equations with nonzero initial conditions or with changing inputs. We want to solve the difference equation

$$1 + \sum_{k=1}^{N} a_k y(n - k) = \sum_{m=0}^{M} b_m x(n - m), \ n \geq 0$$

subject to these initial conditions:

$$\{y(i), i = -1, \ldots, -N\} \quad \text{and} \quad \{x(i), i = -1, \ldots, -M\}.$$

We now demonstrate this solution using an example.

☐ **EXAMPLE 4.14**   Solve

$$y(n) - \frac{3}{2}y(n - 1) + \frac{1}{2}y(n - 2) = x(n), \ n \geq 0$$

where

$$x(n) = \left(\frac{1}{4} \right)^n u(n)$$

subject to $y(-1) = 4$ and $y(-2) = 10$.

Solution   Taking the one-sided z-transform of both sides of the difference equation, we obtain

$$Y^+(z) - \frac{3}{2} \left[y(-1) + z^{-1}Y^+(z) \right] + \frac{1}{2} \left[y(-2) + z^{-1}y(-1) + z^{-2}Y^+(z) \right] = \frac{1}{1 - \frac{1}{4}z^{-1}}$$

Substituting the initial conditions and rearranging,

$$Y^+(z) \left[1 - \frac{3}{2}z^{-1} + \frac{1}{2}z^{-2} \right] = \frac{1}{1 - \frac{1}{4}z^{-1}} + (1 - 2z^{-1})$$

or

$$Y^+(z) = \frac{\frac{1}{1 - \frac{1}{4}z^{-1}}}{1 - \frac{3}{2}z^{-1} + \frac{1}{2}z^{-2}} + \frac{1 - 2z^{-1}}{1 - \frac{3}{2}z^{-1} + \frac{1}{2}z^{-2}} \qquad (4.27)$$

Finally,

$$Y^+(z) = \frac{2 - \frac{9}{4}z^{-1} + \frac{1}{2}z^{-2}}{\left(1 - \frac{1}{2}z^{-1}\right)\left(1 - z^{-1}\right)\left(1 - \frac{1}{4}z^{-1}\right)}$$

Using the partial fraction expansion, we obtain

$$Y^+(z) = \frac{1}{1 - \frac{1}{2}z^{-1}} + \frac{\frac{2}{3}}{1 - z^{-1}} + \frac{\frac{1}{3}}{1 - \frac{1}{4}z^{-1}} \qquad (4.28)$$

After inverse transformation the solution is

$$y(n) = \left[\left(\frac{1}{2}\right)^n + \frac{2}{3} + \frac{1}{3}\left(\frac{1}{4}\right)^n\right] u(n) \quad \square \qquad (4.29)$$

Forms of the Solutions The above solution is the *complete response* of the difference equation. It can be expressed in several forms.

- Homogeneous and particular parts:

$$y(n) = \underbrace{\left[\left(\frac{1}{2}\right)^n + \frac{2}{3}\right] u(n)}_{\text{homogeneous part}} + \underbrace{\frac{1}{3}\left(\frac{1}{4}\right)^n u(n)}_{\text{particular part}}$$

The homogeneous part is due to the *system poles* and the particular part is due to the *input poles*.

- Transient and steady-state responses:

$$y(n) = \underbrace{\left[\frac{1}{3}\left(\frac{1}{4}\right)^n + \left(\frac{1}{2}\right)^n\right] u(n)}_{\text{transient response}} + \underbrace{\frac{2}{3}u(n)}_{\text{steady-state response}}$$

The transient response is due to poles that are *inside* the unit circle, while the steady-state response is due to poles that are *on* the unit circle. Note that when the poles are *outside* the unit circle, the response is termed an *unbounded* response.

- Zero-input (or initial condition) and zero-state responses:
In equation (4.27) $Y^+(z)$ has two parts. The first part can be interpreted as

$$Y_{ZS}(z) = H(z)X(z)$$

while the second part as

$$Y_{ZI}(z) = H(z)X_{IC}(z)$$

where $X_{IC}(z)$ can be thought of as an equivalent *initial-condition input* that generates the same output Y_{ZI} as generated by the initial conditions. In this example $x_{IC}(n)$ is

$$x_{IC}(n) = \{1, -2\}$$
$$\uparrow$$

Now taking the inverse z-transform of each part of (4.27), we write the complete response as

$$y(n) = \underbrace{\left[\frac{1}{3}\left(\frac{1}{4}\right)^n - 2\left(\frac{1}{2}\right)^n + \frac{8}{3}\right]u(n)}_{\text{zero-state response}} + \underbrace{\left[3\left(\frac{1}{2}\right)^n - 2\right]u(n)}_{\text{zero-input response}}$$

From this example it is clear that each part of the complete solution is, in general, a different function and emphasizes a different aspect of system analysis.

MATLAB
IMPLEMEN-
TATION

In Chapter 2 we used the `filter` function to solve the difference equation, given its coefficients and an input. This function can also be used to find the complete response when initial conditions are given. In this form the `filter` function is invoked by

```
y = filter(b,a,x,xic)
```

where `xic` is an equivalent initial-condition input array. To find the complete response in Example 4.14, we will use

```
>> n = [0:7]; x = (1/4).^n; xic = [1, -2];
>> format long
>> y1 = filter(b,a,x,xic)
y1 =
  Columns 1 through 4
    2.00000000000000   1.25000000000000   0.93750000000000   0.79687500000000
  Columns 5 through 8
    0.73046875000000   0.69824218750000   0.68237304687500   0.67449951171875
>> y2 = (1/3)*(1/4).^n+(1/2).^n+(2/3)*ones(1,8) % Matlab Check
y2 =
  Columns 1 through 4
    2.00000000000000   1.25000000000000   0.93750000000000   0.79687500000000
  Columns 5 through 8
    0.73046875000000   0.69824218750000   0.68237304687500   0.67449951171875
```

which agrees with the response given in (4.29). In Example 4.14 we computed $x_{IC}(n)$ analytically. However, in practice, and especially for large-order difference equations, it is tedious to determine $x_{IC}(n)$ analytically. MATLAB provides a function called `filtic`, which is available only in the Signal Processing toolbox (version 2.0b or later). It is invoked by

```
xic = filtic(b,a,Y,X)
```

in which `b` and `a` are the filter coefficient arrays and `Y` and `X` are the initial-condition arrays from the initial conditions on $y(n)$ and $x(n)$, respectively, in the form

$$Y = [y(-1), \ y(-2), \dots, \ y(-N)]$$
$$X = [x(-1), \ x(-2), \dots, \ x(-M)]$$

If $x(n) = 0, \quad n \le -1$ then X need not be specified in the `filtic` function. In Example 4.14 we could have used

```
>> Y = [4, 10];
>> xic = filtic(b,a,Y)
Y =
     1    -2
```

to determine $x_{IC}(n)$.

☐ **EXAMPLE 4.15**   Solve the difference equation

$$y(n) = \frac{1}{3}\left[x(n) + x(n-1) + x(n-2)\right] + 0.95y(n-1) - 0.9025y(n-2), \quad n \ge 0$$

where $x(n) = \cos(\pi n/3)u(n)$ and

$$y(-1) = -2, \ y(-2) = -3; \quad x(-1) = 1, \ x(-2) = 1$$

First determine the solution analytically and then by using MATLAB.

Solution   Taking a one-sided z-transform of the difference equation

$$Y^+(z) = \frac{1}{3}\left[X^+(z) + x(-1) + z^{-1}X^+(z) + x(-2) + z^{-1}x(-1) + z^{-2}X^+(z)\right]$$

$$+ 0.95\left[y(-1) + z^{-1}Y^+(z)\right] - 0.9025\left[y(-2) + z^{-1}y(-1) + z^{-2}Y^+(z)\right]$$

and substituting the initial conditions, we obtain

$$Y^+(z) = \frac{\frac{1}{3} + \frac{1}{3}z^{-1} + \frac{1}{3}z^{-2}}{1 - 0.95z^{-1} + 0.9025z^{-2}}X^+(z) + \frac{1.4742 + 2.1383z^{-1}}{1 - 0.95z^{-1} + 0.9025z^{-2}}$$

Clearly, $x_{IC}(n) = [1.4742, 2.1383]$. Now substituting $X^+(z) = \dfrac{1 - 0.5z^{-1}}{1 - z^{-1} + z^{-2}}$ and simplifying, we will obtain $Y^+(z)$ as a rational function. This simplification and further partial fraction expansion can be done using MATLAB as shown below.

```
>> b = [1,1,1]/3; a = [1,-0.95,0.9025];
>> Y = [-2,-3]; X = [1,1];
>> xic=filtic(b,a,Y,X)
xic =
    1.4742    2.1383
>> bxplus = [1,-0.5]; axplus = [1,-1,1]; % X(z) transform coeff.
>> ayplus = conv(a,axplus) % Denominator of Yplus(z)
ayplus =
    1.0000   -1.9500    2.8525   -1.8525    0.9025
>> byplus = conv(b,bxplus)+conv(xic,axplus)   % Numerator of Yplus(z)
byplus =
    1.8075    0.8308   -0.4975    1.9717
>> [R,p,C] = residuez(byplus,ayplus)
R =
   0.0584 + 3.9468i   0.0584 - 3.9468i   0.8453 + 2.0311i   0.8453 - 2.0311i
p =
   0.5000 - 0.8660i   0.5000 + 0.8660i   0.4750 + 0.8227i   0.4750 - 0.8227i
C =
    []
>> Mp = abs(p), Ap = angle(p)/pi % Polar form
Mp =
    1.0000    1.0000    0.9500    0.9500
Ap =
   -0.3333    0.3333    0.3333   -0.3333
```

Hence

$$Y^{+}(z) = \frac{1.8076 + 0.8308z^{-1} - 0.4975z^{-2} + 1.9717z^{-3}}{1 - 1.95z^{-1} + 2.8525z^{-2} - 1.8525z^{-3} + 0.9025z^{-4}}$$

$$= \frac{0.0584 + j3.9468}{1 - e^{-j\pi/3}z^{-1}} + \frac{0.0584 - j3.9468}{1 - e^{j\pi/3}z^{-1}}$$

$$+ \frac{0.8453 + j2.0311}{1 - 0.95e^{j\pi/3}z^{-1}} + \frac{0.8453 - j2.0311}{1 - 0.95e^{-j\pi/3}z^{-1}}$$

Now from Table 4.1

$$y(n) = (0.0584 + j3.9468)\, e^{-j\pi n/3} + (0.0584 - j3.9468)\, e^{j\pi n/3}$$

$$+ (0.8453 + j2.031)(0.95)^{n}\, e^{j\pi n/3} + (0.8453 - j2.031)(0.95)^{n}\, e^{-j\pi n/3}$$

$$= 0.1169 \cos(\pi n/3) + 7.8937 \sin(\pi n/3)$$

$$+ (0.95)^{n}\, [1.6906 \cos(\pi n/3) - 4.0623 \sin(\pi n/3)], \quad n \ge 0$$

The first two terms of $y(n)$ correspond to the steady-state response, as well as to the particular response, while the last two terms are the transient response (and homogeneous response) terms.

To solve this example using MATLAB, we will need the **filtic** function, which we have already used to determine the $x_{IC}(n)$ sequence. The solution will be a numerical one. Let us determine the first 8 samples of $y(n)$.

Chapter 4 ■ THE z-TRANSFORM

```
>> n = [0:7]; x = cos(pi*n/3);
>> y = filter(b,a,x,xic)
y =
  Columns 1 through 4
    1.80750000000000    4.35545833333333    2.83975000000000  -1.56637197916667
  Columns 5 through 8
   -4.71759442187500   -3.40139732291667    1.35963484230469    5.02808085078841
% Matlab Verification
>> A=real(2*R(1)); B=imag(2*R(1)); C=real(2*R(3)); D=imag(2*R(4));
>> y=A*cos(pi*n/3)+B*sin(pi*n/3)+((0.95).^n).*(C*cos(pi*n/3)+D*sin(pi*n/3))
y =
  Columns 1 through 4
    1.80750000000048    4.35545833333359    2.83974999999978  -1.56637197916714
  Columns 5 through 8
   -4.71759442187528   -3.40139732291648    1.35963484230515    5.02808085078871
```

$\square$

PROBLEMS

---■---

P4.1 Determine the z-transform of the following sequences using definition (4.1). Indicate the region of convergence for each sequence and verify the z-transform expression using MATLAB.

a. $x(n) = \{3, 2, 1, -2, -3\}$
         ↑

b. $x(n) = (0.8)^n u(n-2)$

c. $x(n) = \left(\frac{4}{3}\right)^n u(1-n)$

d. $x(n) = 2^{-|n|} + \left(\frac{1}{3}\right)^{|n|}$

e. $x(n) = (n+1)(3)^n u(n)$

P4.2 Determine the z-transform of the following sequences using the z-transform table and the z-transform properties. Express $X(z)$ as a rational function in z^{-1}. Verify your results using MATLAB. Indicate the region of convergence in each case and provide a pole-zero plot.

a. $x(n) = 2\delta(n-2) + 3u(n-3)$

b. $x(n) = \left(\frac{1}{3}\right)^n u(n-2) + (0.9)^{n-3} u(n)$

c. $x(n) = n\sin\left(\frac{\pi n}{3}\right) u(n) + (0.9)^n u(n-2)$

d. $x(n) = \left(\frac{1}{2}\right)^n \cos\left(\frac{\pi n}{4} - 45°\right) u(n-1)$

e. $x(n) = (n-3)\left(\frac{1}{4}\right)^{n-2} \cos\left\{\frac{\pi}{2}(n-1)\right\} u(n)$

P4.3 The z-transform of $x(n)$ is $X(z) = \left(1 + 2z^{-1}\right)$, $|z| \neq 0$. Find the z-transforms of the following sequences, and indicate their regions of convergence.

a. $x_1(n) = x(3-n) + x(n-3)$

b. $x_2(n) = \left(1 + n + n^2\right) x(n)$

c. $x_3(n) = \left(\frac{1}{2}\right)^n x(n-2)$

d. $x_4(n) = x(n+2) * x(n-2)$

e. $x_5(n) = \cos(\pi n/2) x^*(n)$

P4.4 Repeat Problem 4.3 if

$$X(z) = \frac{1 + z^{-1}}{1 + \frac{5}{6}z^{-1} + \frac{1}{6}z^{-2}}; \quad |z| > \frac{1}{2}$$

P4.5 The inverse z-transform of $X(z)$ is $x(n) = \left(\frac{1}{2}\right)^n u(n)$. Using the z-transform properties, determine the sequences in each of the following cases.

a. $X_1(z) = \frac{z-1}{z}X(z)$

b. $X_2(z) = zX\left(z^{-1}\right)$

c. $X_3(z) = 2X(3z) + 3X(z/3)$

d. $X_4(z) = X(z)X\left(z^{-1}\right)$

e. $X_5(z) = z^2\frac{dX(z)}{dz}$

P4.6 If sequences $x_1(n)$, $x_2(n)$, and $x_3(n)$ are related by $x_3(n) = x_1(n) * x_2(n)$, then

$$\sum_{n=-\infty}^{\infty} x_3(n) = \left(\sum_{n=-\infty}^{\infty} x_1(n)\right)\left(\sum_{n=-\infty}^{\infty} x_2(n)\right)$$

a. Prove the above result by substituting the definition of convolution in the left-hand side.

b. Prove the above result using the convolution property.

c. Verify the above result using MATLAB by choosing any two random sequences $x_1(n)$ and $x_2(n)$.

P4.7 Determine the results of the following polynomial operations using MATLAB.

a. $X_1(z) = \left(1 - 2z^{-1} + 3z^{-2} - 4z^{-3}\right)\left(4 + 3z^{-1} - 2z^{-2} + z^{-3}\right)$

b. $X_2(z) = \left(z^2 - 2z + 3 + 2z^{-1} + z^{-2}\right)\left(z^3 - z^{-3}\right)$

c. $X_3(z) = \left(1 + z^{-1} + z^{-2}\right)^3$

d. $X_4(z) = X_1(z)X_2(z) + X_3(z)$

e. $X_5(z) = \left(z^{-1} - 3z^{-3} + 2z^{-5} + 5z^{-7} - z^{-9}\right)\left(z + 3z^2 + 2z^3 + 4z^4\right)$

P4.8 The `deconv` function is useful in dividing two causal sequences. Write a MATLAB function `deconv_m` to divide two noncausal sequences (similar to the `conv` function). The format of this function should be

```
function [p,np,r,nr] = deconv_m(b,nb,a,na)
% Modified deconvolution routine for noncausal sequences
% function [p,np,r,nr] = deconv_m(b,nb,a,na)
%
%   p = polynomial part of support np1 <= n <= np2
% np = [np1, np2]
%   r = remainder part of support nr1 <= n <= nr2
% nr = [nr1, nr2]
%   b = numerator polynomial of support nb1 <= n <= nb2
% nb = [nb1, nb2]
%   a = denominator polynomial of support na1 <= n <= na2
% na = [na1, na2]
%
```

Check your function on the following operation:

$$\frac{z^2 + z + 1 + z^{-1} + z^{-2} + z^{-3}}{z + 2 + z^{-1}} = \left(z - 1 + 2z^{-1} - 2z^{-2}\right) + \frac{3z^{-2} + 3z^{-3}}{z + 2 + z^{-1}}$$

P4.9 Determine the following inverse z-transforms using the partial fraction expansion method.

a. $X_1(z) = \left(1 - z^{-1} - 4z^{-2} + 4z^{-3}\right) / \left(1 - \frac{11}{4}z^{-1} + \frac{13}{8}z^{-2} - \frac{1}{4}z^{-3}\right)$. The sequence is right-sided.

b. $X_2(z) = \left(1 - z^{-1} - 4z^{-2} + 4z^{-3}\right) / \left(1 - \frac{11}{4}z^{-1} + \frac{13}{8}z^{-2} - \frac{1}{4}z^{-3}\right)$. The sequence is absolutely summable.

c. $X_3(z) = \left(z^3 - 3z^2 + 4z + 1\right) / \left(z^3 - 4z^2 + z - 0.16\right)$. The sequence is left-sided.

d. $X_4(z) = z / \left(z^3 + 2z^2 + 1.25z + 0.25\right)$, $|z| > 1$.

e. $X_5(z) = z / \left(z^2 - 0.25\right)^2$, $|z| < 0.5$.

P4.10 Suppose $X(z)$ is given as follows:

$$X(z) = \frac{2 + 3z^{-1}}{1 - z^{-1} + 0.81z^{-2}}, \quad |z| > 0.9$$

a. Determine $x(n)$ in a form that contains no complex numbers.

b. Using MATLAB, find the first 20 samples of $x(n)$ and compare them with your answer in the above part.

P4.11 For the linear and time-invariant systems described by the impulse responses below, determine (i) the system function representation, (ii) the difference equation representation, (iii) the pole-zero plot, and (iv) the output $y(n)$ if the input is $x(n) = \left(\frac{1}{4}\right)^n u(n)$.

a. $h(n) = 2\left(\frac{1}{2}\right)^n u(n)$

b. $h(n) = n\left(\frac{1}{3}\right)^n u(n) + \left(-\frac{1}{4}\right)^n u(n)$

c. $h(n) = 3(0.9)^n \cos(\pi n/4 + \pi/3) u(n + 1)$

d. $h(n) = n\left[u(n) - u(n - 10)\right]$

e. $h(n) = \left[2 - \sin(\pi n)\right] u(n)$

P4.12 For the linear and time-invariant systems described by the system functions below, determine (i) the impulse response representation, (ii) the difference equation representation, (iii) the pole-zero plot, and (iv) the output $y(n)$ if the input is $x(n) = 3\cos(\pi n/3) u(n)$.

a. $H(z) = (z + 1) / (z - 0.5)$, causal system.

b. $H(z) = \left(1 + z^{-1} + z^{-2}\right) / \left(1 + 0.5z^{-1} - 0.25z^{-2}\right)$, stable system.

c. $H(z) = \left(z^2 - 1\right) / (z - 3)^2$, anticausal system.

d. $H(z) = \frac{z}{z - 0.25} + \frac{1 - 0.5z^{-1}}{1 + 2z^{-1}}$, stable system.

e. $H(z) = \left(1 + z^{-1} + z^{-2}\right)^2$, stable system.

P4.13 For the linear, causal, and time-invariant systems described by the difference equations below, determine (i) the impulse response representation, (ii) the system function representation, (iii) the pole-zero plot, and (iv) the output $y(n)$ if the input is $x(n) = 2(0.9)^n u(n)$.

a. $y(n) = [x(n) + 2x(n-1) + x(n-3)]/4$

b. $y(n) = x(n) + 0.5x(n-1) - 0.5y(n-1) + 0.25y(n-2)$

c. $y(n) = 2x(n) + 0.9y(n-1)$

d. $y(n) = -0.45x(n) - 0.4x(n-1) + x(n-2) + 0.4y(n-1) + 0.45y(n-2)$

e. $y(n) = \sum_{m=0}^{4}(0.8)^m x(n-m) - \sum_{\ell=1}^{4}(0.9)^\ell y(n-\ell)$

P4.14 The output sequence $y(n)$ in Problem 4.13 is the total response. For each of the systems given in that problem, separate $y(n)$ into (i) the homogeneous part, (ii) the particular part, (iii) the transient response, and (iv) the steady-state response.

P4.15 A stable system has the following pole-zero locations:

$$z_1 = j, \quad z_2 = -j, \quad p_1 = -\frac{1}{2} + j\frac{1}{2}, \quad p_2 = -\frac{1}{2} - j\frac{1}{2}$$

It is also known that the frequency response function $H(e^{j\omega})$ evaluated at $\omega = 0$ is equal to 0.8; that is,

$$H(e^{j0}) = 0.8$$

a. Determine the system function $H(z)$ and indicate its region of convergence.

b. Determine the difference equation representation.

c. Determine the steady-state response $y_{ss}(n)$ if the input is $x(n) = \frac{1}{\sqrt{2}}\sin\left(\frac{\pi n}{2}\right)u(n)$.

d. Determine the transient response $y_{tr}(n)$ if the input is $x(n) = \frac{1}{\sqrt{2}}\sin\left(\frac{\pi n}{2}\right)u(n)$.

P4.16 A digital filter is described by the difference equation

$$y(n) = x(n) + x(n-1) + 0.9y(n-1) - 0.81y(n-2)$$

a. Using the `freqz` function, plot the magnitude and phase of the frequency response of the above filter. Note the magnitude and phase at $\omega = \pi/3$ and at $\omega = \pi$.

b. Generate 200 samples of the signal $x(n) = \sin(\pi n/3) + 5\cos(\pi n)$ and process through the filter. Compare the steady-state portion of the output to $x(n)$. How are the amplitudes and phases of two sinusoids affected by the filter?

P4.17 Solve the following difference equation for $y(n)$ using the one-sided z-transform approach.

$$y(n) = 0.5y(n-1) + 0.25y(n-2) + x(n), \quad n \geq 0; \quad y(-1) = 1, \, y(-2) = 2$$

$$x(n) = (0.8)^n u(n)$$

Generate the first 20 samples of $y(n)$ using MATLAB and compare them with your answer.

P4.18 Solve the difference equation for $y(n)$, $\quad n \geq 0$

$$y(n) - 0.4y(n-1) - 0.45y(n-2) = 0.45x(n) + 0.4x(n-1) - x(n-2)$$

driven by the input $x(n) = 2 + \left(\frac{1}{2}\right)^n u(n)$ and subject to

$$y(-1) = 0, \, y(-2) = 3; \, x(-1) == x(-2) = 2$$

Decompose the solution $y(n)$ into (i) transient response, (ii) steady-state response, (iii) zero input response, and (iv) zero-state response.

P4.19 A causal, linear, and time-invariant system is given by the following difference equation:

$$y(n) = y(n-1) + y(n-2) + x(n-1)$$

a. Find the system function $H(z)$ for this system.

b. Plot the poles and zeros of $H(z)$ and indicate the region of convergence (ROC).

c. Find the unit sample response $h(n)$ of this system.

d. Is this system stable? If the answer is yes, justify it. If the answer is no, find a stable unit sample response that satisfies the difference equation.

P4.20 Determine the zero-state response of the system

$$y(n) = \tfrac{1}{4}y(n-1) + x(n) + 3x(n-1), \quad n \geq 0; \quad y(-1) = 2$$

to the input

$$x(n) = e^{j\pi n/4}u(n)$$

What is the steady-state response of the system?

5 *THE DISCRETE FOURIER TRANSFORM*

In Chapters 3 and 4 we studied transform-domain representations of discrete signals. The discrete-time Fourier transform provided the frequency-domain (ω) representation for absolutely summable sequences. The z-transform provided a generalized frequency-domain (z) representation for arbitrary sequences. These transforms have two features in common. First, the transforms are defined for infinite-length sequences. Second, and the most important, they are functions of continuous variables (ω or z). From the numerical computation viewpoint (or from MATLAB's viewpoint), these two features are troublesome because one has to evaluate *infinite sums* at *uncountably infinite* frequencies. To use MATLAB, we have to truncate sequences and then evaluate the expressions at finitely many points. This is what we did in many examples in the two previous chapters. The evaluations were obviously approximations to the exact calculations. In other words, the discrete-time Fourier transform and the z-transform are not *numerically computable* transforms.

Therefore we turn our attention to a numerically computable transform. It is obtained by sampling the discrete-time Fourier transform in the frequency domain (or the z-transform on the unit circle). We develop this transform by first analyzing periodic sequences. From Fourier analysis we know that a periodic function (or sequence) can always be represented by a linear combination of harmonically related complex exponentials (which is a form of sampling). This gives us the *Discrete Fourier Series* (or DFS) representation. Since the sampling is in the frequency domain, we study the effects of sampling in the time domain and the issue of reconstruction in the z-domain. We then extend the DFS to *finite-duration sequences*, which leads to a new transform, called the *Discrete Fourier Transform* (or DFT). The DFT avoids the two problems mentioned above and is a numerically computable transform that is suitable for computer implementation. We study its properties and its use in system analysis in detail. The numerical computation of the DFT for long sequences is prohibitively time consuming. Therefore several algorithms have been developed to efficiently compute the DFT. These are collectively called fast Fourier transform (or FFT) algorithms. We will study two such algorithms in detail.

THE DISCRETE FOURIER SERIES

In Chapter 2 we defined the periodic sequence by $\tilde{x}(n)$, satisfying the condition

$$\tilde{x}(n) = \tilde{x}(n + kN), \quad \forall n, k \tag{5.1}$$

where N is the fundamental period of the sequence. From Fourier analysis we know that the periodic functions can be synthesized as a linear combination of complex exponentials whose frequencies are multiples (or harmonics) of the fundamental frequency (which in our case is $2\pi/N$). From the frequency-domain periodicity of the discrete-time Fourier transform, we conclude that there are a finite number of harmonics; the frequencies are $\left\{ \frac{2\pi}{N} k, \quad k = 0, 1, \ldots, N - 1 \right\}$. Therefore a periodic sequence $\tilde{x}(n)$ can be expressed as

$$\tilde{x}(n) = \frac{1}{N} \sum_{k=0}^{N-1} \tilde{X}(k) e^{j \frac{2\pi}{N} kn}, \quad n = 0, \pm 1, \ldots, \tag{5.2}$$

where $\{\tilde{X}(k), \quad k = 0, \pm 1, \ldots, \}$ are called the discrete Fourier series coefficients, which are given by

$$\tilde{X}(k) = \sum_{n=0}^{N-1} \tilde{x}(n) e^{-j \frac{2\pi}{N} nk}, \quad k = 0, \pm 1, \ldots, \tag{5.3}$$

Note that $\tilde{X}(k)$ is itself a (complex-valued) periodic sequence with fundamental period equal to N, that is,

$$\tilde{X}(k + N) = \tilde{X}(k) \tag{5.4}$$

The pair of equations (5.3) and (5.2) taken together is called the discrete Fourier series representation of periodic sequences. Using $W_N \stackrel{\triangle}{=} e^{-j \frac{2\pi}{N}}$ to

---

denote the complex exponential term, we express (5.3) and (5.2) as

$$\tilde{X}(k) \triangleq \text{DFS}\left[\tilde{x}(n)\right] = \sum_{n=0}^{N-1} \tilde{x}(n)W_N^{nk} \qquad \text{: Analysis or a} \\ \text{DFS equation}$$

$$\tilde{x}(n) \triangleq \text{IDFS}\left[\tilde{X}(k)\right] = \frac{1}{N}\sum_{k=0}^{N-1} \tilde{X}(k)W_N^{-nk} \qquad \text{: Synthesis or an inverse} \\ \text{DFS equation}$$

(5.5)

□ **EXAMPLE 5.1** Find DFS representation of the periodic sequence given below:

$$\tilde{x}(n) = \{\ldots, 0, 1, 2, 3, 0, 1, 2, 3, 0, 1, 2, 3, \ldots\}$$
$$\uparrow$$

Solution
The fundamental period of the above sequence is $N = 4$. Hence $W_4 = e^{-j\frac{2\pi}{4}} = -j$. Now

$$\tilde{X}(k) = \sum_{n=0}^{3} \tilde{x}(n)W_4^{nk}, \quad k = 0, \pm 1, \pm 2, \ldots$$

Hence

$$\tilde{X}(0) = \sum_{0}^{3} \tilde{x}(n)W_4^{0 \cdot n} = \sum_{0}^{3} \tilde{x}(n) = \tilde{x}(0) + \tilde{x}(1) + \tilde{x}(2) + \tilde{x}(3) = 6$$

Similarly,

$$\tilde{X}(1) = \sum_{0}^{3} \tilde{x}(n)W_4^{n} = \sum_{0}^{3} \tilde{x}(n)(-j)^{n} = (-2 + 2j)$$

$$\tilde{X}(2) = \sum_{0}^{3} \tilde{x}(n)W_4^{2n} = \sum_{0}^{3} \tilde{x}(n)(-j)^{2n} = -2$$

$$\tilde{X}(3) = \sum_{0}^{3} \tilde{x}(n)W_4^{3n} = \sum_{0}^{3} \tilde{x}(n)(-j)^{3n} = (-2 - 2j) \qquad \square$$

MATLAB IMPLEMEN-TATION
A careful look at (5.5) reveals that the DFS is a numerically computable representation. It can be implemented in many ways. To compute each sample $\tilde{X}(k)$, we can implement the summation as a `for...end` loop. To compute all DFS coefficients would require another `for...end` loop. This will result in a nested two `for...end` loop implementation. This is clearly inefficient in MATLAB. An efficient implementation in MATLAB

would be to use a matrix-vector multiplication for each of the relations in (5.5). We have used this approach earlier in implementing a numerical approximation to the discrete-time Fourier transform. Let $\tilde{\mathbf{x}}$ and $\tilde{\mathbf{X}}$ denote column vectors corresponding to the primary periods of sequences $x(n)$ and $X(k)$, respectively. Then (5.5) is given by

$$\tilde{\mathbf{X}} = \mathbf{W}_N \tilde{\mathbf{x}}$$
$$\tilde{\mathbf{x}} = \frac{1}{N} \mathbf{W}_N^* \tilde{\mathbf{X}} \tag{5.6}$$

where the matrix $\mathbf{W}_N$ is given by

$$\mathbf{W}_N \triangleq \left[W_N^{kn} {}_{0 \leq k, n \leq N-1} \right] = \begin{matrix} \\ k \\ \downarrow \end{matrix} \overset{n \longrightarrow}{\begin{bmatrix} 1 & 1 & \cdots & 1 \\ 1 & W_N^1 & \cdots & W_N^{(N-1)} \\ \vdots & \vdots & \ddots & \vdots \\ 1 & W_N^{(N-1)} & \cdots & W_N^{(N-1)^2} \end{bmatrix}} \tag{5.7}$$

The matrix $\mathbf{W}_N$ is a square matrix and is called a *DFS matrix*. The following MATLAB function dfs implements the above procedure.

```
function [Xk] = dfs(xn,N)
% Computes Discrete Fourier Series Coefficients
% ----------------------------------------------
% [Xk] = dfs(xn,N)
% Xk = DFS coeff. array over 0 <= k <= N-1
% xn = One period of periodic signal over 0 <= n <= N-1
%  N = Fundamental period of xn
%
n = [0:1:N-1];              % row vector for n
k = [0:1:N-1];              % row vecor for k
WN = exp(-j*2*pi/N);        % Wn factor
nk = n'*k;                  % creates a N by N matrix of nk values
WNnk = WN .^ nk;            % DFS matrix
Xk = xn * WNnk;             % row vector for DFS coefficients
```

The DFS in Example 5.1 can be computed using MATLAB as

```
>> xn = [0,1,2,3]; N = 4;
>> Xk = dfs(xn,N)
Xk =
   6.0000              -2.0000 + 2.0000i  -2.0000 - 0.0000i  -2.0000 - 2.0000i
```

The following `idfs` function implements the synthesis equation.

```
function [xn] = idfs(Xk,N)
% Computes Inverse Discrete Fourier Series
% -----------------------------------------
% [xn] = idfs(Xk,N)
% xn = One period of periodic signal over 0 <= n <= N-1
% Xk = DFS coeff. array over 0 <= k <= N-1
%  N = Fundamental period of Xk
%
n = [0:1:N-1];                  % row vector for n
k = [0:1:N-1];                  % row vecor for k
WN = exp(-j*2*pi/N);            % Wn factor
nk = n'*k;                      % creates a N by N matrix of nk values
WNnk = WN .^ (-nk);             % IDFS matrix
xn = (Xk * WNnk)/N;             % row vector for IDFS values
```

Caution: The above functions are efficient approaches of implementing (5.5) in MATLAB. They are not computationally efficient, especially for large N. We will deal with this problem later in this chapter.

☐ **EXAMPLE 5.2** A periodic "square wave" sequence is given by

$$\tilde{x}(n) = \begin{cases} 1, & mN \le n \le mN + L - 1 \\ 0, & mN + L \le n \le (m+1)N - 1 \end{cases}; \quad m = 0, \pm 1, \pm 2, \dots$$

where N is the fundamental period and L/N is the duty cycle.

    a. Determine an expression for $|\tilde{X}(k)|$ in terms of L and N.
    b. Plot the magnitude $|\tilde{X}(k)|$ for $L = 5$, $N = 20$; $L = 5$, $N = 40$; $L = 5$, $N = 60$; and $L = 7$, $N = 60$.
    c. Comment on the results.

Solution A plot of this sequence for $L = 5$ and $N = 20$ is shown in Figure 5.1.

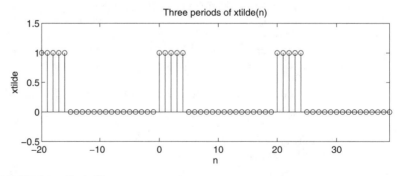

FIGURE 5.1 *Periodic square wave sequence*

a. By applying the analysis equation (5.3),

$$\tilde{X}(k) = \sum_{n=0}^{N-1} \tilde{x}(n) e^{-j\frac{2\pi}{N}nk} = \sum_{n=0}^{L-1} e^{-j\frac{2\pi}{N}nk} = \sum_{n=0}^{L-1} \left(e^{-j\frac{2\pi}{N}k} \right)^n$$

$$= \begin{cases} L, & k = 0, \pm N, \pm 2N, \ldots \\ \dfrac{1 - e^{-j2\pi Lk/N}}{1 - e^{-j2\pi k/N}}, & \text{otherwise} \end{cases}$$

The last step follows from the sum of the geometric terms formula (2.5) in Chapter 2. The last expression can be simplified to

$$\frac{1 - e^{-j2\pi Lk/N}}{1 - e^{-j2\pi k/N}} = \frac{e^{-j\pi Lk/N}}{e^{-j\pi k/N}} \frac{e^{j\pi Lk/N} - e^{-j\pi Lk/N}}{e^{j\pi k/N} - e^{-j\pi k/N}}$$

$$= e^{-j\pi(L-1)k/N} \frac{\sin{(\pi kL/N)}}{\sin{(\pi k/N)}}$$

or the magnitude of $\tilde{X}(k)$ is given by

$$\left| \tilde{X}(k) \right| = \begin{cases} L, & k = 0, \pm N, \pm 2N, \ldots \\ \left| \dfrac{\sin{(\pi kL/N)}}{\sin{(\pi k/N)}} \right|, & \text{otherwise} \end{cases}$$

b. MATLAB script for $L = 5$ and $N = 20$ is given below.

```
>> L = 5; N = 20; k = [-N/2:N/2];          % Sq wave parameters
>> xn = [ones(1,L), zeros(1,N-L)];          % Sq wave x(n)
>> Xk = dfs(xn,N);                          % DFS
>> magXk = abs([Xk(N/2+1:N) Xk(1:N/2+1)]); % DFS magnitude
>> subplot(2,2,1); stem(k,magXk); axis([-N/2,N/2,-0.5,5.5])
>> xlabel('k'); ylabel('Xtilde(k)')
>> title('DFS of SQ. wave: L=5, N=20')
```

The plots for the above and all other cases are shown in Figure 5.2. Note that since $\tilde{X}(k)$ is periodic, the plots are shown from $-N/2$ to $N/2$.

c. Several interesting observations can be made from plots in Figure 5.2. The envelopes of the DFS coefficients of square waves look like "sinc" functions. The amplitude at $k = 0$ is equal to L, while the zeros of the functions are at multiples of N/L, which is the reciprocal of the duty cycle. We will study these functions later in this chapter. □

RELATION TO THE z-TRANSFORM

Let $x(n)$ be a finite-duration sequence of duration N such that

$$x(n) = \begin{cases} \text{Nonzero}, & 0 \leq n \leq N - 1 \\ 0, & \text{elsewhere} \end{cases} \tag{5.8}$$

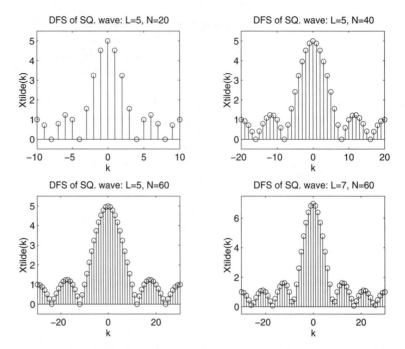

FIGURE 5.2 *The DFS plots of a periodic square wave for various L and N.*

Then we can find its z-transform:

$$X(z) = \sum_{n=0}^{N-1} x(n)z^{-n} \tag{5.9}$$

Now we construct a periodic sequence $\tilde{x}(n)$ by periodically repeating $x(n)$ with period N, that is,

$$x(n) = \begin{cases} \tilde{x}(n), & 0 \le n \le N-1 \\ 0, & \text{elsewhere} \end{cases} \tag{5.10}$$

The DFS of $\tilde{x}(n)$ is given by

$$\tilde{X}(k) = \sum_{n=0}^{N-1} \tilde{x}(n)e^{-j\frac{2\pi}{N}nk} = \sum_{n=0}^{N-1} x(n)\left[e^{j\frac{2\pi}{N}k}\right]^{-n} \tag{5.11}$$

Comparing it with (5.9), we have

$$\tilde{X}(k) = X(z)\big|_{z=e^{j\frac{2\pi}{N}k}} \tag{5.12}$$

which means that the DFS $\tilde{X}(k)$ represents N evenly spaced samples of the z-transform $X(z)$ around the unit circle.

Since $x(n)$ in (5.8) is of finite duration of length N, it is also absolutely summable. Hence its DTFT exists and is given by

$$X(e^{j\omega}) = \sum_{n=0}^{N-1} x(n)e^{-j\omega n} = \sum_{n=0}^{N-1} \tilde{x}(n)e^{-j\omega n} \qquad (5.13)$$

Comparing (5.13) with (5.11), we have

$$\tilde{X}(k) = X(e^{j\omega})\Big|_{\omega=\frac{2\pi}{N}k} \qquad (5.14)$$

Let

$$\omega_1 \triangleq \frac{2\pi}{N} \quad \text{and} \quad \omega_k \triangleq \frac{2\pi}{N}k = k\omega_1$$

then the DFS $X(k) = X(e^{j\omega_k}) = X(e^{jk\omega_1})$, which means that the DFS is obtained by *evenly sampling* the DTFT at $\omega_1 = \frac{2\pi}{N}$ intervals. From (5.12) and (5.14) we observe that the DFS representation gives us a sampling mechanism in the frequency domain which, in principle, is similar to sampling in the time domain. The interval $\omega_1 = \frac{2\pi}{N}$ is the *sampling interval* in the frequency domain. It is also called the *frequency resolution* because it tells us how close are the frequency samples (or measurements).

☐   **EXAMPLE 5.3**   Let $x(n) = \{0, 1, 2, 3\}$.
                  ↑

      a. Compute its discrete-time Fourier transform $X(e^{j\omega})$.
      b. Sample $X(e^{j\omega})$ at $k\omega_1 = \frac{2\pi}{4}k$, $k = 0, 1, 2, 3$ and show that it is equal to $\tilde{X}(k)$ in Example 5.1.

Solution

The sequence $x(n)$ is not periodic but is of finite duration.

      a. The discrete-time Fourier transform is given by

$$X(e^{j\omega}) = \sum_{n=-\infty}^{\infty} x(n)e^{-j\omega n} = e^{-j\omega} + 2e^{-j2\omega} + 3e^{-j3\omega}$$

      b. Sampling at $k\omega_1 = \frac{2\pi}{4}k$, $k = 0, 1, 2, 3$, we obtain

$$X(e^{j0}) = 1 + 2 + 3 = 6 = \tilde{X}(0)$$

$$X(e^{j2\pi/4}) = e^{-j2\pi/4} + 2e^{-j4\pi/4} + 3e^{-j6\pi/4} = -2 + 2j = \tilde{X}(1)$$

$$X(e^{j4\pi/4}) = e^{-j4\pi/4} + 2e^{-j8\pi/4} + 3e^{-j12\pi/4} = 2 = \tilde{X}(2)$$

$$X(e^{j6\pi/4}) = e^{-j6\pi/4} + 2e^{-j12\pi/4} + 3e^{-j18\pi/4} = -2 - 2j = \tilde{X}(3)$$

as expected.   ☐

SAMPLING AND RECONSTRUCTION IN THE z-DOMAIN

Let $x(n)$ be an arbitrary absolutely summable sequence, which may be of infinite duration. Its z-transform is given by

$$X(z) = \sum_{m=-\infty}^{\infty} x(m)z^{-m}$$

and we assume that the ROC of $X(z)$ includes the unit circle. We sample $X(z)$ on the unit circle at equispaced points separated in angle by $\omega_1 = 2\pi/N$ and call it a DFS sequence,

$$\tilde{X}(k) \overset{\triangle}{=} X(z)\big|_{z=e^{j\frac{2\pi}{N}k}}, \qquad k = 0, \pm 1, \pm 2, \ldots \qquad \textbf{(5.15)}$$

$$= \sum_{m=-\infty}^{\infty} x(m)e^{-j\frac{2\pi}{N}km} = \sum_{m=-\infty}^{\infty} x(m)W_N^{km}$$

which is periodic with period N. Finally, we compute the IDFS of $\tilde{X}(k)$,

$$\tilde{x}(n) = \text{IDFS}\big[\tilde{X}(k)\big]$$

which is also periodic with period N. Clearly, there must be a relationship between the arbitrary $x(n)$ and the periodic $\tilde{x}(n)$. This is an important issue. In order to compute the inverse DTFT or the inverse z-transform numerically, we must deal with a finite number of samples of $X(z)$ around the unit circle. Therefore we must know the effect of such sampling on the time-domain sequence. This relationship is easy to obtain.

$$\tilde{x}(n) = \frac{1}{N}\sum_{k=0}^{N-1}\tilde{X}(k)W_N^{-kn} \qquad \text{(from (5.2))}$$

$$= \frac{1}{N}\sum_{k=0}^{N-1}\left\{\sum_{m=-\infty}^{\infty} x(m)W_N^{km}\right\}W_N^{-kn} \qquad \text{(from (5.15))}$$

or

$$\tilde{x}(n) = \sum_{m=-\infty}^{\infty} x(m)\underbrace{\frac{1}{N}\sum_{0}^{N-1}W_N^{-k(n-m)}}_{=\begin{cases}1, & n-m=rN \\ 0, & \text{elsewhere}\end{cases}} = \sum_{m=-\infty}^{\infty} x(m)\sum_{r=-\infty}^{\infty}\delta(n-m-rN)$$

$$= \sum_{r=-\infty}^{\infty}\sum_{m=-\infty}^{\infty} x(m)\delta(n-m-rN)$$

or

$$\tilde{x}(n) = \sum_{r=-\infty}^{\infty} x(n - rN) = \cdots + x(n+N) + x(n) + x(n-N) + \cdots \quad \textbf{(5.16)}$$

which means that when we sample $X(z)$ on the unit circle, we obtain a periodic sequence in the time domain. This sequence is a linear combination of the original $x(n)$ and its infinite replicas, each shifted by multiples of $\pm N$. This is illustrated in Example 5.5. From (5.16) we observe that if $x(n) = 0$ for $n < 0$ and $n \geq N$, then there will be no overlap or aliasing in the time domain. Hence we should be able to recognize and recover $x(n)$ from $\tilde{x}(n)$, that is,

$$x(n) = \tilde{x}(n) \text{ for } 0 \leq n \leq (N-1)$$

or

$$x(n) = \tilde{x}(n)\mathcal{R}_N(n) = \begin{cases} \tilde{x}(n), & 0 \leq n \leq N-1 \\ 0, & \text{else} \end{cases}$$

where $\mathcal{R}_N(n)$ is called a *rectangular window* of length N. Therefore we have the following theorem.

■ **THEOREM 1**   *Frequency Sampling*

If $x(n)$ is time-limited (i.e., of finite duration) to $[0, N-1]$, then N samples of $X(z)$ on the unit circle determine $X(z)$ for all z.

☐ **EXAMPLE 5.4**   Let $x_1(n) = \{6, 5, 4, 3, 2, 1\}$. Its DTFT $X_1(e^{j\omega})$ is sampled at
$\uparrow$

$$\omega_k = \frac{2\pi k}{4}, \quad k = 0, \pm 1, \pm 2, \pm 3, \ldots$$

to obtain a DFS sequence $\tilde{X}_2(k)$. Determine the sequence $\tilde{x}_2(n)$, which is the inverse DFS of $\tilde{X}_2(k)$.

Solution   Without computing the DTFT, the DFS, or the inverse DFS, we can evaluate $\tilde{x}_2(n)$ by using the aliasing formula (5.16).

$$\tilde{x}_2(n) = \sum_{r=-\infty}^{\infty} x_1(n - 4r)$$

Thus $x(4)$ is aliased into $x(0)$, and $x(5)$ is aliased into $x(1)$. Hence

$$\tilde{x}_2(n) = \{\ldots, 8, 6, 4, 3, 8, 6, 4, 3, 8, 6, 4, 3, \ldots\}$$
$\uparrow$
        ☐

Let $x(n) = (0.7)^n u(n)$. Sample its z-transform on the unit circle with $N = 5$, 10, 20, 50 and study its effect on the time domain.

Solution   From Table 4.1 the z-transform of $x(n)$ is

$$X(z) = \frac{1}{1 - 0.7z^{-1}} = \frac{z}{z - 0.7}, \quad |z| > 0.7$$

We can now use MATLAB to implement the sampling operation

$$\tilde{X}(k) = X(z)|_{z=e^{j2\pi k/N}}, \quad k = 0, \pm1, \pm2, \ldots$$

and the inverse DFS computation to determine the corresponding time-domain sequence. The MATLAB script for $N = 5$ is shown below.

```
>> N = 5; k = 0:1:N-1;                          % sample index
>> wk = 2*pi*k/N; zk = exp(j*wk);               % samples of z
>> Xk = (zk)./(zk-0.7);                          % DFS as samples of X(z)
>> xn = real(idfs(Xk,N));                        % IDFS
>> xtilde = xn'* ones(1,8); xtilde = (xtilde(:))'; % Periodic sequence
>> subplot(2,2,1); stem(0:39,xtilde);axis([0,40,-0.1,1.5])
>> xlabel('n'); ylabel('xtilde(n)'); title('N=5')
```

The plots in Figure 5.3 clearly demonstrate the aliasing in the time domain, especially for $N = 5$ and $N = 10$. For large values of N the tail end of $x(n)$

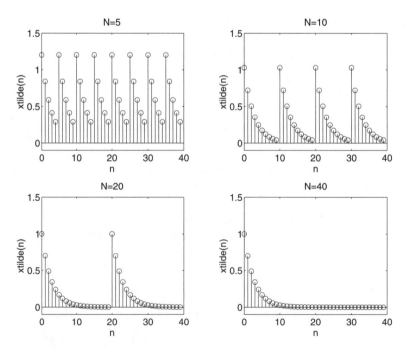

FIGURE 5.3   *Plots in Example 5.5*

is sufficiently small to result in any appreciable amount of aliasing in practice. Such information is useful in effectively truncating an infinite-duration sequence prior to taking its transform. $\qquad\square$

RECONSTRUCTION FORMULA

Let $x(n)$ be time-limited to $[0, N - 1]$. Then from Theorem 1 we should be able to recover the z-transform $X(z)$ using its samples $\tilde{X}(k)$. This is given by

$$X(z) = \mathcal{Z}\left[x(n)\right] = \mathcal{Z}\left[\tilde{x}(n)\mathcal{R}_N(n)\right]$$

$$= \mathcal{Z}[\mathrm{IDFS}\{ \underbrace{\tilde{X}(k)}_{\text{samples of } X(z)} \}\mathcal{R}_N(n)]$$

The above approach results in the z-domain reconstruction formula.

$$X(z) = \sum_0^{N-1} x(n)z^{-n} = \sum_0^{N-1} \tilde{x}(n)z^{-n}$$

$$= \sum_0^{N-1}\left\{\frac{1}{N}\sum_0^{N-1}\tilde{X}(k)W_N^{-kn}\right\}z^{-n}$$

$$= \frac{1}{N}\sum_{k=0}^{N-1}\tilde{X}(k)\left\{\sum_0^{N-1}W_N^{-kn}z^{-n}\right\}$$

$$= \frac{1}{N}\sum_{k=0}^{N-1}\tilde{X}(k)\left\{\sum_0^{N-1}\left(W_N^{-k}z^{-1}\right)^n\right\}$$

$$= \frac{1}{N}\sum_{k=0}^{N-1}\tilde{X}(k)\left\{\frac{1 - W_N^{-kN}z^{-N}}{1 - W_N^{-k}z^{-1}}\right\}$$

Since $W_N^{-kN} = 1$, we have

$$X(z) = \frac{1 - z^{-N}}{N}\sum_{k=0}^{N-1}\frac{\tilde{X}(k)}{1 - W_N^{-k}z^{-1}} \tag{5.17}$$

THE DTFT INTERPOLATION FORMULA

The reconstruction formula (5.17) can be specialized for the discrete-time Fourier transform by evaluating it on the unit circle $z = e^{j\omega}$. Then

$$X(e^{j\omega}) = \frac{1 - e^{-j\omega N}}{N}\sum_{k=0}^{N-1}\frac{\tilde{X}(k)}{1 - e^{j2\pi k/N}e^{-j\omega}}$$

$$= \sum_{k=0}^{N-1}\tilde{X}(k)\frac{1 - e^{-j\omega N}}{N\left\{1 - e^{j2\pi k/N}e^{-j\omega}\right\}}$$

Consider

$$\frac{1 - e^{-j\omega N}}{N\left\{1 - e^{j2\pi k/N}e^{-j\omega}\right\}} = \frac{1 - e^{-j(\omega - \frac{2\pi k}{N})N}}{N\left\{1 - e^{-j(\omega - \frac{2\pi k}{N})}\right\}}$$

$$= \frac{e^{-j\frac{N}{2}(\omega - \frac{2\pi k}{N})}}{e^{-\frac{1}{2}j(\omega - \frac{2\pi k}{N})}}\left\{\frac{\sin\left[(\omega - \frac{2\pi k}{N})\frac{N}{2}\right]}{N\sin\left[(\omega - \frac{2\pi k}{N})\frac{1}{2}\right]}\right\}$$

Let

$$\Phi(\omega) \triangleq \frac{\sin(\frac{\omega N}{2})}{N\sin(\frac{\omega}{2})}e^{-j\omega(\frac{N-1}{2})} : \text{an interpolating polynomial} \qquad (5.18)$$

Then

$$X(e^{j\omega}) = \sum_{k=0}^{N-1}\tilde{X}(k)\Phi\left(\omega - \frac{2\pi k}{N}\right) \qquad (5.19)$$

This is the DTFT interpolation formula to reconstruct $X(e^{j\omega})$ from its samples $\tilde{X}(k)$. Since $\Phi(0) = 1$, we have that $X(e^{j2\pi k/N}) = \tilde{X}(k)$, which means that the interpolation is exact at sampling points. Recall the time-domain interpolation formula (3.33) for analog signals:

$$x_a(t) = \sum_{n=-\infty}^{\infty}x(n)\,\text{sinc}\,[F_s(t - nT_s)] \qquad (5.20)$$

The DTFT interpolating formula (5.19) looks similar.

However, there are some differences. First, the time-domain formula (5.20) reconstructs an arbitrary *nonperiodic* analog signal, while the frequency-domain formula (5.19) gives us a periodic waveform. Second, in (5.19) we use a $\frac{\sin(Nx)}{N\sin x}$ interpolation function instead of our more familiar $\frac{\sin x}{x}$ (sinc) function. Therefore the $\Phi(\omega)$ function is sometimes called a *digital sinc* function, which itself is periodic. This is the function we observed in Example 5.2.

MATLAB
IMPLEMEN-
TATION

The interpolation formula (5.19) suffers the same fate as that of (5.20) while trying to implement it in practice. One has to generate several interpolating polynomials (5.18) and perform their linear combinations to obtain the discrete-time Fourier transform $X(e^{j\omega})$ from its computed samples $\tilde{X}(k)$. Furthermore, in MATLAB we have to evaluate (5.19) on a finer grid over $0 \le \omega \le 2\pi$. This is clearly an inefficient approach. Another approach is to use the cubic spline interpolation function as an efficient approximation to (5.19). This is what we did to implement (5.20) in Chapter 3. However, there is an alternate and efficient approach based on the DFT, which we will study in the next section.

THE DISCRETE FOURIER TRANSFORM

The discrete Fourier series provided us a mechanism for numerically computing the discrete-time Fourier transform. It also alerted us to a potential problem of aliasing in the time domain. Mathematics dictates that the sampling of the discrete-time Fourier transform result in a periodic sequence $\tilde{x}(n)$. But most of the signals in practice are not periodic. They are likely to be of finite duration. How can we develop a numerically computable Fourier representation for such signals? Theoretically, we can take care of this problem by defining a periodic signal whose primary shape is that of the finite-duration signal and then using the DFS on this periodic signal. Practically, we define a new transform called the *Discrete Fourier Transform* (DFT), which is the primary period of the DFS. This DFT is the ultimate numerically computable Fourier transform for arbitrary finite-duration sequences.

First we define a finite-duration sequence $x(n)$ that has N samples over $0 \leq n \leq N-1$ as an *N-point sequence*. Let $\tilde{x}(n)$ be a periodic signal of period N, created using the N-point sequence $x(n)$; that is, from (5.19)

$$\tilde{x}(n) = \sum_{r=-\infty}^{\infty} x(n-rN)$$

This is a somewhat cumbersome representation. Using the modulo-N operation on the argument we can simplify it to

$$\tilde{x}(n) = x(n \bmod N) \tag{5.21}$$

A simple way to interpret this operation is the following: if the argument n is between 0 and $N-1$, then leave it as it is; otherwise add or subtract multiples of N from n until the result is between 0 and $N-1$. Note carefully that (5.21) is valid only if the length of $x(n)$ is N or less. Furthermore, we use the following convenient notation to denote the modulo-N operation.

$$x((n))_N \overset{\triangle}{=} x(n \bmod N) \tag{5.22}$$

Then the compact relationships between $x(n)$ and $\tilde{x}(n)$ are

$$
\begin{aligned}
\tilde{x}(n) &= x((n))_N && \text{(Periodic extension)} \\
x(n) &= \tilde{x}(n)\mathcal{R}_N(n) && \text{(Window operation)}
\end{aligned}
\tag{5.23}
$$

The `rem(n,N)` function in MATLAB determines the remainder after dividing n by N. This function can be used to implement our modulo-N

operation when $n \geq 0$. When $n < 0$, we need to modify the result to obtain correct values. This is shown below in the `m=mod(n,N)` function.

```
function m = mod(n,N)
% Computes m = (n mod N) index
% ---------------------------
% m = mod(n,N)
m = rem(n,N);
m = m+N;
m = rem(m,N);
```

In this function `n` can be any integer array, and the array `m` contains the corresponding modulo-N values.

From the frequency sampling theorem we conclude that N equispaced samples of the discrete-time Fourier transform $X(e^{j\omega})$ of the N-point sequence $x(n)$ can uniquely reconstruct $X(e^{j\omega})$. These N samples around the unit circle are called the discrete Fourier transform coefficients. Let $\tilde{X}(k) = \text{DFS}\,\tilde{x}(n)$, which is a periodic (and hence of infinite duration) sequence. Its primary interval then is the discrete Fourier transform, which is of finite duration. These notions are made clear in the following definitions. The Discrete Fourier Transform of an N-point sequence is given by

$$X(k) \triangleq \text{DFT}\,[x(n)] = \begin{cases} \tilde{X}(k), & 0 \leq k \leq N-1 \\ 0, & \text{elsewhere} \end{cases} = \tilde{X}(k)\mathcal{R}_N(k)$$

or

$$X(k) = \sum_{n=0}^{N-1} x(n)W_N^{nk}, \quad 0 \leq k \leq N-1 \tag{5.24}$$

Note that the DFT $X(k)$ is also an N-point sequence, that is, it is not defined outside of $0 \leq k \leq N-1$. From (5.23) $\tilde{X}(k) = X((k))_N$; that is, outside the $0 \leq k \leq N-1$ interval only the DFS $\tilde{X}(k)$ is defined, which of course is the periodic extension of $X(k)$. Finally, $X(k) = \tilde{X}(k)\mathcal{R}_N(k)$ means that the DFT $X(k)$ is the primary interval of $\tilde{X}(k)$.

The inverse discrete Fourier transform of an N-point DFT $X(k)$ is given by

$$x(n) \triangleq \text{IDFT}\,[X(k)] = \tilde{x}(n)\mathcal{R}_N(n)$$

or

$$x(n) = \frac{1}{N}\sum_{k=0}^{N-1} X(k)W_N^{-kn}, \quad 0 \leq n \leq N-1 \tag{5.25}$$

Once again $x(n)$ is not defined outside $0 \leq n \leq N - 1$. The extension of $x(n)$ outside this range is $\tilde{x}(n)$.

It is clear from the discussions at the top of this section that the DFS is practically equivalent to the DFT when $0 \leq n \leq N - 1$. Therefore the implementation of the DFT can be done in a similar fashion. If $x(n)$ and $X(k)$ are arranged as column vectors $\mathbf{x}$ and $\mathbf{X}$, respectively, then from (5.24) and (5.25) we have

$$\mathbf{X} = \mathbf{W}_N \mathbf{x}$$
$$\mathbf{x} = \frac{1}{N} \mathbf{W}_N^* \mathbf{X}$$

(5.26)

where $\mathbf{W}_N$ is the matrix defined in (5.7) and will now be called a *DFT matrix*. Hence the earlier `dfs` and `idfs` MATLAB functions can be renamed as the `dft` and `idft` functions to implement the discrete Fourier transform computations.

```
function [Xk] = dft(xn,N)
% Computes Discrete Fourier Transform
% -----------------------------------
% [Xk] = dft(xn,N)
% Xk = DFT coeff. array over 0 <= k <= N-1
% xn = N-point finite-duration sequence
%  N = Length of DFT
%
n = [0:1:N-1];                  % row vector for n
k = [0:1:N-1];                  % row vecor for k
WN = exp(-j*2*pi/N);            % Wn factor
nk = n'*k;                      % creates a N by N matrix of nk values
WNnk = WN .^ nk;                % DFT matrix
Xk = xn * WNnk;                 % row vector for DFT coefficients

function [xn] = idft(Xk,N)
% Computes Inverse Discrete Transform
% -----------------------------------
% [xn] = idft(Xk,N)
% xn = N-point sequence over 0 <= n <= N-1
% Xk = DFT coeff. array over 0 <= k <= N-1
%  N = length of DFT
%
n = [0:1:N-1];                  % row vector for n
k = [0:1:N-1];                  % row vecor for k
WN = exp(-j*2*pi/N);            % Wn factor
nk = n'*k;                      % creates a N by N matrix of nk values
WNnk = WN .^ (-nk);             % IDFT matrix
xn = (Xk * WNnk)/N;             % row vector for IDFT values
```

□ **EXAMPLE 5.6**   Let $x(n)$ be a 4-point sequence:

$$x(n) = \begin{cases} 1, & 0 \leq n \leq 3 \\ 0, & \text{otherwise} \end{cases}$$

a. Compute the discrete-time Fourier transform $X(e^{j\omega})$ and plot its magnitude and phase.

b. Compute the 4-point DFT of $x(n)$.

Solution **a.** The discrete-time Fourier transform is given by

$$X(e^{j\omega}) = \sum_{0}^{3} x(n)e^{-j\omega n} = 1 + e^{-j\omega} + e^{-j2\omega} + e^{-j3\omega}$$

$$= \frac{1 - e^{-j4\omega}}{1 - e^{-j\omega}} = \frac{\sin(2\omega)}{\sin(\omega/2)} e^{-j3\omega/2}$$

Hence

$$\left| X(e^{j\omega}) \right| = \left| \frac{\sin(2\omega)}{\sin(\omega/2)} \right|$$

and

$$\angle X(e^{j\omega}) = \begin{cases} -\dfrac{3\omega}{2}, & \text{when } \dfrac{\sin(2\omega)}{\sin(\omega/2)} > 0 \\[3mm] -\dfrac{3\omega}{2} \pm \pi, & \text{when } \dfrac{\sin(2\omega)}{\sin(\omega/2)} < 0 \end{cases}$$

The plots are shown in Figure 5.4.

b. Let us denote the 4-point DFT by $X_4(k)$. Then

$$X_4(k) = \sum_{n=0}^{3} x(n)W_4^{nk}; \quad k = 0, 1, 2, 3; \ W_4 = e^{-j2\pi/4} = -j$$

These calculations are similar to those in Example 5.1. We can also use MATLAB to compute this DFT.

```
>> x = [1,1,1,1]; N = 4;
>> X = dft(x,N);
>> magX = abs(X), phaX = angle(X)*180/pi
magX =
     4.0000    0.0000    0.0000    0.0000
phaX =
          0 -134.9810  -90.0000  -44.9979
```

Hence

$$X_4(k) = \{4, 0, 0, 0\}$$
$$\uparrow$$

Note that when the magnitude sample is zero, the corresponding angle is not zero. This is due to a particular algorithm used by MATLAB to compute the

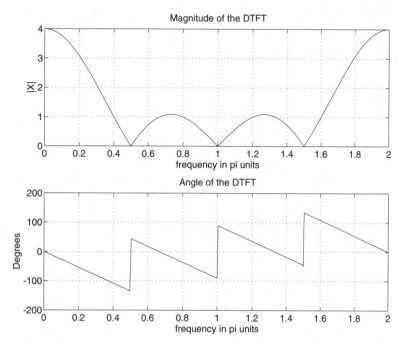

FIGURE 5.4 *The DTFT plots in Example 5.6*

angle part. Generally these angles should be ignored. The plot of DFT values is shown in Figure 5.5. The plot of $X(e^{j\omega})$ is also shown as a dashed line for comparison. From the plot in Figure 5.5 we observe that X_4 correctly gives 4 samples of $X(e^{j\omega})$, but it has only one nonzero sample. Is this surprising? By looking at the 4-point $x(n)$, which contains all 1's, one must conclude that its periodic extension is

$$\tilde{x}(n) = 1, \ \forall n$$

which is a constant (or a DC) signal. This is what is predicted by the DFT $X_4(k)$, which has a nonzero sample at $k = 0$ (or $\omega = 0$) and has no values at other frequencies. □

□ **EXAMPLE 5.7** How can we obtain other samples of the DTFT $X(e^{j\omega})$?

Solution It is clear that we should sample at dense (or finer) frequencies; that is, we should increase N. Suppose we take twice the number of points, or $N = 8$ instead of 4. This we can achieve by treating $x(n)$ as an 8-point sequence by *appending* 4 zeros.

$$x(n) = \{1, 1, 1, 1, 0, 0, 0, 0\}$$
$$\uparrow$$

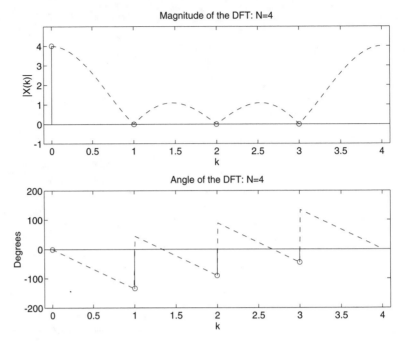

FIGURE 5.5 *The DFT plots of Example 5.6*

This is a very important operation called a *zero-padding operation*. This operation is necessary in practice to obtain a *dense spectrum* of signals as we shall see. Let $X_8(k)$ be an 8-point DFT, then

$$X_8(k) = \sum_{n=0}^{7} x(n)W_8^{nk}; \quad k = 0, 1, \dots, 7; \; W_8 = e^{-j\pi/4}$$

In this case the frequency resolution is $\omega_1 = 2\pi/8 = \pi/4$.

```
>> x = [1,1,1,1, zeros(1,4)]; N = 8;
>> X = dft(x,N);
>> magX = abs(X), phaX = angle(X)*180/pi
magX =
    4.0000    2.6131    0.0000    1.0824    0.0000    1.0824    0.0000    2.6131
phaX =
         0  -67.5000 -134.9810  -22.5000  -90.0000   22.5000  -44.9979   67.5000
```

Hence

$$X_8(k) = \{4, \; 2.6131e^{-j67.5°}, \; 0, \; 1.0824e^{-j22.5°}, \; 0, \; 1.0824e^{j22.5°},$$
$$\underset{\uparrow}{}$$
$$0, \; 2.6131e^{j67.5°}\}$$

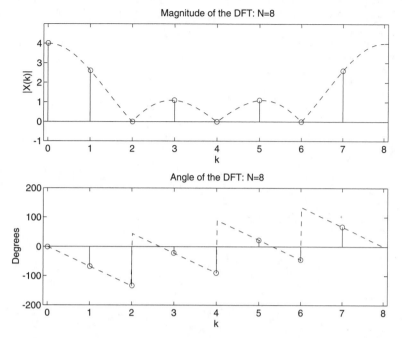

FIGURE 5.6 *The DFT plots of Example 5.7: $N = 8$*

which is shown in Figure 5.6. Continuing further, if we treat $x(n)$ as a 16-point sequence by padding 12 zeros, such that

$$x(n) = \{1, 1, 1, 1, 0, 0, 0, 0, 0, 0, 0, 0, 0, 0, 0, 0\}$$
$$\uparrow$$

then the frequency resolution is $\omega_1 = 2\pi/16 = \pi/8$ and $W_{16} = e^{-j\pi/8}$. Therefore we get a more dense spectrum with spectral samples separated by $\pi/8$. The sketch of $X_{16}(k)$ is shown in Figure 5.7. $\qquad\square$

Comments: Based on the last two examples there are several comments that we can make.

1. Zero-padding is an operation in which more zeros are appended to the original sequence. The resulting longer DFT provides closely spaced samples of the discrete-time Fourier transform of the original sequence. In MATLAB zero-padding is implemented using the `zeros` function.

2. In Example 5.6 all we needed to accurately plot the discrete-time Fourier transform $X(e^{j\omega})$ of $x(n)$ was $X_4(k)$, the 4-point DFT. This is because $x(n)$ had only 4 nonzero samples, so we could have used the interpolation formula (5.19) on $X_4(k)$ to obtain $X(e^{j\omega})$. However, in practice, it is easier to obtain $X_8(k)$ and $X_{16}(k)$, and so on, to *fill in* the values of $X(e^{j\omega})$ rather than using the interpolation formula. This approach can

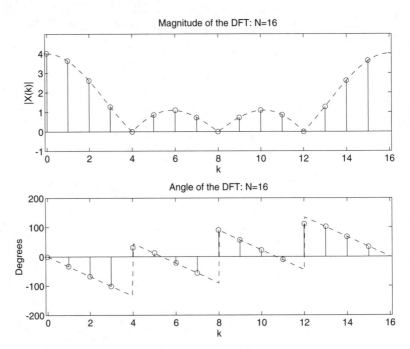

FIGURE 5.7 *The DFT plots of Example 5.7:* $N = 16$

be made even more efficient using fast Fourier transform algorithms to compute the DFT.

3. The zero-padding gives us a *high-density spectrum* and provides a better displayed version for plotting. But it does not give us a *high-resolution spectrum* because no new information is added to the signal; only additional zeros are added in the data.

4. To get a high-resolution spectrum, one has to obtain more data from the experiment or observations (see Example 5.8 below). There are also other advanced methods that use additional side information or non-linear techniques.

☐   **EXAMPLE 5.8**   To illustrate the difference between the high-density spectrum and the high-resolution spectrum, consider the sequence

$$x(n) = \cos{(0.48\pi n)} + \cos{(0.52\pi n)}$$

We want to determine its spectrum based on the finite number of samples.

    a. Determine and plot the discrete-time Fourier transform of $x(n)$, $\quad 0 \le n \le 10$.
    b. Determine and plot the discrete-time Fourier transform of $x(n)$, $\quad 0 \le n \le 100$.

Solution
We could determine analytically the discrete-time Fourier transform in each case, but MATLAB is a good vehicle to study these problems.

a. We can first determine the 10-point DFT of $x(n)$ to obtain an estimate of its discrete-time Fourier transform.

```
>> n = [0:1:99]; x = cos(0.48*pi*n)+cos(0.52*pi*n);
>> n1 = [0:1:9] ;y1 = x(1:1:10);
>> subplot(2,1,1) ;stem(n1,y1); title('signal x(n), 0 <= n <= 9');xlabel('n')
>> Y1 = dft(y1,10); magY1 = abs(Y1(1:1:6));
>> k1 = 0:1:5 ;w1 = 2*pi/10*k1;
>> subplot(2,1,2);plot(w1/pi,magY1);title('Samples of DTFT Magnitude');
>> xlabel('frequency in pi units')
```

The plots in Figure 5.8 show there aren't enough samples to draw any conclusions. Therefore we will pad 90 zeros to obtain a dense spectrum.

```
>> n2 = [0:1:99]; y2 = [x(1:1:10) zeros(1,90)];
>> subplot(2,1,1) ;stem(n2,y2) ;title('signal x(n), 0 <= n <= 9 + 90 zeros');
>> xlabel('n')
>> Y2 =dft(y2,100); magY2 = abs(Y2(1:1:51));
>> k2 = 0:1:50; w2 = 2*pi/100*k2;
>> subplot(2,1,2); plot(w2/pi,magY2); title('DTFT Magnitude');
>> xlabel('frequency in pi units')
```

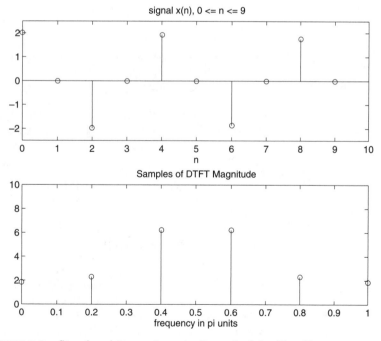

FIGURE 5.8 *Signal and its spectrum in Example 5.8a: $N = 10$*

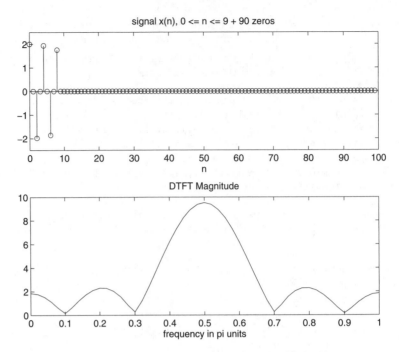

FIGURE 5.9 *Signal and its spectrum in Example 5.8a: $N = 100$*

Now the plot in Figure 5.9 shows that the sequence has a dominant frequency at $\omega = 0.5\pi$. This fact is not supported by the original sequence, which has two frequencies. The zero-padding provided a smoother version of the spectrum in Figure 5.8.

b. To get better spectral information, we will take the first 100 samples of $x(n)$ and determine its discrete-time Fourier transform.

```
>> subplot(2,1,1); stem(n,x);
>> title('signal x(n), 0 <= n <= 99'); xlabel('n')
>> X = dft(x,100); magX = abs(X(1:1:51));
>> k = 0:1:50; w = 2*pi/100*k;
>> subplot(2,1,2); plot(w/pi,magX); title('DTFT Magnitude');
>> xlabel('frequency in pi units')
```

Now the discrete-time Fourier transform plot in Figure 5.10 clearly shows two frequencies, which are very close to each other. This is the high-resolution spectrum of $x(n)$. Note that padding more zeros to the 100-point sequence will result in a smoother rendition of the spectrum in Figure 5.10 but will not reveal any new information. Students are encouraged to verify this. □

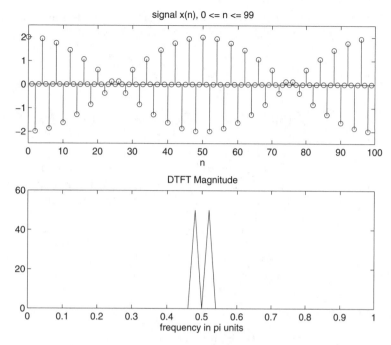

FIGURE 5.10 *Signal and its spectrum in Example 5.8b: $N = 100$*

PROPERTIES OF THE DISCRETE FOURIER TRANSFORM

The DFT properties are derived from those of the DFS because mathematically DFS is the valid representation. We discuss several useful properties, which are given without proof. These properties also apply to the DFS with necessary changes. Let $X(k)$ be an N-point DFT of the sequence $x(n)$. Unless otherwise stated, the N-point DFTs will be used in these properties.

1. **Linearity:** The DFT is a linear transform

$$\text{DFT}\left[ax_1(n) + bx_2(n)\right] = a\,\text{DFT}\left[x_1(n)\right] + b\,\text{DFT}\left[x_2(n)\right] \qquad \textbf{(5.27)}$$

Note: If $x_1(n)$ and $x_2(n)$ have different durations—that is, they are N_1-point and N_2-point sequences, respectively—then choose $N_3 = \max(N_1, N_2)$ and proceed by taking N_3-point DFTs.

2. **Circular folding:** If an N-point sequence is folded, then the result $x(-n)$ would not be an N-point sequence, and it would not be possible

to compute its DFT. Therefore we use the modulo-N operation on the argument $(-n)$ and define folding by

$$x\left((-n)\right)_N = \begin{cases} x(0), & n = 0 \\ x(N-n), & 1 \leq n \leq N-1 \end{cases} \tag{5.28}$$

This is called a *circular folding*. To visualize it, imagine that the sequence $x(n)$ is wrapped around a circle in the counterclockwise direction so that indices $n = 0$ and $n = N$ overlap. Then $x((-n))_N$ can be viewed as a clockwise wrapping of $x(n)$ around the circle; hence the name circular folding. In MATLAB the circular folding can be achieved by x=x(mod(-n,N)+1). Note that the arguments in MATLAB begin with 1. Then its DFT is given by

$$\text{DFT}\left[x\left((-n)\right)_N\right] = X\left((-k)\right)_N = \begin{cases} X(0), & k = 0 \\ X(N-k), & 1 \leq k \leq N-1 \end{cases} \tag{5.29}$$

◻ **EXAMPLE 5.9**   Let $x(n) = 10\,(0.8)^n$, $\quad 0 \leq n \leq 10$.

      a. Determine and plot $x\left((-n)\right)_{11}$.
      b. Verify the circular folding property.

Solution   **a. MATLAB Script**

```
>> n = 0:100; x = 10*(0.8) .^ n;
>> y = x(mod(-n,11)+1);
>> subplot(2,1,1); stem(n,x); title('Original sequence')
>> xlabel('n'); ylabel('x(n)');
>> subplot(2,1,2); stem(n,y); title('Circularly folded sequence')
>> xlabel('n'); ylabel('x(-n mod 10)');
```

The plots in Figure 5.11 show the effect of circular folding.

      b. MATLAB Script

```
>> X = dft(x,11); Y = dft(y,11);
>> subplot(2,2,1); stem(n,real(X));
>> title('Real{DFT[x(n)]}'); xlabel('k');
>> subplot(2,2,2); stem(n,imag(X));
>> title('Imag{DFT[x(n)]}'); xlabel('k');
>> subplot(2,2,3); stem(n,real(Y));
>> title('Real{DFT[x((-n))11]}'); xlabel('k');
>> subplot(2,2,4); stem(n,imag(Y));
>> title('Imag{DFT[x((-n))11]}'); xlabel('k');
```

The plots in Figure 5.12 verify the property.   ◻

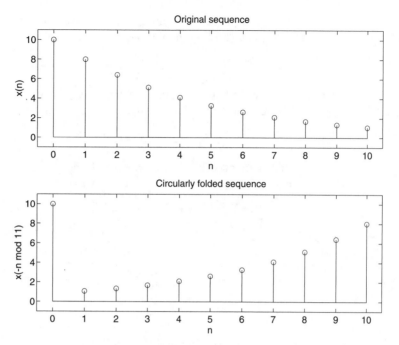

FIGURE 5.11 *Circular folding in Example 5.9a*

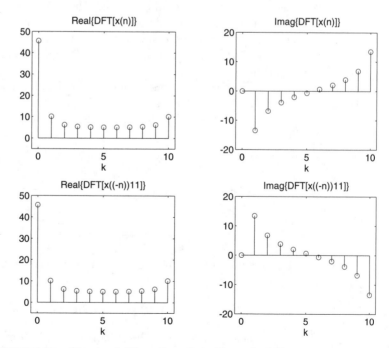

FIGURE 5.12 *Circular folding property in Example 5.9b*

3. **Conjugation:** Similar to the above property we have to introduce the circular folding in the frequency domain.

$$\text{DFT}\,[x^*(n)] = X^*\,((-k))_N \qquad (5.30)$$

4. **Symmetry properties for real sequences:** Let $x(n)$ be a real-valued N-point sequence. Then $x(n) = x^*(n)$. Using the above property,

$$X(k) = X^*\,((-k))_N \qquad (5.31)$$

This symmetry is called a *circular conjugate symmetry*. It further implies that

$$\text{Re}\,[X(k)] = \text{Re}\,[X\,((-k))_N] \qquad\Longrightarrow \text{Circular-even sequence}$$
$$\text{Im}\,[X(k)] = -\,\text{Im}\,[X\,((N-k))_N] \Longrightarrow \text{Circular-odd sequence}$$
$$|X(k)| = |X\,((-k))_N| \qquad\qquad \Longrightarrow \text{Circular-even sequence}$$
$$\angle X(k) = -\angle X\,((-k))_N \qquad\quad \Longrightarrow \text{Circular-odd sequence}$$

$$(5.32)$$

Comments: 1. Observe the magnitudes and angles of the various DFTs in Examples 5.6 and 5.7. They do satisfy the above circular symmetries. These symmetries are different than the usual even and odd symmetries. To visualize this, imagine that the DFT samples are arranged around a circle so that the indices $k = 0$ and $k = N$ overlap; then the samples will be symmetric with respect to $k = 0$, which justifies the name circular symmetry.

2. The corresponding symmetry for the DFS coefficients is called the *periodic conjugate symmetry*.

3. Since these DFTs have symmetry, one needs to compute $X(k)$ only for

$$k = 0, 1, \ldots, \frac{N}{2}; \quad N \text{ even}$$

or for

$$k = 0, 1, \ldots, \frac{N-1}{2}; \quad N \text{ odd}$$

This results in about 50% savings in computation as well as in storage.

4. From (5.30)

$$X(0) = X^*((-0))_N = X^*(0)$$

which means that the DFT coefficient at $k = 0$ must be a real number. But $k = 0$ means that the frequency $\omega_k = k\omega_1 = 0$, which is the DC frequency. Hence the DC coefficient for a real-valued $x(n)$ must be a real number. In addition, if N is even, then $N/2$ is also an integer. Then from (5.32)

$$X\left(\frac{N}{2}\right) = X^*\left(\left(-\frac{N}{2}\right)\right)_N = X^*\left(\frac{N}{2}\right)$$

which means that even the $k = N/2$ component is also real-valued. This component is called the *Nyquist component* since $k = N/2$ means that the frequency $\omega_{N/2} = (N/2)(2\pi/N) = \pi$, which is the digital Nyquist frequency.

The real-valued signals can also be decomposed into their even and odd components, $x_e(n)$ and $x_o(n)$, respectively, as discussed in Chapter 2. However, these components are not N-point sequences and therefore we cannot take their N-point DFTs. Hence we define a new set of components using the circular folding discussed above. These are called *circular-even* and *circular-odd* components defined by

$$x_{ec}(n) \triangleq \frac{1}{2}[x(n) + x((-n))_N] = \begin{cases} x(0), & n = 0 \\ \frac{1}{2}[x(n) + x(N-n)], & 1 \le n \le N-1 \end{cases}$$

$$x_{oc}(n) \triangleq \frac{1}{2}[x(n) - x((-n))_N] = \begin{cases} 0, & n = 0 \\ \frac{1}{2}[x(n) - x(N-n)], & 1 \le n \le N-1 \end{cases}$$

(5.33)

Then

$$\text{DFT}[x_{ec}(n)] = \text{Re}[X(k)] = \text{Re}[X((-k))_N]$$
$$\text{DFT}[x_{oc}(n)] = \text{Im}[X(k)] = \text{Im}[X((-k))_N]$$

(5.34)

Implication: If $x(n)$ is real and circular-even, then its DFT is also real and circular-even. Hence only the first $0 \le n \le N/2$ coefficients are necessary for complete representation.

Using (5.33), it is easy to develop a function to decompose an N-point sequence into its circular-even and circular-odd components. The following `circevod` function uses the `mod` function given earlier to implement the $n \,\text{MOD}\, N$ operation.

```
function [xec, xoc] = circevod(x)
% signal decomposition into circular-even and circular-odd parts
% ----------------------------------------------------------------
% [xec, xoc] = circevod(x)
%
if any(imag(x) ~= 0)
        error('x is not a real sequence')
```

```
end
N = length(x); n = 0:(N-1);
xec = 0.5*(x + x(mod(-n,N)+1));
xoc = 0.5*(x - x(mod(-n,N)+1));
```

☐   **EXAMPLE 5.10**   Let $x(n) = 10\,(0.8)^n$, $0 \le n \le 10$ as in Example 5.9.

     a. Decompose and plot the $x_{ec}(n)$ and $x_{oc}(n)$ components of $x(n)$.
     b. Verify the property in (5.34).

Solution   **a.** MATLAB Script _____

```
>> n = 0:10; x = 10*(0.8) .^ n;
>> [xec,xoc] = circevod(x);
>> subplot(2,1,1); stem(n,xec); title('Circular-even component')
>> xlabel('n'); ylabel('xec(n)'); axis([-0.5,10.5,-1,11])
>> subplot(2,1,2); stem(n,xoc); title('Circular-odd component')
>> xlabel('n'); ylabel('xoc(n)'); axis([-0.5,10.5,-4,4])
```

The plots in Figure 5.13 show the circularly symmetric components of $x(n)$.

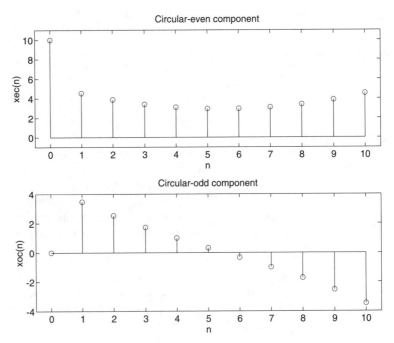

FIGURE 5.13   *Circular-even and circular-odd components of the sequence in Example 5.10a*

```
>> X = dft(x,11); Xec = dft(xec,11); Xoc = dft(xoc,11);
>> subplot(2,2,1); stem(n,real(X)); axis([-0.5,10.5,-5,50])
>> title('Real{DFT[x(n)]}'); xlabel('k');
>> subplot(2,2,2); stem(n,imag(X)); axis([-0.5,10.5,-20,20])
>> title('Imag{DFT[x(n)]}'); xlabel('k');
>> subplot(2,2,3); stem(n,real(Xec)); axis([-0.5,10.5,-5,50])
>> title('DFT[xec(n)]'); xlabel('k');
>> subplot(2,2,4); stem(n,imag(Xoc)); axis([-0.5,10.5,-20,20])
>> title('DFT[xoc(n)]'); xlabel('k');
```

From the plots in Figure 5.14 we observe that the DFT of $x_{ec}(n)$ is the same as the real part of $X(k)$ and that the DFT of $x_{oc}(n)$ is the same as the imaginary part of $X(k)$. □

A similar property for complex-valued sequences is explored in Exercise 5.10.

5. **Circular shift of a sequence:** If an N-point sequence is shifted in either direction, then the result is no longer between $0 \leq n \leq N - 1$. Therefore we first convert $x(n)$ into its periodic extension $\tilde{x}(n)$, and then shift it by m samples to obtain

$$\tilde{x}(n - m) = x\left((n - m)\right)_N \qquad (5.35)$$

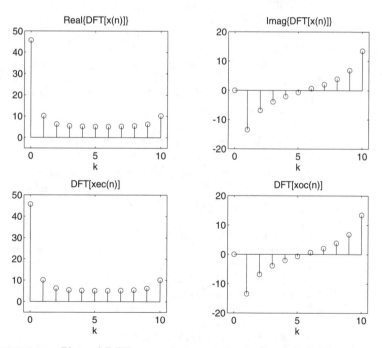

FIGURE 5.14 *Plots of DFT symmetry properties in Example 5.10b*

This is called a *periodic shift* of $\tilde{x}(n)$. The periodic shift is then converted into an N-point sequence. The resulting sequence

$$\tilde{x}(n-m)\mathcal{R}_N(n) = x\left((n-m)\right)_N \mathcal{R}_N(n) \tag{5.36}$$

is called the *circular shift* of $x(n)$. Once again to visualize this, imagine that the sequence $x(n)$ is wrapped around a circle. Now rotate the circle by k samples and unwrap the sequence from $0 \leq n \leq N-1$. Its DFT is given by

$$\text{DFT}\left[x\left((n-m)\right)_N \mathcal{R}_N(n)\right] = W_N^{km} X(k) \tag{5.37}$$

□   **EXAMPLE 5.11**   Let $x(n) = 10\,(0.8)^n$, $0 \leq n \leq 10$ be an 11-point sequence.

    a. Sketch $x\left((n+4)\right)_{11} R_{11}(n)$, that is, a circular shift by 4 samples toward the left.

    b. Sketch $x\left((n-3)\right)_{15} R_{15}(n)$, that is, a circular shift by 3 samples toward the right, where $x(n)$ is assumed to be a 15-point sequence.

Solution

We will use a step-by-step graphical approach to illustrate the circular shifting operation. This approach shows the periodic extension $\tilde{x}(n) = x\left((n)\right)_N$ of $x(n)$, followed by a linear shift in $\tilde{x}(n)$ to obtain $\tilde{x}(n-m) = x\left((n-m)\right)_N$, and finally truncating $\tilde{x}(n-m)$ to obtain the circular shift.

    a. Figure 5.15 shows four sequences. The top-left shows $x(n)$, the bottom-left shows $\tilde{x}(n)$, the top-right shows $\tilde{x}(n+4)$, and finally the bottom-right shows $x\left((n+4)\right)_{11} R_{11}(n)$. Note carefully that as samples move out of the $[0, N-1]$ window in one direction, they reappear from the opposite direction. This is the meaning of the circular shift, and it is different from the linear shift.

    b. In this case the sequence $x(n)$ is treated as a 15-point sequence by padding 4 zeros. Now the circular shift will be different than when $N = 11$. This is shown in Figure 5.16. In fact the circular shift $x\left((n-3)\right)_{15}$ looks like a linear shift $x(n-3)$.   □

To implement a circular shift, we do not have to go through the periodic shift as shown in Example 5.11. It can be implemented directly in two ways. In the first approach, the modulo-N operation can be used on the argument $(n-m)$ in the time domain. This is shown below in the `cirshftt` function.

```
function y = cirshftt(x,m,N)
% Circular shift of m samples wrt size N in sequence x: (time domain)
% ------------------------------------------------------------------
% [y] = cirshftt(x,m,N)
% y = output sequence containing the circular shift
% x = input sequence of length <= N
% m = sample shift
% N = size of circular buffer
%   Method: y(n) = x((n-m) mod N)
```

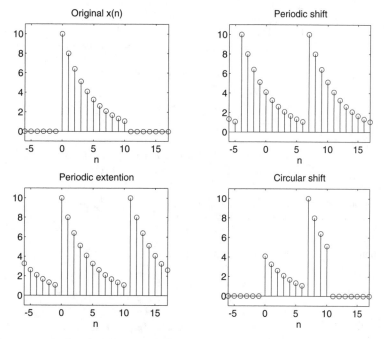

FIGURE 5.15 *Graphical interpretation of circular shift, $N = 11$*

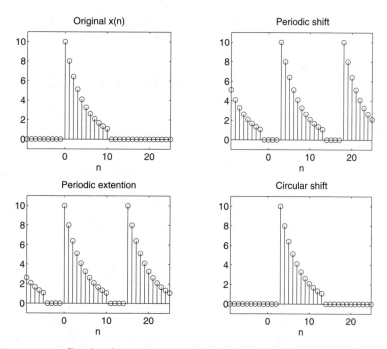

FIGURE 5.16 *Graphical interpretation of circular shift, $N = 15$*

```
% Check for length of x
if length(x) > N
        error('N must be >= the length of x')
end
x = [x zeros(1,N-length(x))];
n = [0:1:N-1];
n = mod(n-m,N);
y = x(n+1);
```

In the second approach, the property 5.37 can be used in the frequency domain. This is explored in Exercise 5.12.

□   **EXAMPLE 5.12**   Given an 11-point sequence $x(n) = 10\,(0.8)^n$, $\quad 0 \le n \le 10$, determine and plot $x\,((n-6))_{15}$.

Solution

MATLAB Script _____
```
>> n = 0:10; x = 10*(0.8) .^ n;
>> y = cirshftt(x,6,15);
>> n = 0:14; x = [x, zeros(1,4)];
>> subplot(2,1,1); stem(n,x); title('Original sequence')
>> xlabel('n'); ylabel('x(n)');
>> subplot(2,1,2); stem(n,y);
>> title('Circularly shifted sequence, N=15')
>> xlabel('n'); ylabel('x((n-6) mod 15)');
```
_____

The results are shown in Figure 5.17.   □

6. **Circular shift in the frequency domain:** This property is a dual of the above property given by

$$\text{DFT}\left[W_N^{-\ell n} x(n)\right] = X\,((k-\ell))_N\, R_N(k) \qquad \textbf{(5.38)}$$

7. **Circular convolution:** A linear convolution between two N-point sequences will result in a longer sequence. Once again we have to restrict our interval to $0 \le n \le N - 1$. Therefore instead of linear shift, we should consider the circular shift. A convolution operation that contains a circular shift is called the *circular convolution* and is given by

$$x_1(n)\,\text{\textcircled{N}}\,x_2(n) = \sum_{m=0}^{N-1} x_1(m)x_2\,((n-m))_N, \quad 0 \le n \le N-1 \quad \textbf{(5.39)}$$

Note that the circular convolution is also an N-point sequence. It has a structure similar to that of a linear convolution. The differences are in the summation limits and in the N-point circular shift. Hence it depends on N and is also called an *N-point circular convolution*. Therefore the

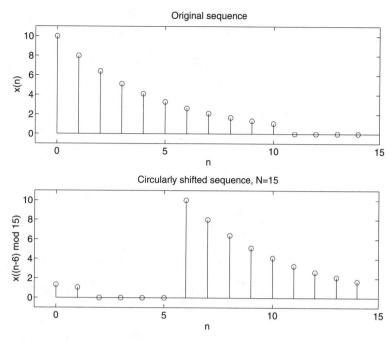

Original sequence

Circularly shifted sequence, N=15

FIGURE 5.17 *Circularly shifted sequence in Example 5.12*

use of the notation $\textcircled{N}$ is appropriate. The DFT property for the circular convolution is

$$\text{DFT}\left[x_1(n) \,\textcircled{N}\, x_2(n)\right] = X_1(k) \cdot X_2(k) \qquad \textbf{(5.40)}$$

An alternate interpretation of this property is that when we multiply two N-point DFTs in the frequency domain, we get the circular convolution (and not the usual linear convolution) in the time domain.

☐   **EXAMPLE 5.13**   Let $x_1(n) = \{1, 2, 2\}$ and $x_2(n) = \{1, 2, 3, 4\}$. Compute the 4-point circular convolution $x_1(n) \,\textcircled{4}\, x_2(n)$.

Solution   Note that $x_1(n)$ is a 3-point sequence, hence we will have to pad one zero to make it a 4-point sequence before we perform the circular convolution. We will solve this problem in the time domain as well as in the frequency domain. In the time domain we will use the mechanism of circular convolution, while in the frequency domain we will use the DFTs.

     • *Time-domain approach:* The 4-point circular convolution is given by

$$x_1(n) \,\textcircled{4}\, x_2(n) = \sum_{m=0}^{3} x_1(m)\, x_2\left(\left(n-m\right)\right)_4$$

Thus we have to create a circularly folded and shifted sequence $x_2\left((n-m)\right)_N$ for each value of n, multiply it sample-by-sample with $x_1(m)$, add the samples to obtain the circular convolution value for that n, and then repeat the procedure for $0 \leq n \leq 3$. Consider

$$x_1(m) = \{1,\ 2,\ 2,\ 0\} \qquad \text{and} \qquad x_2(m) = \{1,\ 2,\ 3,\ 4\}$$

for $n = 0$

$$\sum_{m=0}^{3} x_1(m) \cdot x_2\left((0-m)\right)_5 = \sum_{m=0}^{3} [\{1,\ 2,\ 2,\ 0\} \cdot \{1,\ 4,\ 3,\ 2\}]$$

$$= \sum_{m=0}^{3} \{1,\ 8,\ 6,\ 0\} = 15$$

for $n = 1$

$$\sum_{m=0}^{3} x_1(m) \cdot x_2\left((1-m)\right)_5 = \sum_{m=0}^{3} [\{1,\ 2,\ 2,\ 0\} \cdot \{2,\ 1,\ 4,\ 3\}]$$

$$= \sum_{m=0}^{3} \{2,\ 2,\ 8,\ 0\} = 12$$

for $n = 2$

$$\sum_{m=0}^{3} x_1(m) \cdot x_2\left((2-m)\right)_5 = \sum_{m=0}^{3} [\{1,\ 2,\ 2,\ 0\} \cdot \{3,\ 2,\ 1,\ 4\}]$$

$$= \sum_{m=0}^{3} \{3,\ 4,\ 2,\ 0\} = 9$$

for $n = 3$

$$\sum_{m=0}^{3} x_1(m) \cdot x_2\left((3-m)\right)_5 = \sum_{m=0}^{3} [\{1,\ 2,\ 2,\ 0\} \cdot \{4,\ 3,\ 2,\ 1\}]$$

$$= \sum_{m=0}^{3} \{4,\ 6,\ 4,\ 0\} = 14$$

Hence

$$x_1(n)\ \textcircled{4}\ x_2(n) = \{15,\ 12,\ 9,\ 14\}$$

- *Frequency-domain approach:* In this approach we first compute 4-point DFTs of $x_1(n)$ and $x_2(n)$, multiply them sample-by-sample, and then take the inverse DFT of the result to obtain the circular convolution.

<u>DFT of $x_1(n)$</u>

$$x_1(n) = \{1, 2, 2, 0\} \Longrightarrow X_1(k) = \{5, \ -1 - j2, \ 1, \ -1 + j2\}$$

<u>DFT of $x_2(n)$</u>

$$x_2(n) = \{1, 2, 3, 4\} \Longrightarrow X_2(k) = \{10, \ -2 + j2, \ -2, \ -2 - j2\}$$

Now

$$X1(k) \cdot X2(k) = \{50, \ 6 + j2, \ -2, \ 6 - j2\}$$

Finally after IDFT,

$$x_1(n) \ \textcircled{4} \ x_2(n) = \{15, \ 12, \ 9, \ 14\}$$

which is the same as before. $\qquad\qquad\qquad\qquad\qquad\qquad\qquad\qquad\qquad$ $\square$

Similar to the circular shift implementation, we can implement the circular convolution in a number of different ways. The simplest approach would be to implement (5.39) literally by using the `cirshftt` function and requiring two nested `for...end` loops. Obviously, this is not efficient. Another approach is to generate a sequence $x((n-m))_N$ for each n in $[0, N-1]$ as rows of a matrix and then implement (5.39) as a matrix-vector multiplication similar to our `dft` function. This would require one `for...end` loop. The following `circonvt` function incorporates these steps.

```
function y = circonvt(x1,x2,N)
% N-point circular convolution between x1 and x2: (time-domain)
% -------------------------------------------------------------
% [y] = circonvt(x1,x2,N)
% y = output sequence containing the circular convolution
% x1 = input sequence of length N1 <= N
% x2 = input sequence of length N2 <= N
% N = size of circular buffer
% Method: y(n) = sum (x1(m)*x2((n-m) mod N))
% Check for length of x1
if length(x1) > N
        error('N must be >= the length of x1')
end
% Check for length of x2
if length(x2) > N
        error('N must be >= the length of x2')
end
x1=[x1 zeros(1,N-length(x1))];
x2=[x2 zeros(1,N-length(x2))];
m = [0:1:N-1];
x2 = x2(mod(-m,N)+1);
H = zeros(N,N);
```

```
for n = 1:1:N
H(n,:) = cirshftt(x2,n-1,N);
end
y = x1*H';
```

The third approach would be to implement the frequency-domain operation (5.40) using the `dft` function. This is explored in Exercise 5.15.

☐ **EXAMPLE 5.14** Let us use MATLAB to perform the circular convolution in Example 5.13.

Solution The sequences are $x_1(n) = \{1, 2, 2\}$ and $x_2(n) = \{1, 2, 3, 4\}$.

```
>> x1 = [1,2,2]; x2 = [1,2,3,4];
>> y = circonvt(x1, x2, 4)
y =
    15    12    9    14
```

Hence

$$x_1(n) \,\, \textcircled{4} \,\, x_2(n) = \{15, \ 12, \ 9, \ 14\}$$

as before.   ☐

☐ **EXAMPLE 5.15** In this example we will study the effect of N on the circular convolution. Obviously, $N \geq 4$; otherwise there will be a time-domain aliasing for $x_2(n)$. We will use the same two sequences from Example 5.13.

    a. Compute $x_1(n) \,\, \textcircled{5} \,\, x_2(n)$.
    b. Compute $x_1(n) \,\, \textcircled{6} \,\, x_2(n)$.
    c. Comment on the results.

Solution The sequences are $x_1(n) = \{1, 2, 2\}$ and $x_2(n) = \{1, 2, 3, 4\}$. Even though the sequences are the same as in Example 5.14, we should expect different results for different values of N. This is not the case with the linear convolution, which is unique, given two sequences.

    a. 5-point circular convolution:

```
>> x1 = [1,2,2]; x2 = [1,2,3,4];
>> y = circonvt(x1, x2, 5)
y =
    9    4    9    14    14
```

Hence

$$x_1(n) \,\, \textcircled{5} \,\, x_2(n) = \{9, \ 4, \ 9, \ 14, \ 14\}$$

b. 6-point circular convolution:

```
>> x1 = [1,2,2]; x2 = [1,2,3,4];
>> y = circonvt(x1, x2, 6)
y =
       1     4     9    14    14     8
```

Hence

$$x_1(n) \;\textcircled{6}\; x_2(n) = \{1,\ 4,\ 9,\ 14,\ 14,\ 8\}$$

c. A careful observation of 4-, 5-, and 6-point circular convolutions from this and the previous example indicates some unique features. Clearly, an N-point circular convolution is an N-point sequence. However, some samples in these convolutions have the same values, while other values can be obtained as a sum of samples in other convolutions. For example, the first sample in the 5-point convolution is a sum of the first and the last sample of the 6-point convolution. The linear convolution between $x_1(n)$ and $x_2(n)$ is given by

$$x_1(n) * x_2(n) = \{1,\ 4,\ 9,\ 14,\ 14,\ 8\}$$

which is equivalent to the 6-point circular convolution. These and other issues are explored in the next section. □

8. **Multiplication:** This is the dual of the circular convolution property. It is given by

$$\mathrm{DFT}\left[x_1(n) \cdot x_2(n)\right] = \frac{1}{N} X_1(k) \;\textcircled{N}\; X_2(k) \tag{5.41}$$

in which the circular convolution is performed in the frequency domain. The MATLAB functions developed for circular convolution can also be used here since $X_1(k)$ and $X_2(k)$ are also N-point sequences.

9. **Parseval's relation:** This relation computes the energy in the frequency domain.

$$E_x = \sum_{n=0}^{N-1} |x(n)|^2 = \frac{1}{N} \sum_{k=0}^{N-1} |X(k)|^2 \tag{5.42}$$

The quantity $\frac{|X(k)|^2}{N}$ is called the *energy spectrum* of finite-duration sequences. Similarly, for periodic sequences, the quantity $\left|\frac{\tilde{X}(k)}{N}\right|^2$ is called the *power spectrum*.

LINEAR CONVOLUTION USING THE DFT
━━━━━■━━━━━

One of the most important operations in linear systems is the linear convolution. In fact FIR filters are generally implemented in practice using this linear convolution. On the other hand, the DFT is a practical approach for implementing linear system operations in the frequency domain. As we shall see later, it is also an efficient operation in terms of computations. However, there is one problem. The DFT operations result in a circular convolution (something that we do not desire), not in a linear convolution that we want. Now we shall see how to use the DFT to perform a linear convolution (or equivalently, how to make a circular convolution identical to the linear convolution). We alluded to this problem in Example 5.15.

Let $x_1(n)$ be an N_1-point sequence and let $x_2(n)$ be an N_2-point sequence. Define the linear convolution of $x_1(n)$ and $x_2(n)$ by $x_3(n)$, that is,

$$x_3(n) = x_1(n) * x_2(n) \tag{5.43}$$

$$= \sum_{k=-\infty}^{\infty} x_1(k)x_2(n-k) = \sum_{0}^{N_1-1} x_1(k)x_2(n-k)$$

Then $x_3(n)$ is a $(N_1 + N_2 - 1)$-point sequence. If we choose $N = \max(N_1, N_2)$ and compute an N-point circular convolution $x_1(n) \, \text{Ⓝ} \, x_2(n)$, then we get an N-point sequence, which obviously is different from $x_3(n)$. This observation also gives us a clue. Why not choose $N = N_1 + N_2 - 1$ and perform an $(N_1 + N_2 - 1)$-point circular convolution? Then at least both of these convolutions will have an equal number of samples.

Therefore let $N = N_1 + N_2 - 1$ and let us treat $x_1(n)$ and $x_2(n)$ as N-point sequences. Define the N-point circular convolution by $x_4(n)$.

$$x_4(n) = x_1(n) \, \text{Ⓝ} \, x_2(n) \tag{5.44}$$

$$= \left[\sum_{m=0}^{N-1} x_1(m)x_2((n-k))_N \right] \mathcal{R}_N(n)$$

$$= \left[\sum_{m=0}^{N-1} x_1(m) \sum_{r=-\infty}^{\infty} x_2(n-k-rN) \right] \mathcal{R}_N(n)$$

$$= \left[\sum_{r=-\infty}^{\infty} \underbrace{\sum_{m=0}^{N_1-1} x_1(m)x_2(n-k-rN)}_{x_3(n-rN)} \right] \mathcal{R}_N(n)$$

$$= \left[\sum_{r=-\infty}^{\infty} x_3(n - rN) \right] \mathcal{R}_N(n) \qquad \text{using (5.43)}$$

This analysis shows that, in general, the circular convolution is an aliased version of the linear convolution. We observed this fact in Example 5.15. Now since $x_3(n)$ is an $N = (N_1 + N_2 - 1)$-point sequence, we have

$$x_4(n) = x_3(n); \quad 0 \le n \le (N - 1)$$

which means that there is no aliasing in the time domain.

Conclusion: If we make both $x_1(n)$ and $x_2(n)$ $N = N_1 + N_2 - 1$ point sequences by padding an appropriate number of zeros, then the circular convolution is identical to the linear convolution.

☐ **EXAMPLE 5.16** Let $x_1(n)$ and $x_2(n)$ be the two 4-point sequences given below.

$$x_1(n) = \{1, \ 2, \ 2, \ 1\}, \quad x_2(n) = \{1, \ -1, \ -1, \ 1\}$$

    a. Determine their linear convolution $x_3(n)$.
    b. Compute the circular convolution $x_4(n)$ so that it is equal to $x_3(n)$.

Solution   We will use MATLAB to do this problem.

    a. MATLAB Script_____

```
>> x1 = [1,2,2,1]; x2=[1,-1,-1,1];
>> x3 = conv(x1,x2)
x3 =    1     1    -1    -2    -1     1     1
```

Hence the linear convolution $x_3(n)$ is a 7-point sequence given by

$$x_3(n) = \{1, 1, -1, -2, -1, 1, 1\}$$

    b. We will have to use $N \ge 7$. Choosing $N = 7$, we have

```
>> x4 = circonvt(x1,x2,7)
x4 =    1     1    -1    -2    -1     1     1
```

Hence

$$x_4 = \{1, 1, -1, -2, -1, 1, 1\} = x_3(n) \qquad \square$$

ERROR ANALYSIS   In order to use the DFT for linear convolution, we must choose N properly. However, in practice it may not be possible to do so, especially when N is very large and there is a limit on memory. Then an error will be introduced when N is chosen less than the required value to perform the circular convolution. We want to compute this error, which is useful in practice. Obviously, $N \geq \max(N_1, N_2)$. Therefore let

$$\max(N_1, N_2) \leq N < (N_1 + N_2 - 1)$$

Then from our previous analysis (5.44)

$$x_4(n) = \left[\sum_{r=-\infty}^{\infty} x_3(n - rN)\right] \mathcal{R}_N(n)$$

Let an error $e(n)$ be given by

$$e(n) \triangleq x_4(n) - x_3(n)$$

$$= \left[\sum_{r \neq 0} x_3(n - rN)\right] \mathcal{R}_N(n)$$

Since $N \geq \max(N_1, N_2)$, only two terms corresponding to $r = \pm 1$ remain in the above summation. Hence

$$e(n) = [x_3(n - N) + x_3(n + N)] \mathcal{R}_N(n)$$

Generally, $x_1(n)$ and $x_2(n)$ are causal sequences. Then $x_3(n)$ is also causal, which means that

$$x_3(n - N) = 0; \quad 0 \leq n \leq N - 1$$

Therefore

$$e(n) = x_3(n + N), \quad 0 \leq n \leq N - 1 \qquad \textbf{(5.45)}$$

This is a simple yet important relation. It implies that when $\max(N_1, N_2) \leq N < (N_1 + N_2 - 1)$, the error value at n is the same as the linear convolution value computed N samples away. Now the linear convolution will be zero after $(N_1 + N_2 - 1)$ samples. This means that the first few samples of the circular convolution are in error, while the remaining ones are the correct linear convolution values.

☐ **EXAMPLE 5.17**   Consider the sequences $x_1(n)$ and $x_2(n)$ from the previous example. Evaluate circular convolutions for $N = 6, 5$, and 4. Verify the error relations in each case.

Solution   Clearly, the linear convolution $x_3(n)$ is still the same.

$$x_3(n) = \{1, 1, -1, -2, -1, 1, 1\}$$

When $N = 6$, we obtain a 6-point sequence.

$$x_4(n) = x_1(n) \; \textcircled{6} \; x_2(n) = \{2, 1, -1, -2, -1, 1\}$$

Therefore

$$e(n) = \{2, 1, -1, -2, -1, 1\} - \{1, 1, -1, -2, -1, 1\}, \quad 0 \le n \le 5$$

$$= \{1, 0, 0, 0, 0, 0\}$$

$$= x_3(n+6)$$

as expected. When $N = 5$, we obtain a 5-point sequence,

$$x_4(n) = x_1(n) \; \textcircled{5} \; x_2(n) = \{2, 2, -1, -2, -1\}$$

and

$$e(n) = \{2, 2, -1, -2, -1\} - \{1, 1, -1, -2, -1\}, \quad 0 \le n \le 4$$

$$= \{1, 1, 0, 0, 0\}$$

$$= x_3(n+5)$$

Finally, when $N = 4$, we obtain a 4-point sequence,

$$x_4(n) = x_1(n) \; \textcircled{4} \; x_2(n) = \{0, 2, 0, -2\}$$

and

$$e(n) = \{0, 2, 0, -2\} - \{1, 1, -1, -2\}, \quad 0 \le n \le 3$$

$$= \{-1, 1, 1, 0\}$$

$$= x_3(n+4)$$

The last case of $N = 4$ also provides the useful observation given below. $\qquad \square$

Observation: When $N = \max(N_1, N_2)$ is chosen for circular convolution, then the first $(M - 1)$ samples are in error (i.e., different from the linear convolution), where $M = \min(N_1, N_2)$. This result is useful in implementing long convolutions in the form of block processing.

BLOCK CON-VOLUTIONS

When we want to filter an input sequence that is being received continuously, such as a speech signal from a microphone, then for practical purposes we can think of this sequence as an infinite-length sequence. If we want to implement this filtering operation as an FIR filter in which the linear convolution is computed using the DFT, then we experience some practical problems. We will have to compute a large DFT, which is generally impractical. Furthermore, output samples are not available until all input samples are processed. This introduces an unacceptably large amount of delay. Therefore we have to segment the infinite-length input sequence into smaller sections (or blocks), process each section using the DFT, and finally assemble the output sequence from the outputs of each section. This procedure is called a *block convolution* (or block processing) operation.

Let us assume that the sequence $x(n)$ is sectioned into N-point sequences and that the impulse response of the filter is an M-point sequence, where $M < N$. Then from the above observation we note that the N-point circular convolution between the input block and the impulse response will yield a block output sequence in which the first $(M-1)$ samples are not the correct output values. If we simply partition $x(n)$ into nonoverlapping sections, then the resulting output sequence will have intervals of incorrect samples. To correct this problem, we can partition $x(n)$ into sections, each overlapping with the previous one by exactly $(M-1)$ samples, save the last $(N-M+1)$ output samples, and finally concatenate these outputs into a sequence. To correct for the first $(M-1)$ samples in the first output block, we set the first $(M-1)$ samples in the first input block to zero. This procedure is called an *overlap-save* method of block convolutions. Clearly, when $N \gg M$, this method is more efficient. We illustrate it using a simple example.

☐ **EXAMPLE 5.18**   Let $x(n) = (n+1)$, $0 \le n \le 9$ and $h(n) = \{1, 0, -1\}$. Implement the overlap-save method using $N = 6$ to compute $y(n) = x(n) * h(n)$.
$$\uparrow$$

Solution   Since $M = 3$, we will have to overlap each section with the previous one by two samples. Now $x(n)$ is a 10-point sequence, and we will need $(M-1) = 2$ zeros in the beginning. Since $N = 6$, we will need 3 sections. Let the sections be

$$x_1(n) = \{0, 0, 1, 2, 3, 4\}$$

$$x_2(n) = \{3, 4, 5, 6, 7, 8\}$$

$$x_3(n) = \{7, 8, 9, 10, 0, 0\}$$

Note that we have to pad $x_3(n)$ by two zeros since $x(n)$ runs out of values at $n = 9$. Now we will compute the 6-point circular convolution of each section with $h(n)$.

$$y_1 = x_1(n) \,\textcircled{6}\, h(n) = \{-3, -4, 1, 2, 2, 2\}$$

$$y_2 = x_2(n) \,\textcircled{6}\, h(n) = \{-4, -4, 2, 2, 2, 2\}$$

$$y_3 = x_3(n) \,\textcircled{6}\, h(n) = \{7, 8, 2, 2, -9, -10\}$$

Noting that the first two samples are to be discarded, we assemble the output $y(n)$ as

$$y(n) = \{1, 2, 2, 2, 2, 2, 2, 2, 2, 2, -9, -10\}$$
$$\uparrow$$

The linear convolution is given by

$$x(n) * h(n) = \{1, 2, 2, 2, 2, 2, 2, 2, 2, 2, -9, -10\}$$
$$\uparrow$$

which agrees with the overlap-save method.   □

MATLAB
IMPLEMEN-
TATION

Using the above example as a guide, we can develop a MATLAB function to implement the overlap-save method for a very long input sequence $x(n)$. The key step in this function is to obtain a proper indexing for the segmentation. Given $x(n)$ for $n \geq 0$, we have to set the first $(M - 1)$ samples to zero to begin the block processing. Let this augmented sequence be

$$\hat{x}(n) \triangleq \underbrace{\{0, 0, \ldots, 0}_{(M-1) \text{ zeros}}, x(n)\}, \quad n \geq 0$$

and let $L = N - M + 1$, then the kth block $x_k(n)$, $0 \leq n \leq N - 1$, is given by

$$x_k(n) = \hat{x}(m); \quad kL \leq m \leq kL + N - 1, \ k \geq 0, \ 0 \leq n \leq N - 1$$

The total number of blocks is given by

$$K = \left\lfloor \frac{N_x + M - 2}{L} \right\rfloor + 1$$

where N_x is the length of $x(n)$ and $\lfloor \cdot \rfloor$ is the truncation operation. Now each block can be circularly convolved with $h(n)$ using the `circonvt` function developed earlier to obtain

$$y_k(n) = x_k(n) \ \textcircled{N} \ h(n)$$

Finally, discarding the first $(M - 1)$ samples from each $y_k(n)$ and concatenating the remaining samples, we obtain the linear convolution $y(n)$. This procedure is incorporated in the following `ovrlpsav` function.

```
function [y] = ovrlpsav(x,h,N)
% Overlap-Save method of block convolution
% ----------------------------------------
% [y] = ovrlpsav(x,h,N)
% y = output sequence
% x = input sequence
% h = impulse response
% N = block length
%
Lenx = length(x); M = length(h);
M1 = M-1; L = N-M1;
```

```
h = [h zeros(1,N-M)];
%
x = [zeros(1,M1), x, zeros(1,N-1)]; % preappend (M-1) zeros
K = floor((Lenx+M1-1)/(L));        % # of blocks
Y = zeros(K+1,N);
% convolution with succesive blocks
for k=0:K
xk = x(k*L+1:k*L+N);
Y(k+1,:) = circonvt(xk,h,N);
end
Y = Y(:,M:N)';                     % discard the first (M-1) samples
y = (Y(:))';                       % assemble output
```

It should be noted that the `ovrlpsav` function as developed here is not the most efficient approach. We will come back to this issue when we discuss the fast Fourier transform.

☐   **EXAMPLE 5.19**   To verify the operation of the `ovrlpsav` function, let us consider the sequences given in Example 5.18.

Solution

MATLAB Script _____
```
>> n = 0:9; x = n+1; h = [1,0,-1]; N = 6;
>> y = ovrlpsav(x,h,N)
y =
     1    2    2    2    2    2    2    2    2    2   -9  -10
```
_____

This is the correct linear convolution as expected.   ☐

There is an alternate method called an *overlap-add* method of block convolutions. In this method the input sequence $x(n)$ is partitioned into nonoverlapping blocks and convolved with the impulse response. The resulting output blocks are overlapped with the subsequent sections and added to form the overall output. This is explored in Exercise 5.20.

THE FAST FOURIER TRANSFORM
■

The DFT (5.24) introduced earlier is the only transform that is discrete in both the time and the frequency domains, and is defined for finite-duration sequences. Although it is a computable transform, the straightforward implementation of (5.24) is very inefficient, especially when the sequence length N is large. In 1965 Cooley and Tukey [4] showed a procedure to substantially reduce the amount of computations involved in the DFT. This led to the explosion of applications of the DFT, including in the digital signal processing area. Furthermore, it also led to the development of other efficient algorithms. All these efficient algorithms are collectively known as fast Fourier transform (FFT) algorithms.

Consider an N-point sequence $x(n)$. Its N-point DFT is given by (5.24) and reproduced here

$$X(k) = \sum_{n=0}^{N-1} x(n) W_N^{nk}, \quad 0 \le k \le N-1 \tag{5.46}$$

where $W_N = e^{-j2\pi/N}$. To obtain one sample of $X(k)$, we need N complex multiplications and $(N-1)$ complex additions. Hence to obtain a complete set of DFT coefficients, we need N^2 complex multiplications and $N(N-1) \simeq N^2$ complex additions. Also one has to store N^2 complex coefficients $\{W_N^{nk}\}$ (or generate internally at an extra cost). Clearly, the number of DFT computations for an N-point sequence depends quadratically on N, which will be denoted by the notation

$$C_N = o\left(N^2\right)$$

For large N, $o\left(N^2\right)$ is unacceptable in practice. Generally, the processing time for one addition is much less than that for one multiplication. Hence from now on we will concentrate on the number of complex multiplications, which itself requires 4 real multiplications and 2 real additions.

Goal of an Efficient Computation In an efficiently designed algorithm the number of computations should be constant per data sample, and therefore the total number of computations should be linear with respect to N.

The quadratic dependence on N can be reduced by realizing that most of the computations (which are done again and again) can be eliminated using the periodicity property

$$W_N^{kn} = W_N^{k(n+N)} = W_N^{(k+N)n}$$

and the symmetry property

$$W_N^{kn+N/2} = -W_N^{kn}$$

of the factor $\{W_N^{nk}\}$.

One algorithm that considers only the periodicity of W_N^{nk} is the Goertzel algorithm. This algorithm still requires $C_N = o(N^2)$ multiplications, but it has certain advantages. This algorithm is described in Chapter 10. We first begin with an example to illustrate the advantages of the symmetry and periodicity properties in reducing the number of computations. We then describe and analyze two specific FFT algorithms that require $C_N = o(N \log N)$ operations. They are the *decimation-in-time* (DIT-FFT) and *decimation-in-frequency* (DIF-FFT) algorithms.

☐   **EXAMPLE 5.20**   Let us discuss the computations of a 4-point DFT and develop an efficient algorithm for its computation.

$$X(k) = \sum_{n=0}^{3} x(n) W_4^{nk}, \quad 0 \le k \le 3; \quad W_4 = e^{-j2\pi/4} = -j$$

Solution   The above computations can be done in the matrix form

$$
\begin{bmatrix} X(0) \\ X(1) \\ X(2) \\ X(3) \end{bmatrix} =
\begin{bmatrix} W_4^0 & W_4^0 & W_4^0 & W_4^0 \\ W_4^0 & W_4^1 & W_4^2 & W_4^3 \\ W_4^0 & W_4^2 & W_4^4 & W_4^6 \\ W_4^0 & W_4^3 & W_4^6 & W_4^9 \end{bmatrix}
\begin{bmatrix} x(0) \\ x(1) \\ x(2) \\ x(3) \end{bmatrix}
$$

which requires 16 complex multiplications.

Efficient Approach: Using periodicity,

$$W_4^0 = W_4^4 = 1 \quad ; \quad W_4^1 = W_4^9 = -j$$
$$W_4^2 = W_4^6 = -1 \quad ; \quad W_4^3 = j$$

and substituting in the above matrix form, we get

$$
\begin{bmatrix} X(0) \\ X(1) \\ X(2) \\ X(3) \end{bmatrix} =
\begin{bmatrix} 1 & 1 & 1 & 1 \\ 1 & -j & -1 & j \\ 1 & -1 & 1 & -1 \\ 1 & j & -1 & -j \end{bmatrix}
\begin{bmatrix} x(0) \\ x(1) \\ x(2) \\ x(3) \end{bmatrix}
$$

Using symmetry, we obtain

$$X(0) = x(0) + x(1) + x(2) + x(3) = \underbrace{[x(0) + x(2)]}_{g_1} + \underbrace{[x(1) + x(3)]}_{g_2}$$

$$X(1) = x(0) - jx(1) - x(2) + jx(3) = \underbrace{[x(0) - x(2)]}_{h_1} - j\underbrace{[x(1) - x(3)]}_{h_2}$$

$$X(2) = x(0) - x(1) + x(2) - x(3) = \underbrace{[x(0) + x(2)]}_{g_1} - \underbrace{[x(1) + x(3)]}_{g_2}$$

$$X(3) = x(0) + jx(1) - x(2) - jx(3) = \underbrace{[x(0) - x(2)]}_{h_1} + j\underbrace{[x(1) - x(3)]}_{h_2}$$

Hence an efficient algorithm is

$$
\begin{array}{c|c}
\text{Step 1} & \text{Step 2} \\
g_1 = x(0) + x(2) & X(0) = g_1 + g_2 \\
g_2 = x(1) + x(3) & X(1) = h_1 - jh_2 \\
h_1 = x(0) - x(2) & X(2) = g_1 - g_2 \\
h_2 = x(1) - x(3) & X(3) = h_1 + jh_2
\end{array}
\tag{5.47}
$$

which requires only 2 complex multiplications, which is a considerably smaller number, even for this simple example. A signal flowgraph structure for this algorithm is given in Figure 5.18.

An Interpretation: This efficient algorithm (5.47) can be interpreted differently. First, a 4-point sequence $x(n)$ is divided into two 2-point sequences, which are arranged into column vectors as given below.

$$
\left[\begin{bmatrix} x(0) \\ x(2) \end{bmatrix}, \begin{bmatrix} x(1) \\ x(3) \end{bmatrix} \right] = \begin{bmatrix} x(0) & x(1) \\ x(2) & x(3) \end{bmatrix}
$$

Second, a smaller 2-point DFT of each column is taken.

$$
\mathbf{W}_2 \begin{bmatrix} x(0) & x(1) \\ x(2) & x(3) \end{bmatrix} = \begin{bmatrix} 1 & 1 \\ 1 & -1 \end{bmatrix} \begin{bmatrix} x(0) & x(1) \\ x(2) & x(3) \end{bmatrix}
$$

$$
= \begin{bmatrix} x(0) + x(2) & x(1) + x(3) \\ x(0) - x(2) & x(1) - x(3) \end{bmatrix} = \begin{bmatrix} g_1 & g_2 \\ h_1 & h_2 \end{bmatrix}
$$

Then each element of the resultant matrix is multiplied by $\{W_4^{pq}\}$, where p is the row index and q is the column index; that is, the following *dot-product* is performed:

$$
\begin{bmatrix} 1 & 1 \\ 1 & -j \end{bmatrix} .* \begin{bmatrix} g_1 & g_2 \\ h_1 & h_2 \end{bmatrix} = \begin{bmatrix} g_1 & g_2 \\ h_1 & -jh_2 \end{bmatrix}
$$

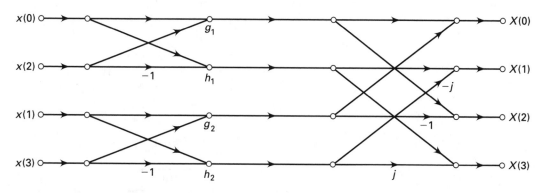

FIGURE 5.18 *Signal flowgraph in Example 5.20*

Finally, two more smaller 2-point DFTs are taken of *row vectors*.

$$
\begin{bmatrix} g_1 & g_2 \\ h_1 & -jh_2 \end{bmatrix} \mathbf{W}_2 = \begin{bmatrix} g_1 & g_2 \\ h_1 & -jh_2 \end{bmatrix} \begin{bmatrix} 1 & 1 \\ 1 & -1 \end{bmatrix} = \begin{bmatrix} g_1 + g_2 & g_1 - g_2 \\ h_1 - jh_2 & h_1 + jh_2 \end{bmatrix}
$$

$$
= \begin{bmatrix} X(0) & X(2) \\ X(1) & X(3) \end{bmatrix}
$$

Although this interpretation seems to have more multiplications than the efficient algorithm, it does suggest a systematic approach of computing a larger DFT based on smaller DFTs. □

DIVIDE-AND-COMBINE APPROACH

To reduce the DFT computation's quadratic dependence on N, one must choose a composite number $N = LM$ since

$$
L^2 + M^2 \ll N^2 \text{ for large } N
$$

Now divide the sequence into M smaller sequences of length L, take M smaller L-point DFTs, and then combine these into a larger DFT using L smaller M-point DFTs. This is the essence of the divide-and-combine approach. Let $N = LM$, then the indices n and k in (5.46) can be written as

$$
\begin{aligned}
n &= M\ell + m, & 0 \le \ell \le L-1, & \quad 0 \le m \le M-1 \\
k &= p + Lq, & 0 \le p \le L-1, & \quad 0 \le q \le M-1
\end{aligned} \tag{5.48}
$$

and write sequences $x(n)$ and $X(k)$ as arrays $x(\ell, m)$ and $X(p, q)$, respectively. Then (5.46) can be written as

$$
X(p, q) = \sum_{m=0}^{M-1} \sum_{\ell=0}^{L-1} x(\ell, m) W_N^{(M\ell+m)(p+Lq)} \tag{5.49}
$$

$$
= \sum_{m=0}^{M-1} \left\{ W_N^{mp} \left[\sum_{\ell=0}^{L-1} x(\ell, m) W_N^{M\ell p} \right] \right\} W_N^{Lmq}
$$

$$
= \sum_{m=0}^{M-1} \left\{ W_N^{mp} \underbrace{\left[\sum_{\ell=0}^{L-1} x(\ell, m) W_L^{\ell p} \right]}_{L\text{-point DFT}} \right\} W_M^{mq}
$$

$$
\underbrace{\phantom{= \sum_{m=0}^{M-1} \left\{ W_N^{mp} \left[\sum_{\ell=0}^{L-1} x(\ell, m) W_L^{\ell p} \right] \right\} W_M^{mq}}}_{M\text{-point DFT}}
$$

Hence (5.49) can be implemented as a three-step procedure:

1. First, we compute the L-point DFT array

$$F(p, m) = \sum_{\ell=0}^{L-1} x(\ell, m) W_L^{\ell p}; \quad 0 \le p \le L - 1 \qquad (5.50)$$

for each of the columns $m = 0, \ldots, M - 1$.

2. Second, we modify $F(p, m)$ to obtain another array.

$$G(p, m) = W_N^{pm} F(p, m), \quad \begin{matrix} 0 \le p \le L - 1 \\ 0 \le m \le M - 1 \end{matrix} \qquad (5.51)$$

The factor W_N^{pm} is called a *twiddle* factor.

3. Finally, we compute the M-point DFTs

$$X(p, q) = \sum_{m=0}^{M-1} G(p, m) W_M^{mq} \quad 0 \le q \le M - 1 \qquad (5.52)$$

for each of the rows $p = 0, \ldots, L - 1$.

The total number of complex multiplications for this approach can now be given by

$$C_N = ML^2 + N + LM^2 < o\left(N^2\right) \qquad (5.53)$$

This procedure can be further repeated if M or L are composite numbers. Clearly, the most efficient algorithm is obtained when N is a highly composite number, that is, $N = R^\nu$. Such algorithms are called *radix-R* FFT algorithms. When $N = R_1^{\nu_1} R_2^{\nu_2} \cdots$, then such decompositions are called *mixed-radix* FFT algorithms. The one most popular and easily programmable algorithm is the radix-2 FFT algorithm.

RADIX-2 FFT ALGORITHM Let $N = 2^\nu$; then we choose $M = 2$ and $L = N/2$ and divide $x(n)$ into two $N/2$-point sequences according to (5.48) as

$$\begin{matrix} g_1(n) = x(2n) \\ g_2(n) = x(2n + 1) \end{matrix}; \quad 0 \le n \le \frac{N}{2} - 1$$

The sequence $g_1(n)$ contains even-ordered samples of $x(n)$, while $g_2(n)$ contains odd-ordered samples of $x(n)$. Let $G_1(k)$ and $G_2(k)$ be $N/2$-point DFTs of $g_1(n)$ and $g_2(n)$, respectively. Then (5.49) reduces to

$$X(k) = G_1(k) + W_N^k G_2(k), \quad 0 \le k \le N - 1 \qquad (5.54)$$

This is called a *merging formula*, which combines two $N/2$-point DFTs into one N-point DFT. The total number of complex multiplications reduces to

$$C_N = \frac{N^2}{2} + N = o\left(N^2/2\right)$$

This procedure can be repeated again and again. At each stage the sequences are decimated and the smaller DFTs combined. This decimation ends after ν stages when we have N one-point sequences, which are also one-point DFTs. The resulting procedure is called the *decimation-in-time* FFT (DIT-FFT) algorithm, for which the total number of complex multiplications is

$$C_N = N\nu = N\log_2 N$$

Clearly, if N is large, then C_N is approximately linear in N, which was the goal of our efficient algorithm. Using additional symmetries, C_N can be reduced to $\frac{N}{2}\log_2 N$. The signal flowgraph for this algorithm is shown in Figure 5.19 for $N = 8$.

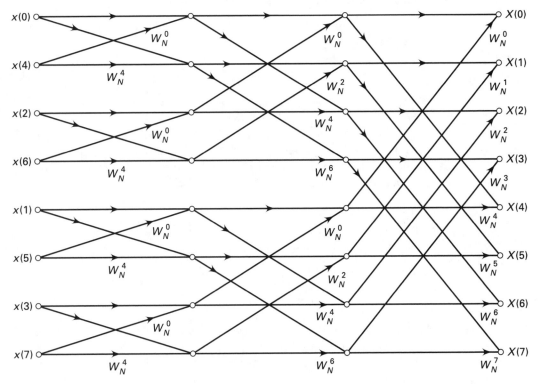

FIGURE 5.19 *Decimation-in-time FFT structure for $N = 8$*

In an alternate approach we choose $L = 2$, $M = N/2$ and follow the steps in (5.49). Note that the initial DFTs are 2-point DFTs, which contain no complex multiplications. From (5.50)

$$F(0, m) = x(0, m) + x(1, m)W_2^0$$
$$= x(n) + x(n + N/2), \ 0 \le n \le N/2$$
$$F(1, m) = x(0, m) + x(1, m)W_2^1$$
$$= x(n) - x(n + N/2), \ 0 \le n \le N/2$$

and from (5.51)

$$G(0, m) = F(0, m)W_N^0$$
$$= x(n) + x(n + N/2), \ 0 \le n \le N/2$$
$$G(1, m) = F(1, m)W_N^m$$
$$= [x(n) - x(n + N/2)] W_N^n, \ 0 \le n \le N/2$$

(5.55)

Let $G(0, m) = d_1(n)$ and $G(1, m) = d_2(n)$ for $0 \le n \le N/2 - 1$ (since they can be considered as time-domain sequences); then from (5.52) we have

$$X(0, q) = \quad X(2q) \quad = D_1(q)$$
$$X(1, q) = X(2q + 1) = D_2(q)$$

(5.56)

This implies that the DFT values $X(k)$ are computed in a decimated fashion. Therefore this approach is called a *decimation-in-frequency* FFT (DIF-FFT) algorithm. Its signal flowgraph is a transposed structure of the DIT-FFT structure, and its computational complexity is also equal to $\frac{N}{2} \log_2 N$.

MATLAB IMPLEMEN-TATION

MATLAB provides a function called `fft` to compute the DFT of a vector x. It is invoked by `X = fft(x,N)`, which computes the N-point DFT. If the length of x is less than N, then x is padded with zeros. If the argument N is omitted, then the length of the DFT is the length of x. If x is a matrix, then `fft(x,N)` computes the N-point DFT of each column of x.

This `fft` function is written in machine language and not using MATLAB commands (i.e., it is not available as a .m file). Therefore it executes very fast. It is written as a mixed-radix algorithm. If N is a power of two, then a high-speed radix-2 FFT algorithm is employed. If N is not a power of two, then N is decomposed into prime factors and a slower mixed-radix FFT algorithm is used. Finally, if N is a prime number, then the `fft` function is reduced to the raw DFT algorithm.

The inverse DFT is computed using the `ifft` function, which has the same characteristics as `fft`.

☐ **EXAMPLE 5.21** In this example we will study the execution time of the `fft` function for $1 \leq N \leq 2048$. This will reveal the divide-and-combine strategy for various values of N.

Solution To determine the execution time, MATLAB provides two functions. The `clock` function provides the instantaneous clock reading, while the `etime(t1,t2)` function computes the elapsed time between two time marks `t1` and `t2`. To determine the execution time, we will generate random vectors from length 1 through 2048, compute their FFTs, and save the computation time in an array. Finally, we will plot this execution time versus N.

MATLAB Script _____
```
>> Nmax = 2048;
>> fft_time=zeros(1,Nmax);
>> for n=1:1:Nmax
>>     x=rand(1,n);
>>     t=clock;fft(x);fft_time(n)=etime(clock,t);
>> end
>> n=[1:1:Nmax];
>> plot(n,fft_time,'.')
>> xlabel('N');ylabel('Time in Sec.')
>> title('FFT execution times')
```
_____

The plot of the execution times is shown in Figure 5.20. This plot is very informative. The points in the plot do not show one clear function but appear to group themselves into various trends. The uppermost group depicts a $o(N^2)$ dependence on N, which means that these values must be prime numbers between 1 and 2048 for which the FFT algorithm defaults to the DFT algorithm. Similarly, there are groups corresponding to the $o\left(N^2/2\right)$, $o\left(N^2/3\right)$, $o\left(N^2/4\right)$, and so on, dependencies for which the number N has fewer decompositions. The last group shows the (almost linear) $o\left(N \log N\right)$ dependence, which is for $N = 2^\nu, 0 \leq \nu \leq 11$. For these values of N, the radix-2 FFT algorithm is used. For all other values, a mixed-radix FFT algorithm is employed. This shows that the divide-and-combine strategy is very effective when N is highly composite. For example, the execution time is 0.16 second for $N = 2048$, 2.48 seconds for $N = 2047$, and 46.96 seconds for $N = 2039$. ☐

The MATLAB functions developed previously in this chapter should now be modified by substituting the `fft` function in place of the `dft` function. From the above example care must be taken to use a highly composite N. A good practice is to choose $N = 2^\nu$ unless a specific situation demands otherwise.

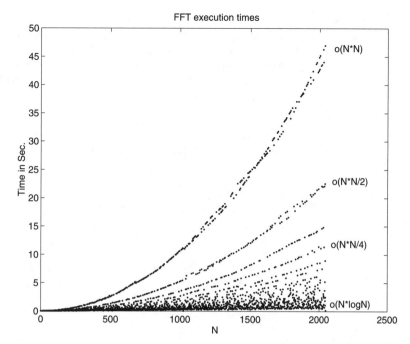

FIGURE 5.20 *FFT execution times for* $1 <= N <= 2048$

FAST CONVO-LUTIONS

The `conv` function in MATLAB is implemented using the `filter` function (which is written in C) and is very efficient for smaller values of N (< 50). For larger values of N it is possible to speed up the convolution using the FFT algorithm. This approach uses the circular convolution to implement the linear convolution, and the FFT to implement the circular convolution. The resulting algorithm is called a *fast convolution* algorithm. In addition, if we choose $N = 2^\nu$ and implement the radix-2 FFT, then the algorithm is called a *high-speed convolution*. Let $x_1(n)$ be a N_1-point sequence and $x_2(n)$ be a N_2-point sequence; then for high-speed convolution N is chosen to be

$$N = 2^{\lceil \log_2(N_1 + N_2 - 1) \rceil} \tag{5.57}$$

where $\lceil x \rceil$ is the smallest integer greater than x (also called a *ceiling* function). The linear convolution $x_1(n) * x_2(n)$ can now be implemented by two N-point FFTs, one N-point IFFT, and one N-point dot-product.

$$x_1(n) * x_2(n) = \text{IFFT}\left[\text{FFT}\left[x_1(n)\right] \cdot \text{FFT}\left[x_2(n)\right]\right] \tag{5.58}$$

For large values of N, (5.58) is faster than the time-domain convolution as we see in the following example.

EXAMPLE 5.22   To demonstrate the effectiveness of the high-speed convolution, let us compare the execution times of two approaches. Let $x_1(n)$ be an L-point uniformly distributed random number between $[0, 1]$, and let $x_2(n)$ be an L-point Gaussian random sequence with mean 0 and variance 1. We will determine the average execution times for $1 \leq L \leq 150$, in which the average is computed over the 100 realizations of random sequences.

Solution

MATLAB Script

```
conv_time = zeros(1,150); fft_time = zeros(1,150);
%
for L = 1:150
    tc = 0; tf=0;
    N = 2*L-1; nu = ceil(log10(NI)/log10(2)); N = 2^nu;
    for I=1:100
        h = randn(1,L);
        x = rand(1,L);
        t0 = clock; y1 = conv(h,x); t1=etime(clock,t0);
        tc = tc+t1;
        t0 = clock; y2 = ifft(fft(h,N).*fft(x,N)); t2=etime(clock,t0);
        tf = tf+t2;
    end
%
    conv_time(L)=tc/100;
    fft_time(L)=tf/100;
end
%
n = 1:150; subplot(1,1,1);
plot(n(25:150),conv_time(25:150),n(25:150),fft_time(25:150))
```

Figure 5.21 shows the linear convolution and the high-speed convolution times for $25 \leq L \leq 150$. It should be noted that these times are affected by the computing platform used to execute the MATLAB script. The plot in Figure 5.21 was obtained on a 33-MHz 486 computer. It shows that for low values of L the linear convolution is faster. The crossover point appears to be $L = 50$, beyond which the linear convolution time increases exponentially, while the high-speed convolution time increases fairly linearly. Note that since $N = 2^\nu$, the high-speed convolution time is constant over a range on L.   □

HIGH-SPEED BLOCK CON-VOLUTIONS   Earlier we discussed a block convolution algorithm called the overlap-and-save method (and its companion the overlap-and-add method), which is used to convolve a very large sequence with a relatively smaller sequence. The MATLAB function `ovrlpsav` developed in that section uses the DFT to implement the linear convolution. We can now replace the DFT by the radix-2 FFT algorithm to obtain a *high-speed* overlap-and-save algo-

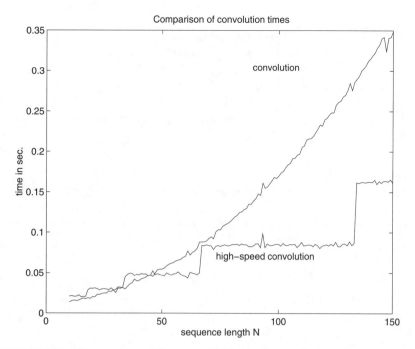

FIGURE 5.21 *Comparison of linear and high-speed convolution times*

rithm. To further reduce the computations, the FFT of the shorter (fixed) sequence can be computed only once. The following `hsolpsav` function shows this algorithm.

```
function [y] = hsolpsav(x,h,N)
% High-speed Overlap-Save method of block convolutions using FFT
% --------------------------------------------------------------
% [y] = hsolpsav(x,h,N)
% y = output sequence
% x = input sequence
% h = impulse response
% N = block length (must be a power of two)
%
N = 2^(ceil(log10(N)/log10(2)));
Lenx = length(x); M = length(h);
M1 = M-1; L = N-M1;
h = fft(h,N);
%
x = [zeros(1,M1), x, zeros(1,N-1)];
K = floor((Lenx+M1-1)/(L)); % # of blocks
Y = zeros(K+1,N);
```

```
for k=0:K
xk = fft(x(k*L+1:k*L+N));
Y(k+1,:) = real(ifft(xk.*h));
end
Y = Y(:,M:N)'; y = (Y(:))';
```

A similar modification can be done to the overlap-and-add algorithm.

PROBLEMS

P5.1 Determine the DFS coefficients of the following periodic sequences using the DFS definition, and verify by using MATLAB.

a. $\tilde{x}_1(n) = \{2, 0, 2, 0\}$, $N = 4$

b. $\tilde{x}_2(n) = \{0, 0, 1, 0, 0\}$, $N = 5$

c. $\tilde{x}_3(n) = \{3, -3, 3, -3\}$, $N = 4$

d. $\tilde{x}_4(n) = \{j, j, -j, -j\}$, $N = 4$

e. $\tilde{x}_5(n) = \{1, j, j, 1\}$, $N = 4$

P5.2 Determine the periodic sequences, given the following periodic DFS coefficients. First use the IDFS definition and then verify using MATLAB.

a. $\tilde{X}_1(k) = \{5, -2j, 3, 2j\}$, $N = 4$

b. $\tilde{X}_2(k) = \{4, -5, 3, -5\}$, $N = 4$

c. $\tilde{X}_3(k) = \{1, 2, 3, 4, 5\}$, $N = 5$

d. $\tilde{X}_4(k) = \{0, 0, 2, 0\}$, $N = 4$

e. $\tilde{X}_5(k) = \{0, j, -2j, -j\}$, $N = 4$

P5.3 Let $\tilde{x}_1(n)$ be periodic with fundamental period $N = 50$, where one period is given by

$$\tilde{x}_1(n) = \begin{cases} ne^{-0.3n}, & 0 \le n \le 25 \\ 0, & 26 \le n \le 49 \end{cases}$$

and let $\tilde{x}_2(n)$ be periodic with fundamental period $N = 100$, where one period is given by

$$\tilde{x}_2(n) = \begin{cases} ne^{-0.3n}, & 0 \le n \le 25 \\ 0, & 26 \le n \le 99 \end{cases}$$

These two periodic sequences differ in their periodicity but otherwise have equal nonzero samples.

a. Find the DFS $\tilde{X}_1(k)$ of $\tilde{x}_1(n)$ and plot samples (using the **stem** function) of its magnitude and angle versus k.

b. Find the DFS $\tilde{X}_2(k)$ of $\tilde{x}_2(n)$ and plot samples of its magnitude and angle versus k.

c. What is the difference between the above two DFS plots?

P5.4 Consider the periodic sequence $\tilde{x}_1(n)$ given in Problem 5.3. Let $\tilde{x}_3(n)$ be periodic with period 100, obtained by concatenating two periods of $\tilde{x}_1(n)$, that is,

$$\tilde{x}_3(n) = [\tilde{x}_1(n), \tilde{x}_1(n)]_{\text{PERIODIC}}$$

Clearly, $\tilde{x}_3(n)$ is different from $\tilde{x}_2(n)$ of Problem 3 even though both of them are periodic with period 100.

a. Find the DFS $\tilde{X}_3(k)$ of $\tilde{x}_3(n)$ and plot samples of its magnitude and angle versus k.

b. What effect does the periodicity doubling have on the DFS?

c. Generalize the above result to M-fold periodicity. In particular, show that if

$$\tilde{x}_M(n) = \left[\underbrace{\tilde{x}_1(n), \ldots, \tilde{x}_1(n)}_{M \text{ times}} \right]_{\text{PERIODIC}}$$

then

$$\tilde{X}_M(Mk) = M\tilde{X}_1(k), \quad k = 0, 1, \ldots, N-1$$
$$\tilde{X}_M(k) = 0, \qquad\quad k \neq 0, M, \ldots, MN$$

P5.5 Let $X(e^{j\omega})$ be the DTFT of a 10-point sequence:

$$x(n) = \{2, 5, 3, -4, -2, 6, 0, -3, -3, 2\}$$

a. Let

$$y_1(n) = \overset{\text{3-point}}{\text{IDFS}} \left[X(e^{j0}), X(e^{j2\pi/3}), X(e^{j4\pi/3}) \right]$$

Determine $y_1(n)$ using the frequency sampling theorem. Verify your answer using MATLAB.

b. Let

$$y_2(n) = \overset{\text{20-point}}{\text{IDFS}} \left[X(e^{j0}), X(e^{j2\pi/20}), X(e^{j4\pi/3}), \ldots, X(e^{j2\pi(19)/20}) \right]$$

Determine $y_2(n)$ using the frequency sampling theorem. Verify your answer using MATLAB.

P5.6 A 12-point sequence is $x(n)$ defined as

$$x(n) = \{1, 2, 3, 4, 5, 6, 6, 5, 4, 3, 2, 1\}$$

a. Determine the DFT $X(k)$ of $x(n)$. Plot (using the stem function) its magnitude and phase.

b. Plot the magnitude and phase of the DTFT $X(e^{j\omega})$ of $x(n)$ using MATLAB.

c. Verify that the above DFT is the sampled version of $X(e^{j\omega})$. It might be helpful to combine the above two plots in one graph using the hold function.

d. Is it possible to reconstruct the DTFT $X(e^{j\omega})$ from the DFT $X(k)$? If possible, give the necessary interpolation formula for reconstruction. If not possible, state why this reconstruction cannot be done.

P5.7 Plot the DTFT magnitudes of the following sequences using the DFT as a computation tool. Make an educated guess about the length N so that your plots are meaningful.

a. $x_1(n) = 2\cos(0.2\pi n)[u(n) - u(n-10)]$

b. $x_2(n) = \sin(0.45\pi n)\sin(0.55\pi n)$, $\quad 0 \leq n \leq 50$

c. $x_3(n) = 3(2)^n$, $\quad -10 \leq n \leq 10$

d. $x_4(n) = (-0.5)^n$, $\quad -10 \leq n \leq 10$

e. $x_5(n) = 5\left(0.9e^{j\pi/4}\right)^n u(n)$

P5.8 Let $H(e^{j\omega})$ be the frequency response of a real, causal discrete-time LTI system.

a. If

$$\mathrm{Re}\left\{H\left(e^{j\omega}\right)\right\} = \sum_{k=0}^{5}(0.5)^k \cos(k\omega)$$

determine the impulse response $h(n)$ analytically. Verify your answer using IDFT as a computation tool. Choose the length N judiciously.

b. If

$$\mathrm{Im}\left\{H\left(e^{j\omega}\right)\right\} = \sum_{\ell=0}^{5} 2\ell \sin(\ell\omega) \qquad \text{and} \qquad \int_{-\pi}^{\pi} H(e^{j\omega})\, d\omega = 0$$

determine the impulse response $h(n)$ analytically. Verify your answer using IDFT as a computation tool. Again choose the length N judiciously.

P5.9 Let $X(k)$ denote the N-point DFT of an N-point sequence $x(n)$. The DFT $X(k)$ itself is an N-point sequence.

a. If the DFT of $X(k)$ is computed to obtain another N-point sequence $x_1(n)$, show that

$$x_1(n) = Nx((n))_N, \quad 0 \leq n \leq N-1$$

b. Using the above property, design a MATLAB function to implement an N-point circular folding operation $x_2(n) = x_1((-n))_N$. The format should be

```
x2 = circfold(x1,N)
% Circular folding using DFT
% x2 = circfold(x1,N)
% x2 = circularly folded output sequence
% x1 = input sequence of length <= N
%  N = circular buffer length
```

c. Determine the circular folding of the following sequence:

$$x_1(n) = \{1, 2, 3, 4, 5, 6, 6, 5, 4, 3, 2, 1\}$$

P5.10 Complex-valued N-point sequences are decomposed into N-point even and odd sequences using the following relations:

$$x_{ec}(n) \triangleq \frac{1}{2}\left[x(n) + x^*((-n))_N\right]$$

$$x_{oc}(n) \triangleq \frac{1}{2}\left[x(n) - x^*((-n))_N\right]$$

Then

$$\mathrm{DFT}\left[x_{ec}(n)\right] = \mathrm{Re}\left[X(k)\right] = \mathrm{Re}\left[X((-k))_N\right]$$

$$\mathrm{DFT}\left[x_{oc}(n)\right] = j\,\mathrm{Im}\left[X(k)\right] = j\,\mathrm{Im}\left[X((-k))_N\right]$$

a. Prove the above property analytically.

b. Modify the `circevod` function developed in the chapter so that it can be used for complex-valued sequences.

c. Verify the above symmetry property and your MATLAB function on the following sequence.

$$x(n) = \left(0.9e^{j\pi/3}\right)^n \left[u(n) - u(n-20)\right]$$

P5.11 The first five values of the 8-point DFT of a real-valued sequence $x(n)$ are given by

$$\{0.25, 0.125 - j0.3, 0, 0.125 - j0.06, 0.5\}$$

Determine the DFT of each of the following sequences using properties.

a. $x_1(n) = x((2-n))_8$

b. $x_2(n) = x((n+5))_{10}$

c. $x_3(n) = x^2(n)$

d. $x_4(n) = x(n) \,\textcircled{8}\, x((-n))_8$

e. $x_5(n) = x(n)\, e^{j\pi n/4}$

P5.12 If $X(k)$ is the DFT of an N-point complex-valued sequence

$$x(n) = x_R(n) + jx_I(n)$$

where $x_R(n)$ and $x_I(n)$ are the real and imaginary parts of $x(n)$, then

$$X_R(k) \triangleq \mathrm{DFT}\left[x_R(n)\right] = X_{ec}(k)$$

$$jX_I(k) \triangleq \mathrm{DFT}\left[x_I(n)\right] = X_{oc}(k)$$

where $X_{ec}(k)$ and $X_{oc}(k)$ are the circular-even and circular-odd components of $X(k)$ as defined in Problem 5.10.

a. Prove the above property analytically.

b. This property can be used to compute the DFTs of two real-valued N-point sequences using one N-point DFT operation. Specifically, let $x_1(n)$ and $x_2(n)$ be two N-point sequences. Then we can form a complex-valued sequence

$$x(n) = x_1(n) + jx_2(n)$$

and use the above property. Develop a MATLAB function to implement this approach with the following format.

```
function [X1,X2] = real2dft(x1,x2,N)
% DFTs of two real sequences
% [X1,X2] = real2dft(x1,x2,N)
%   X1 = n-point DFT of x1
%   X2 = n-point DFT of x2
%   x1 = sequence of length <= N
%   x2 = sequence of length <= N
%    N = length of DFT
```

c. Compute the DFTs of the following two sequences:

$$x(n) = \cos(0.25\pi n), \; x(n) = \sin(0.75\pi n); \quad 0 \le n \le 63$$

P5.13 Using the frequency-domain approach, develop a MATLAB function to determine a circular shift $x((n-m))_N$, given an N_1-point sequence $x(n)$, where $N_1 \le N$. Your function should have the following format.

```
function y = cirshftf(x,m,N)
%
%function y=cirshftf(x,m,N)
%
%  Circular shift of m samples wrt size N in sequence x: (freq domain)
%  -----------------------------------------------------------------
%        y : output sequence containing the circular shift
%        x : input sequence of length <= N
%        m : sample shift
%        N : size of circular buffer
%
%  Method: y(n) = idft(dft(x(n))*WN^(mk))
%
% If m is a scalar then y is a sequence (row vector)
% If m is a vector then y is a matrix, each row is a circular shift
% in x corresponding to entries in vecor m
% M and x should not be matrices
```

Verify your function on the following sequence

$$x_1(n) = 11 - n, \quad 0 \le n \le 10$$

with $m = 10$ and $N = 15$.

P5.14 Using the analysis and synthesis equations of the DFT, show that

$$\sum_{n=0}^{N-1} |x(n)|^2 = \frac{1}{N} \sum_{k=0}^{N-1} |X(k)|^2$$

This is commonly referred to as a Parseval's relation for the DFT. Verify this relation by using MATLAB on the sequence in Problem 5.9.

P5.15 Using the frequency domain approach, develop a MATLAB function to implement the circular convolution operation between two sequences. The format of the sequence should be

```
function x3 = circonvf(x1,x2,N)
% Circular convolution in the frequency domain
%   x3 = circonvf(x1,x2,N)
%   x3 = convolution result of length N
%   x1 = sequence of length <= N
%   x2 = sequence of length <= N
%    N = length of circular buffer
```

P5.16 The `circonvt` function developed in this chapter implements the circular convolution as a matrix-vector multiplication. The matrix corresponding to the circular shifts $\left\{ x\left((n-m) \right)_N ; \quad 0 \le n \le N-1 \right\}$ has an interesting structure. This matrix is called a *circulant* matrix, which is a special case of the Toeplitz matrix introduced in Chapter 2.

a. Consider the sequences given in Example 5.13. Express $x_1(n)$ as a column vector $\mathbf{x}_1$ and $x_2\left((n-m) \right)_N$ as a matrix $\mathbf{X}_2$ with rows corresponding to $n = 0, 1, 2, 3$. Characterize this matrix $\mathbf{X}_2$. Can it completely be described by its first row (or column)?

b. Determine the circular convolution as $\mathbf{X}_2 \mathbf{x}_1$ and verify your calculations.

P5.17 Develop a MATLAB function to construct a circulant matrix $\mathbf{C}$, given an N-point sequence $x(n)$. Use the `cirshftf` function developed in Problem 5.13. Your subroutine function should have the following format.

```
function [C] = circulnt(x,N)
% Circulant Matrix from an N-point sequence
% [C] = circulnt(x,N)
% C = circulant matrix of size NxN
% x = sequence of length <= N
% N = size of circulant matrix
```

Using this function, modify the circular convolution function `circonvt` discussed in the chapter so that the `for...end` loop is eliminated. Verify your functions on the sequences in Problem 5.16.

P5.18 Compute the N-point circular convolution for the following sequences.

a. $x_1(n) = \{1, 1, 1, 1\}$, $x_2(n) = \cos(\pi n/4)\, \mathcal{R}_N(n)$; $\quad N = 8$

b. $x_1(n) = \cos(2\pi n/N)\, \mathcal{R}_N(n)$, $x_2(n) = \sin(2\pi n/N)\, \mathcal{R}_N(n)$; $\quad N = 32$

c. $x_1(n) = (0.8)^n\, \mathcal{R}_N(n)$, $x_2(n) = (-0.8)^n\, \mathcal{R}_N(n)$; $\quad N = 20$

d. $x_1(n) = n\mathcal{R}_N(n)$, $x_2(n) = (N - n)\, \mathcal{R}_N(n)$; $\quad N = 10$

e. $x_1(n) = \{1, -1, 1, -1\}$, $x_2(n) = \{1, 0, -1, 0\}$; $\quad N = 4$

P5.19 For the following sequences compute (i) the N-point circular convolution $x_3(n) = x_1(n)$ (N) $x_2(n)$, (ii) the linear convolution $x_4(n) = x_1(n) * x_2(n)$, and (iii) the error sequence $e(n) = x_3(n) - x_4(n)$.

a. $x_1(n) = \{1, 1, 1, 1\}$, $x_2(n) = \cos(\pi n/4)\, \mathcal{R}_6(n)$; $\quad N = 8$

b. $x_1(n) = \cos(2\pi n/N)\, \mathcal{R}_{16}(n)$, $x_2(n) = \sin(2\pi n/N)\, \mathcal{R}_{16}(n)$; $\quad N = 32$

c. $x_1(n) = (0.8)^n\, \mathcal{R}_{10}(n)$, $x_2(n) = (-0.8)^n\, \mathcal{R}_{10}(n)$; $\quad N = 15$

d. $x_1(n) = n\mathcal{R}_{10}(n)$, $x_2(n) = (N - n)\, \mathcal{R}_{10}(n)$; $\quad N = 10$

e. $x_1(n) = \{1, -1, 1, -1\}$, $x_2(n) = \{1, 0, -1, 0\}$; $\quad N = 5$

In each case verify that $e(n) = x_4(n + N)$.

P5.20 The overlap-add method of block convolution is an alternative to the overlap-save method. Let $x(n)$ be a long sequence of length ML, where $M, L \gg 1$. Divide $x(n)$ into M segments $\{x_m(n),\ m = 1, \ldots, M\}$, each of length L.

$$x_m(n) = \begin{cases} x(n), & mM \le n \le (m+1)M - 1 \\ 0, & \text{elsewhere} \end{cases} \qquad \text{so that} \qquad x(n) = \sum_{m=0}^{M-1} x_m(n)$$

Let $h(n)$ be an L-point impulse response; then

$$y(n) = x(n) * h(n) = \sum_{m=0}^{M-1} x_m(n) * h(n) = \sum_{m=0}^{M-1} y_m(n); \quad y_m(n) \triangleq x_m(n) * h(n)$$

Clearly, $y_m(n)$ is a $(2L - 1)$-point sequence. In this method we have to save the intermediate convolution results and then properly overlap these before adding to form the final result $y(n)$. To use DFT for this operation, we have to choose $N \ge (2L - 1)$.

a. Develop a MATLAB function to implement the overlap-add method using the circular convolution operation. The format should be

```
function [y] = ovrlpadd(x,h,N)
% Overlap-Add method of block convolution
% [y] = ovrlpadd(x,h,N)
%
% y = output sequence
% x = input sequence
% h = impulse response
% N = block length >= 2*length(h)-1
```

b. Incorporate the radix-2 FFT implementation in the above function to obtain a high-speed overlap-add block convolution routine. Remember to choose $N = 2^\nu$.

c. Verify your functions on the following two sequences:

$$x(n) = \cos(\pi n/500)\, \mathcal{R}_{4000}(n), \quad h(n) = \{1, -1, 1, -1\}$$

P5.21 Given the sequences $x_1(n)$ and $x_2(n)$ shown below:

$$x_1(n) = \{2, 1, 1, 2\}, \quad x_2(n) = \{1, -1, -1, 1\}$$

a. Compute the circular convolution $x_1(n) \,\textcircled{N}\, x_2(n)$ for $N = 4$, 7, and 8.

b. Compute the linear convolution $x_1(n) * x_2(n)$.

c. Using results of calculations, determine the minimum value of N necessary so that linear and circular convolutions are the same on the N-point interval.

d. Without performing the actual convolutions, explain how you could have obtained the result of part c.

P5.22 Let

$$x\left(n\right) = \begin{cases} A\cos\left(2\pi\ell n/N\right), & 0 \le n \le N - 1 \\ 0, & \text{elsewhere} \end{cases} = A\cos\left(2\pi\ell n/N\right)\mathcal{R}_N\left(n\right)$$

where ℓ is an integer. Notice that $x\left(n\right)$ contains *exactly* ℓ periods (or cycles) of the cosine waveform in N samples. This is a windowed cosine sequence containing *no leakage*.

a. Show that the DFT $X\left(k\right)$ is a real sequence given by

$$X\left(k\right) = \frac{AN}{2}\delta\left(k - \ell\right) + \frac{AN}{2}\delta\left(k - N + \ell\right); \quad 0 \le k \le \left(N - 1\right), 0 < \ell < N$$

b. Show that if $\ell = 0$, then the DFT $X\left(k\right)$ is given by

$$X\left(k\right) = AN\delta\left(k\right); \quad 0 \le k \le \left(N - 1\right)$$

c. Explain clearly how the above results should be modified if $\ell < 0$ or $\ell > N$.

d. Verify the results of parts a, b, and c by using the following sequences. Plot the real parts of the DFT sequences using the **stem** function.

(i) $x_1\left(n\right) = 3\cos\left(0.04\pi n\right)\mathcal{R}_{200}\left(n\right)$

(ii) $x_2\left(n\right) = 5\mathcal{R}_{50}\left(n\right)$

(iii) $x_3\left(n\right) = \left[1 + 2\cos\left(0.5\pi n\right) + \cos\left(\pi n\right)\right]\mathcal{R}_{100}\left(n\right)$

(iv) $x_4\left(n\right) = \cos\left(25\pi n/16\right)\mathcal{R}_{64}\left(n\right)$

(v) $x_5\left(n\right) = \left[4\cos\left(0.1\pi n\right) - 3\cos\left(1.9\pi n\right)\right]\mathcal{R}_N\left(n\right)$

P5.23 Let $x\left(n\right) = A\cos\left(\omega_0 n\right)\mathcal{R}_N\left(n\right)$, where ω_0 is a real number.

a. Using the properties of the DFT, show that the real and the imaginary parts of $X\left(k\right)$ are given by

$$X\left(k\right) = X_R\left(k\right) + jX_I\left(k\right)$$

$$X_R\left(k\right) = \left(A/2\right)\cos\left[\frac{\pi\left(N - 1\right)}{N}\left(k - f_0 N\right)\right]\frac{\sin\left[\pi\left(k - f_0 N\right)\right]}{\sin\left[\pi\left(k - f_0 N\right)/N\right]}$$

$$+ \left(A/2\right)\cos\left[\frac{\pi\left(N - 1\right)}{N}\left(k + f_0 N\right)\right]\frac{\sin\left[\pi\left(k - N + f_0 N\right)\right]}{\sin\left[\pi\left(k - N + f_0 N\right)/N\right]}$$

$$X_I\left(k\right) = -\left(A/2\right)\sin\left[\frac{\pi\left(N - 1\right)}{N}\left(k - f_0 N\right)\right]\frac{\sin\left[\pi\left(k - f_0 N\right)\right]}{\sin\left[\pi\left(k - f_0 N\right)/N\right]}$$

$$- \left(A/2\right)\sin\left[\frac{\pi\left(N - 1\right)}{N}\left(k + f_0 N\right)\right]\frac{\sin\left[\pi\left(k - N + f_0 N\right)\right]}{\sin\left[\pi\left(k - N + f_0 N\right)/N\right]}$$

b. The above result implies that the original frequency ω_0 of the cosine waveform has *leaked* into other frequencies that form the harmonics of the time-limited sequence, and hence it is called the leakage property of cosines. It is a natural result due to the fact that band-limited periodic cosines are sampled over noninteger periods. Explain this result using the periodic extension $\tilde{x}(n)$ of $x(n)$ and the result in Problem 5.22 part a.

c. Verify the leakage property using $x(n) = \cos(5\pi n/99)\,\mathcal{R}_{200}(n)$. Plot the real and the imaginary parts of $X(k)$ using the **stem** function.

P5.24 Let

$$x(n) = \begin{cases} A\sin(2\pi\ell n/N), & 0 \le n \le N-1 \\ 0, & \text{elsewhere} \end{cases} = A\sin(2\pi\ell n/N)\,\mathcal{R}_N(n)$$

where ℓ is an integer. Notice that $x(n)$ contains *exactly* ℓ periods (or cycles) of the sine waveform in N samples. This is a windowed sine sequence containing *no leakage*.

a. Show that the DFT $X(k)$ is a purely imaginary sequence given by

$$X(k) = \frac{AN}{2j}\delta(k-\ell) - \frac{AN}{2j}\delta(k-N+\ell); \quad 0 \le k \le (N-1),\ 0 < \ell < N$$

b. Show that if $\ell = 0$, then the DFT $X(k)$ is given by

$$X(k) = 0; \quad 0 \le k \le (N-1)$$

c. Explain clearly how the above results should be modified if $\ell < 0$ or $\ell > N$.

d. Verify the results of parts a, b, and c using the following sequences. Plot the imaginary parts of the DFT sequences using the **stem** function.

(i) $x_1(n) = 3\sin(0.04\pi n)\,\mathcal{R}_{200}(n)$

(ii) $x_2(n) = 5\sin 10\pi n\,\mathcal{R}_{50}(n)$

(iii) $x_3(n) = [2\sin(0.5\pi n) + \sin(\pi n)]\,\mathcal{R}_{100}(n)$

(iv) $x_4(n) = \sin(25\pi n/16)\,\mathcal{R}_{64}(n)$

(v) $x_5(n) = [4\sin(0.1\pi n) - 3\sin(1.9\pi n)]\,\mathcal{R}_N(n)$

P5.25 Let $x(n) = A\sin(\omega_0 n)\,\mathcal{R}_N(n)$, where ω_0 is a real number.

a. Using the properties of the DFT, show that the real and the imaginary parts of $X(k)$ are given by

$$X(k) = X_R(k) + jX_I(k)$$

$$X_R(k) = (A/2)\sin\left[\frac{\pi(N-1)}{N}(k-f_0 N)\right]\frac{\sin[\pi(k-f_0 N)]}{\sin[\pi(k-f_0 N)/N]}$$

$$\quad -(A/2)\sin\left[\frac{\pi(N-1)}{N}(k+f_0 N)\right]\frac{\sin[\pi(k-N+f_0 N)]}{\sin[\pi(k-N+f_0 N)/N]}$$

$$X_I(k) = -(A/2)\cos\left[\frac{\pi(N-1)}{N}(k-f_0 N)\right]\frac{\sin[\pi(k-f_0 N)]}{\sin[\pi(k-f_0 N)/N]}$$

$$\quad +(A/2)\cos\left[\frac{\pi(N-1)}{N}(k+f_0 N)\right]\frac{\sin[\pi(k-N+f_0 N)]}{\sin[\pi(k-N+f_0 N)/N]}$$

b. The above result is the leakage property of sines. Explain it using the periodic extension $\tilde{x}(n)$ of $x(n)$ and the result in Problem 5.24 part a.

c. Verify the leakage property using $x(n) = \sin(5\pi n/99)\, \mathcal{R}_{200}(n)$. Plot the real and the imaginary parts of $X(k)$ using the **stem** function.

P5.26 An analog signal $x_a(t) = 2\sin(4\pi t) + 5\cos(8\pi t)$ is sampled at $t = 0.01n$ for $n = 0, 1, \ldots, N-1$ to obtain an N-point sequence $x(n)$. An N-point DFT is used to obtain an estimate of the magnitude spectrum of $x_a(t)$.

a. From the following values of N, choose the one that will provide the accurate estimate of the spectrum of $x_a(t)$. Plot the real and imaginary parts of the DFT spectrum $|X(k)|$.

   (i) $N = 40$,   (ii) $N = 50$,   (iii) $N = 60$.

b. From the following values of N, choose the one that will provide the least amount of leakage in the spectrum of $x_a(t)$. Plot the real and imaginary parts of the DFT spectrum $|X(k)|$.

   (i) $N = 90$,   (ii) $N = 95$,   (iii) $N = 99$.

P5.27 Using (5.49), determine and draw the signal flowgraph for the $N = 8$ point, radix-2 decimation-in-frequency FFT algorithm. Using this flowgraph, determine the DFT of the sequence

$$x(n) = \cos(\pi n/2), \quad 0 \le n \le 7$$

P5.28 Using (5.49), determine and draw the signal flowgraph for the $N = 16$-point, radix-4 decimation-in-time FFT algorithm. Using this flowgraph, determine the DFT of the sequence

$$x(n) = \cos(\pi n/2), \quad 0 \le n \le 15$$

P5.29 Let $x(n) = \cos(\pi n/99)$, $0 \le n \le (N-1)$ be an N-point sequence. Choose $N = 4^{\nu}$ and determine the execution times in MATLAB for $\nu = 5, 6, \ldots, 10$. Verify that these times are proportional to

$$N \log_4 N$$

DIGITAL FILTER STRUCTURES

In earlier chapters we studied the theory of discrete systems in both the time and frequency domains. We will now use this theory for the processing of digital signals. To process signals, we have to design and implement systems called filters (or spectrum analyzers in some contexts). The filter design issue is influenced by such factors as the type of the filter (i.e., IIR or FIR) or the form of its implementation (structures). Hence before we discuss the design issue, we first concern ourselves with how these filters can be implemented in practice. This is an important concern because different filter structures dictate different design strategies.

As we discussed earlier, IIR filters are characterized by infinite-duration impulse responses. Some of these impulse responses can be modeled by rational system functions or, equivalently, by difference equations. Such filters are termed as auto-regressive moving average (ARMA) or, more generally, as *recursive* filters. Those IIR filters that cannot be so modeled are called *nonrecursive* filters. In DSP, IIR filters generally imply recursive ones because these can be implemented efficiently. Therefore we will always use the term IIR to imply recursive filters. Furthermore, ARMA filters include moving average filters that are FIR filters. However, we will treat FIR filters separately from IIR filters for both design and implementation purposes.

We begin with a description of basic building blocks that are used to describe filter structures. In the remaining sections we briefly describe IIR, FIR, and lattice filter structures, respectively, and provide MATLAB functions to implement these structures.

BASIC ELEMENTS

Since our filters are LTI systems, we need the following three elements to describe digital filter structures. These elements are shown in Figure 6.1.

- **Adder:** This element has two inputs and one output and is shown in Figure 6.1(a). Note that the addition of three or more signals is implemented by successive two-input adders.
- **Multiplier (gain):** This is a single-input, single-output element and is shown in Figure 6.1(b). Note that the multiplication by 1 is understood and hence not explicitly shown.
- **Delay element (shifter or memory):** This element delays the signal passing through it by one sample as shown in Figure 6.1(c). It is implemented by using a shift register.

Using these basic elements, we can now describe various structures of both IIR and FIR filters. MATLAB is a convenient tool in the development of these structures that require operations on polynomials.

IIR FILTER STRUCTURES

The system function of an IIR filter is given by

$$H(z) = \frac{B(z)}{A(z)} = \frac{\sum_{n=0}^{M} b_n z^{-n}}{\sum_{n=0}^{N} a_n z^{-n}} = \frac{b_0 + b_1 z^{-1} + \cdots + b_M z^{-M}}{1 + a_1 z^{-1} + \cdots + a_N z^{-N}}; \quad a_0 = 1 \quad \textbf{(6.1)}$$

where b_n and a_n are the coefficients of the filter. We have assumed without loss of generality that $a_0 = 1$. The order of such an IIR filter is called N if

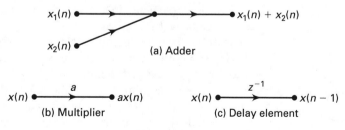

FIGURE 6.1 *Three basic elements*

$a_N \neq 0$. The difference equation representation of an IIR filter is expressed as

$$y(n) = \sum_{m=0}^{M} b_m x(n-m) - \sum_{m=1}^{N} a_m y(n-m) \qquad \textbf{(6.2)}$$

Three different structures can be used to implement an IIR filter:

- **Direct form:** In this form the difference equation (6.2) is implemented directly as given. There are two parts to this filter, namely the moving average part and the recursive part (or equivalently, the numerator and denominator parts). Therefore this implementation leads to two versions: direct form I and direct form II structures.

- **Cascade form:** In this form the system function $H(z)$ in equation (6.1) is factored into smaller second-order sections, called *biquads*. The system function is then represented as a *product* of these biquads. Each biquad is implemented in a direct form, and the entire system function is implemented as a *cascade* of biquad sections.

- **Parallel form:** This is similar to the cascade form, but after factorization, a partial fraction expansion is used to represent $H(z)$ as a *sum* of smaller second-order sections. Each section is again implemented in a direct form, and the entire system function is implemented as a *parallel* network of sections.

We will briefly discuss these forms in this section. IIR filters are generally described using the rational form version (or the direct form structure) of the system function. Hence we will provide MATLAB functions for converting direct form structures to cascade and parallel form structures.

DIRECT FORM As the name suggests, the difference equation (6.2) is implemented as given using delays, multipliers, and adders. For the purpose of illustration, let $M = N = 4$. Then the difference equation is

$$y(n) = b_0 x(n) + b_1 x(n-1) + b_2 x(n-2) + b_3 x(n-3) + b_4 x(n-4)$$
$$- a_1 y(n-1) - a_2 y(n-2) - a_3 y(n-3) - a_4 y(n-4)$$

which can be implemented as shown in Figure 6.2. This block diagram is called *direct form I* structure.

The direct form I structure implements each part of the rational function $H(z)$ separately with a cascade connection between them. The numerator part is a tapped delay line followed by the denominator part, which is a feedback tapped delay line. Thus there are two separate delay lines in this structure, and hence it requires eight delay elements. We can reduce this delay element count or eliminate one delay line by interchanging the order in which the two parts are connected in the cascade.

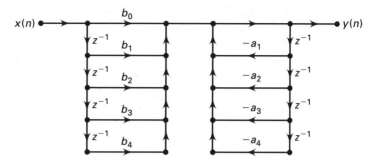

FIGURE 6.2 *Direct form I structure*

Now the two delay lines are close to each other, connected by a unity gain branch. Therefore one delay line can be removed, and this reduction leads to a canonical structure called *direct form II* structure, shown in Figure 6.3. It should be noted that both direct forms are equivalent from the input-output point of view. Internally, however, they have different signals.

MATLAB
IMPLEMEN-
TATION

In MATLAB the direct form structure is described by two row vectors; b containing the $\{b_n\}$ coefficients and a containing the $\{a_n\}$ coefficients. The structure is implemented by the `filter` function, which is discussed in Chapter 2.

CASCADE
FORM

In this form the system function $H(z)$ is written as a product of second-order sections with real coefficients. This is done by factoring the numerator and denominator polynomials into their respective roots and then combining either a complex conjugate root pair or any two real roots into second-order polynomials. In the remainder of this chapter we assume

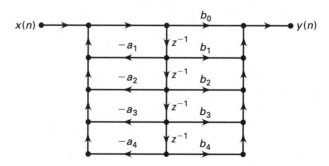

FIGURE 6.3 *Direct form II structure*

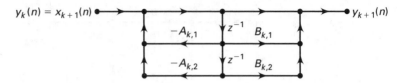

$y_k(n) = x_{k+1}(n)$ $y_{k+1}(n)$

FIGURE 6.4 *Biquad section structure*

that N is an even integer. Then

$$H(z) = \frac{b_0 + b_1 z^{-1} + \cdots + b_N z^{-N}}{1 + a_1 z^{-1} + \cdots + a_N z^{-N}} \tag{6.3}$$

$$= b_0 \frac{1 + \frac{b_1}{b_0} z^{-1} + \cdots + \frac{b_N}{b_0} z^{-N}}{1 + a_1 z^{-1} + \cdots + a_N z^{-N}}$$

$$= b_0 \prod_{k=1}^{K} \frac{1 + B_{k,1} z^{-1} + B_{k,2} z^{-2}}{1 + A_{k,1} z^{-1} + A_{k,2} z^{-2}}$$

where K is equal to $\frac{N}{2}$, and $B_{k,1}$, $B_{k,2}$, $A_{k,1}$, and $A_{k,2}$ are real numbers representing the coefficients of second-order sections. The second-order section

$$H_k(z) = \frac{Y_{k+1}(z)}{Y_k(z)} = \frac{1 + B_{k,1} z^{-1} + B_{k,2} z^{-2}}{1 + A_{k,1} z^{-1} + A_{k,2} z^{-2}}; \quad k = 1, \ldots, K$$

with

$$Y_1(z) = b_0 X(z); \quad Y_{K+1}(z) = Y(z)$$

is called the kth biquad section. The input to the kth biquad section is the output from the $(k-1)$th biquad section, while the output from the kth biquad is the input to the $(k+1)$th biquad. Now each biquad section $H_k(z)$ can be implemented in direct form II as shown in Figure 6.4. The entire filter is then implemented as a cascade of biquads.

As an example, consider $N = 4$. Figure 6.5 shows a cascade form structure for this fourth-order IIR filter.

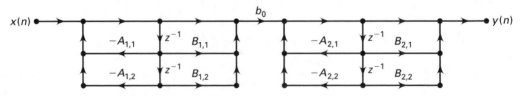

FIGURE 6.5 *Cascade form structure for $N = 4$*

Given the coefficients $\{b_n\}$ and $\{a_n\}$ of the direct form filter, we have to obtain the coefficients b_0, $\{B_{k,i}\}$, and $\{A_{k,i}\}$. This is done by the function dir2cas given below.

```
function [b0,B,A] = dir2cas(b,a);
% DIRECT-form to CASCADE-form conversion (cplxpair version)
% ------------------------------------------------------------
% [b0,B,A] = dir2cas(b,a)
% b0 = gain coefficient
% B = K by 3 matrix of real coefficients containing bk's
% A = K by 3 matrix of real coefficients containing ak's
% b = numerator polynomial coefficients of DIRECT form
% a = denominator polynomial coefficients of DIRECT form

% compute gain coefficient b0
b0 = b(1); b = b/b0;
a0 = a(1); a = a/a0;
b0 = b0/a0;
%
M = length(b); N = length(a);
if N > M
b = [b zeros(1,N-M)];
elseif M > N
a = [a zeros(1,M-N)]; N = M;
else
NM = 0;
end
%
K = floor(N/2); B = zeros(K,3); A = zeros(K,3);
if K*2 == N;
b = [b 0];
a = [a 0];
end
%
broots = cplxpair(roots(b));
aroots = cplxpair(roots(a));
for i=1:2:2*K
Brow = broots(i:1:i+1,:);
Brow = real(poly(Brow));
B(fix((i+1)/2),:) = Brow;
Arow = aroots(i:1:i+1,:);
Arow = real(poly(Arow));
A(fix((i+1)/2),:) = Arow;
end
```

The above function converts the b and a vectors into $K \times 3$ B and A matrices. It begins by computing b_0, which is equal to b_0/a_0 (assuming $a_0 \neq 1$). It then makes the vectors b and a of equal length by zero-padding the shorter vector. This ensures that each biquad has a nonzero numerator and denominator. Next it computes the roots of the $B(z)$ and $A(z)$ polynomials. Using the `cplxpair` function, these roots are ordered in complex conjugate pairs. Now every pair is converted back into a second-order numerator or denominator polynomial using the `poly` function.

The cascade form is implemented using a `casfiltr` function, which is described below. It employs the `filter` function in a loop using the coefficients of each biquad stored in B and A matrices. The input is scaled by b0, and the output of each filter operation is used as an input to the next filter operation. The output of the final filter operation is the overall output.

```
function y = casfiltr(b0,B,A,x);
% CASCADE form realization of IIR and FIR filters
% -----------------------------------------------
% y = casfiltr(b0,B,A,x);
%   y = output sequence
%  b0 = gain coefficient of CASCADE form
%   B = K by 3 matrix of real coefficients containing bk's
%   A = K by 3 matrix of real coefficients containing ak's
%   x = input sequence
%
[K,L] = size(B);
N = length(x);
w = zeros(K+1,N);
w(1,:) = x;
for i = 1:1:K
        w(i+1,:) = filter(B(i,:),A(i,:),w(i,:));
end
y = b0*w(K+1,:);
```

The following MATLAB function, `cas2dir`, converts a cascade form to a direct form. This is a simple operation that involves multiplication of several second-order polynomials. For this purpose the MATLAB function `conv` is used in a loop over K factors.

```
function [b,a] = cas2dir(b0,B,A);
% CASCADE-to-DIRECT form conversion
% --------------------------------
% [b,a] = cas2dir(b0,B,A)
%   b = numerator polynomial coefficients of DIRECT form
%   a = denominator polynomial coefficients of DIRECT form
%  b0 = gain coefficient
%   B = K by 3 matrix of real coefficients containing bk's
%   A = K by 3 matrix of real coefficients containing ak's
%
[K,L] = size(B);
```

```
b = [1];
a = [1];
for i=1:1:K
b=conv(b,B(i,:));
a=conv(a,A(i,:));
end
b = b*b0;
```

□ **EXAMPLE 6.1**   A filter is described by the following difference equation:

$$16y(n) + 12y(n-1) + 2y(n-2) - 4y(n-3) - y(n-4)$$
$$= x(n) - 3x(n-1) + 11x(n-2) - 27x(n-3) + 18x(n-4)$$

Determine its cascade form structure.

Solution

MATLAB Script _____
```
>> b=[1 -3 11 -27 18];
>> a=[16 12 2 -4 -1];
>> [b0,B,A]=dir2cas(b,a)
  b0 = 0.0625
   B =
          1.0000    -0.0000     9.0000
          1.0000    -3.0000     2.0000
   A =
          1.0000  1.0000     0.5000
          1.0000    -0.2500    -0.1250
```
_____

The resulting structure is shown in Figure 6.6. To check that our cascade structure is correct, let us compute the first 8 samples of the impulse response using both forms.

```
>> delta = impseq(0,0,7);
delta =
       1     0     0     0     0     0     0     0
>> format long
>> hcas=casfiltr(b0,B,A,delta)
```

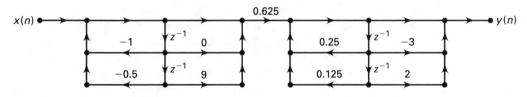

FIGURE 6.6   *Cascade structure in Example 6.1.*

```
hcas =
  Columns 1 through 4
    0.06250000000000   -0.23437500000000    0.85546875000000   -2.28417968750000
  Columns 5 through 8
    2.67651367187500   -1.52264404296875    0.28984069824219    0.49931716918945
>> hdir=filter(b,a,delta)
hdir =
  Columns 1 through 4
    0.06250000000000   -0.23437500000000    0.85546875000000   -2.28417968750000
  Columns 5 through 8
    2.67651367187500   -1.52264404296875    0.28984069824219    0.49931716918945
```
$\square$

PARALLEL FORM

In this form the system function $H(z)$ is written as a sum of second order sections using partial fraction expansion (PFE).

$$H(z) = \frac{B(z)}{A(z)} = \frac{b_0 + b_1 z^{-1} + \cdots + b_M z^{-M}}{1 + a_1 z^{-1} + \cdots + a_N z^{-N}} \tag{6.4}$$

$$= \frac{\hat{b}_0 + \hat{b}_1 z^{-1} + \cdots + \hat{b}_{N-1} z^{1-N}}{1 + a_1 z^{-1} + \cdots + a_N z^{-N}} + \underbrace{\sum_0^{M-N} C_k z^{-k}}_{\text{only if } M \geq N}$$

$$= \sum_{k=1}^{K} \frac{B_{k,0} + B_{k,1} z^{-1}}{1 + A_{k,1} z^{-1} + A_{k,2} z^{-2}} + \underbrace{\sum_0^{M-N} C_k z^{-k}}_{\text{only if } M \geq N}$$

where K is equal to $\frac{N}{2}$, and $B_{k,0}$, $B_{k,1}$, $A_{k,1}$, and $A_{k,2}$ are real numbers representing the coefficients of second-order sections. The second-order section

$$H_k(z) = \frac{Y_{k+1}(z)}{Y_k(z)} = \frac{B_{k,0} + B_{k,1} z^{-1}}{1 + A_{k,1} z^{-1} + A_{k,2} z^{-2}}; \quad k = 1, \ldots, K$$

with

$$Y_k(z) = H_k(z) X(z), \quad Y(z) = \sum Y_k(z), \quad M < N$$

is the kth proper rational biquad section. The filter input is available to all biquad sections as well as to the polynomial section if $M \geq N$ (which is an FIR part). The output from these sections is summed to form the filter output. Now each biquad section $H_k(z)$ can be implemented in direct form II. Due to the summation of subsections, a parallel structure can be built to realize $H(z)$. As an example, consider $M = N = 4$. Figure 6.7 shows a parallel form structure for this fourth-order IIR filter.

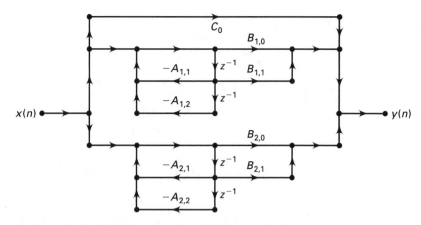

FIGURE 6.7 *Parallel form structure for $N = 4$*

The function `dir2par` given below converts the direct form coefficients $\{b_n\}$ and $\{a_n\}$ into parallel form coefficients $\{B_{k,i}\}$ and $\{A_{k,i}\}$.

```
function [C,B,A] = dir2par(b,a);
% DIRECT-form to PARALLEL-form conversion
% --------------------------------------
% [C,B,A] = dir2par(b,a)
%   C = Polynomial part when length(b) >= length(a)
%   B = K by 2 matrix of real coefficients containing bk's
%   A = K by 3 matrix of real coefficients containing ak's
%   b = numerator polynomial coefficients of DIRECT form
%   a = denominator polynomial coefficients of DIRECT form
%
M = length(b); N = length(a);

[r1,p1,C] = residuez(b,a);
p = cplxpair(p1,10000000*eps);
I = cplxcomp(p1,p);
r = r1(I);

K = floor(N/2); B = zeros(K,2); A = zeros(K,3);
if K*2 == N; %N even, order of A(z) odd, one factor is first order
for i=1:2:N-2
Brow = r(i:1:i+1,:);
Arow = p(i:1:i+1,:);
[Brow,Arow] = residuez(Brow,Arow,[]);
B(fix((i+1)/2),:) = real(Brow);
A(fix((i+1)/2),:) = real(Arow);
end
[Brow,Arow] = residuez(r(N-1),p(N-1),[]);
B(K,:) = [real(Brow) 0]; A(K,:) = [real(Arow) 0];
```

```
else
        for i=1:2:N-1
Brow = r(i:1:i+1,:);
Arow = p(i:1:i+1,:);
[Brow,Arow] = residuez(Brow,Arow,[]);
B(fix((i+1)/2),:) = real(Brow);
A(fix((i+1)/2),:) = real(Arow);
end
end
```

The **dir2cas** function first computes the z-domain partial fraction expansion using the **residuez** function. We need to arrange pole-and-residue pairs into complex conjugate pole-and-residue pairs followed by real pole-and-residue pairs. To do this, the **cplxpair** function from MATLAB can be used; this sorts a complex array into complex conjugate pairs. However, two consecutive calls to this function, one each for pole and residue arrays, will not guarantee that poles and residues will correspond to each other. Therefore a new **cplxcomp** function is developed, which compares two shuffled complex arrays and returns the index of one array, which can be used to rearrange another array.

```
function I = cplxcomp(p1,p2)
%  I = cplxcomp(p1,p2)
% Compares two complex pairs which contain the same scalar elements
%  but (possibly) at differrent indices.  This routine should be
%  used after CPLXPAIR routine for rearranging pole vector and its
%  corresponding residue vector.
%      p2 = cplxpair(p1)
%
I=[];
for j=1:1:length(p2)
    for i=1:1:length(p1)
if (abs(p1(i)-p2(j)) < 0.0001)
   I=[I,i];
        end
    end
end
I=I';
```

After collecting these pole-and-residue pairs, the **dir2cas** function computes the numerator and denominator of the biquads by employing the **residuez** function in the reverse fashion.

These parallel form coefficients are then used in the function **parfiltr**, which implements the parallel form. The **parfiltr** function uses the **filter** function in a loop using the coefficients of each biquad stored in the B and A matrices. The input is first filtered through the FIR part C and stored in the first row of a w matrix. Then the outputs of all biquad filters are computed for the same input and stored as subsequent rows in the w matrix. Finally, all the columns of the w matrix are summed to yield the output.

```
function y = parfiltr(C,B,A,x);
% PARALLEL form realization of IIR filters
% -----------------------------------------
%   [y] = parfiltr(C,B,A,x);
%   y = output sequence
%   C = polynomial (FIR) part when M >= N
%   B = K by 2 matrix of real coefficients containing bk's
%   A = K by 3 matrix of real coefficients containing ak's
%   x = input sequence
%
[K,L] = size(B);
N = length(x);
w = zeros(K+1,N);
w(1,:) = filter(C,1,x);
for i = 1:1:K
        w(i+1,:) = filter(B(i,:),A(i,:),x);
end
y = sum(w);
```

To obtain a direct form from a parallel form, the function par2dir can be used. It computes poles and residues of each proper biquad and combines these into system poles and residues. Another call of the residuez function in reverse order computes the numerator and denominator polynomials.

```
function [b,a] = par2dir(C,B,A);
% PARALLEL-to-DIRECT form conversion
% ---------------------------------
%   [b,a] = par2dir(C,B,A)
%   b = numerator polynomial coefficients of DIRECT form
%   a = denominator polynomial coefficients of DIRECT form
%   C = Polynomial part of PARALLEL form
%   B = K by 2 matrix of real coefficients containing bk's
%   A = K by 3 matrix of real coefficients containing ak's
%
[K,L] = size(A); R = []; P = [];

for i=1:1:K
[r,p,k]=residuez(B(i,:),A(i,:));
R = [R;r]; P = [P;p];
end
[b,a] = residuez(R,P,C);
b = b(:)'; a = a(:)';
```

□   **EXAMPLE 6.2**   Consider the filter given in Example 6.1.

$$16y(n) + 12y(n-1) + 2y(n-2) - 4y(n-3) - y(n-4)$$
$$= x(n) - 3x(n-1) + 11x(n-2) - 27x(n-3) + 18x(n-4)$$

Now determine its parallel form.

Solution

MATLAB Script _____
```
>> b=[1 -3 11 -27 18];
>> a=[16 12 2 -4 -1];
>> [C,B,A]=dir2par(b,a)
C =
   -18
B =
   10.0500    -3.9500
   28.1125   -13.3625
A =
    1.0000     1.0000     0.5000
    1.0000    -0.2500    -0.1250
```

The resulting structure is shown in Figure 6.8. To check our parallel structure, let us compute the first 8 samples of the impulse response using both forms.

```
>> format long; delta = impseq(0,0,7);
>> hpar=parfiltr(C,B,A,delta)
hpar =
  Columns 1 through 4
    0.06250000000000   -0.23437500000000    0.85546875000000   -2.28417968750000
```

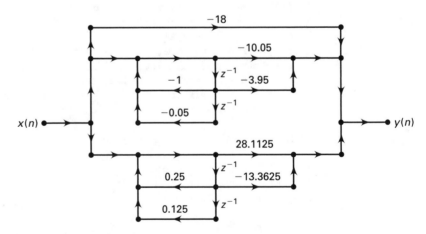

FIGURE 6.8   *Parallel form structure in Example 6.2.*

Columns 5 through 8
```
   2.67651367187500  -1.52264404296875   0.28984069824219   0.49931716918945
>> hdir = filter(b,a,delta)
hdir =
  Columns 1 through 4
   0.06250000000000  -0.23437500000000   0.85546875000000  -2.28417968750000
  Columns 5 through 8
   2.67651367187500  -1.52264404296875   0.28984069824219   0.49931716918945
```
□

□ **EXAMPLE 6.3** What would be the overall direct, cascade, or parallel form if a structure contains a combination of these forms? Consider the block diagram shown in Figure 6.9.

Solution This structure contains a cascade of two parallel sections. The first parallel section contains two biquads, while the second one contains three biquads. We will have to convert each parallel section into a direct form using the `par2dir` function, giving us a cascade of two direct forms. The overall direct form can be computed by convolving the corresponding numerator and denominator polynomials. The overall cascade and parallel forms can now be derived from the direct form.

```
>> C0=0; B1=[2 4;3 1]; A1=[1 1 0.9; 1 0.4 -0.4];
>> B2=[0.5 0.7;1.5 2.5;0.8 1]; A2=[1 -1 0.8;1 0.5 0.5;1 0 -0.5];
>> [b1,a1]=par2dir(C0,B1,A1)
b1 =
    5.0000    8.8000    4.5000   -0.7000
```

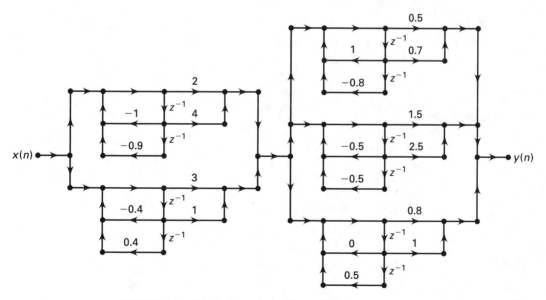

FIGURE 6.9 *Block diagram in Example 6.3*

```
a1 =
   1.0000    1.4000    0.9000   -0.0400   -0.3600
>> [b2,a2]=par2dir(C0,B2,A2)
b2 =
   2.8000    2.5500   -1.5600    2.0950    0.5700   -0.7750
a2 =
   1.0000   -0.5000    0.3000    0.1500    0.0000    0.0500   -0.2000
>> b=conv(b1,b2) % Overall direct form numerator
b =
  Columns 1 through 7
   14.0000   37.3900   27.2400    6.2620   12.4810   11.6605   -5.7215
  Columns 8 through 9
   -3.8865    0.5425
>> a=conv(a1,a2) % Overall direct form denominator
a =
  Columns 1 through 7
    1.0000    0.9000    0.5000    0.0800    0.1400    0.3530   -0.2440
  Columns 8 through 11
   -0.2890   -0.1820   -0.0100    0.0720
>> [b0,Bc,Ac]=dir2cas(b,a) % Overall cascade form
b0 =
   14.0000
Bc =
    1.0000    1.8836    1.1328
    1.0000   -0.6915    0.6719
    1.0000    2.0776    0.8666
    1.0000         0         0
    1.0000   -0.5990    0.0588
Ac =
    1.0000    1.0000    0.9000
    1.0000    0.5000    0.5000
    1.0000   -1.0000    0.8000
    1.0000    1.5704    0.6105
    1.0000   -1.1704    0.3276
>> [C0,Bp,Ap]=dir2par(b,a) % Overall parallel form
C0 = []
Bp =
  -20.4201   -1.6000
   24.1602    5.1448
    2.4570    3.3774
   -0.8101   -0.2382
    8.6129   -4.0439
Ap =
    1.0000    1.0000    0.9000
    1.0000    0.5000    0.5000
    1.0000   -1.0000    0.8000
    1.0000    1.5704    0.6105
    1.0000   -1.1704    0.3276
```

This example shows that by using the MATLAB functions developed in this section, we can probe and construct a wide variety of structures. □

FIR FILTER STRUCTURES
———————————■———————————

A finite-duration impulse response filter has a system function of the form

$$H(z) = b_0 + b_1 z^{-1} + \cdots + b_{M-1} z^{1-M} = \sum_{n=0}^{M-1} b_n z^{-n} \qquad \textbf{(6.5)}$$

Hence the impulse response $h(n)$ is

$$h(n) = \begin{cases} b_n, & 0 \leq n \leq M - 1 \\ 0, & \text{else} \end{cases} \qquad \textbf{(6.6)}$$

and the difference equation representation is

$$y(n) = b_0 x(n) + b_1 x(n-1) + \cdots + b_{M-1} x(n - M + 1) \qquad \textbf{(6.7)}$$

which is a linear convolution of finite support.

The order of the filter is $M - 1$, while the *length* of the filter (which is equal to the number of coefficients) is M. The FIR filter structures are always stable, and they are relatively simple compared to IIR structures. Furthermore, FIR filters can be designed to have a linear-phase response, which is desirable in some applications.

We will consider the following four structures:

- **Direct form:** In this form the difference equation (6.7) is implemented directly as given.
- **Cascade form:** In this form the system function $H(z)$ in (6.5) is factored into second-order factors, which are then implemented in a cascade connection.
- **Linear-phase form:** When an FIR filter has a linear phase response, its impulse response exhibits certain symmetry conditions. In this form we exploit these symmetry relations to reduce multiplications by about half.
- **Frequency sampling form:** This structure is based on the DFT of the impulse response $h(n)$ and leads to a parallel structure. It is also suitable for a design technique based on the sampling of frequency response $H\left(e^{j\omega}\right)$.

We will briefly describe the above four forms along with some examples. The MATLAB function `dir2cas` developed in the previous section is also applicable for the cascade form.

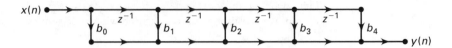

FIGURE 6.10   *Direct form FIR structure*

DIRECT FORM

The difference equation (6.7) is implemented as a tapped delay line since there are no feedback paths. Let $M = 5$ (i.e., a fourth-order FIR filter); then

$$y(n) = b_0 x(n) + b_1 x(n-1) + b_2 x(n-2) + b_3 x(n-3) + b_4 x(n-4)$$

The direct form structure is given in Figure 6.10. Note that since the denominator is equal to unity, there is only one direct form structure.

MATLAB IMPLEMEN-TATION

In MATLAB the direct form FIR structure is described by the row vector b containing the $\{b_n\}$ coefficients. The structure is implemented by the `filter` function, in which the vector a is set to the scalar value 1 as discussed in Chapter 2.

CASCADE FORM

This form is similar to that of the IIR form. The system function $H(z)$ is converted into products of second-order sections with real coefficients. These sections are implemented in direct form and the entire filter as a cascade of second-order sections. From (6.5)

$$H(z) = b_0 + b_1 z^{-1} + \cdots + b_{M-1} z^{-M+1} \tag{6.8}$$

$$= b_0 \left(1 + \frac{b_1}{b_0} z^{-1} + \cdots + \frac{b_{M-1}}{b_0} z^{-M+1} \right)$$

$$= b_0 \prod_{k=1}^{K} \left(1 + B_{k,1} z^{-1} + B_{k,2} z^{-2} \right)$$

where K is equal to $\lfloor \frac{M}{2} \rfloor$, and $B_{k,1}$ and $B_{k,2}$ are real numbers representing the coefficients of second-order sections. For $M = 7$ the cascade form is shown in Figure 6.11.

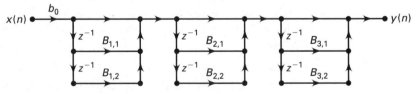

FIGURE 6.11   *Cascade form FIR structure*

Although it is possible to develop a new MATLAB function for the FIR cascade form, we will use our `dir2cas` function by setting the denominator vector **a** equal to 1. Similarly, `cas2dir` can be used to obtain the direct form from the cascade form.

LINEAR-PHASE FORM

For frequency-selective filters (e.g., lowpass filters) it is generally desirable to have a phase response that is a linear function of frequency; that is, we want

$$\angle H(e^{j\omega}) = \beta - \alpha\omega, \quad -\pi < \omega \leq \pi \qquad (6.9)$$

where $\beta = 0$ or $\pm\pi/2$ and α is a constant. For a causal FIR filter with impulse response over $[0, \ M-1]$ interval, the linear-phase condition (6.9) imposes the following symmetry conditions on the impulse response $h(n)$:

$$h(n) = h(M-1-n); \quad \beta = 0, \ 0 \leq n \leq M-1 \qquad (6.10)$$

$$h(n) = -h(M-1-n); \quad \beta = \pm\pi/2, \ 0 \leq n \leq M-1 \qquad (6.11)$$

An impulse response that satisfies (6.10) is called a symmetric impulse response, while that in (6.11) is called an antisymmetric impulse response. These symmetry conditions can now be exploited in a structure called the linear-phase form.

Consider the difference equation given in (6.7) with a symmetric impulse response in (6.10). We have

$$y(n) = b_0 x(n) + b_1 x(n-1) + \cdots + b_1 x(n-M+2) + b_0 x(n-M+1)$$

$$= b_0[x(n) + x(n-M+1)] + b_1[x(n-1) + x(n-M+2)] + \cdots$$

The block diagram implementation of the above difference equation is shown in Figure 6.12 for both odd and even M.

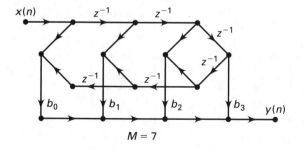
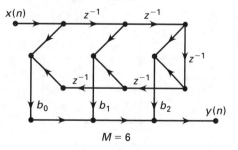

FIGURE 6.12 *Linear phase form FIR structures (symmetric impulse response)*

Clearly, this structure requires 50% fewer multiplications than the direct form. A similar structure can be derived for an antisymmetric impulse response.

MATLAB
IMPLEMEN-
TATION

The linear-phase structure is essentially a direct form drawn differently to save on multiplications. Hence in a MATLAB implementation the linear-phase structure is equivalent to the direct form.

☐ **EXAMPLE 6.4** An FIR filter is given by the system function

$$H(z) = 1 + 16\frac{1}{16}z^{-4} + z^{-8}$$

Determine and draw the direct, linear-phase, and cascade form structures.

    a. Direct form: The difference equation is given by

$$y(n) = x(n) + 16.0625x(n-4) + x(n-8)$$

and the direct form structure is shown in Figure 6.13(a).

    b. Linear-phase form: The difference equation can be written in the form

$$y(n) = [x(n) + x(n-8)] + 16.0625x(n-4)$$

and the resulting structure is shown in Figure 6.13(b).

    c. Cascade form:

```
>> b=[1,0,0,0,16+1/16,0,0,0,1];
>> [b0,B,A] = dir2cas(b,1)
```

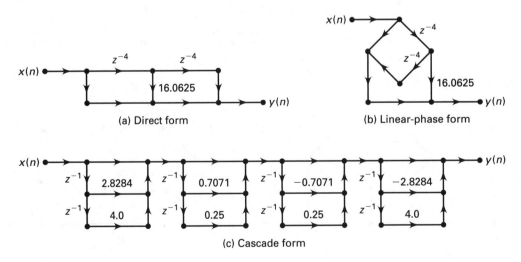

(a) Direct form   (b) Linear-phase form

(c) Cascade form

FIGURE 6.13 *FIR filter structures in Example 6.4*

```
b0 = 1
B =
      1.0000      2.8284      4.0000
      1.0000      0.7071      0.2500
      1.0000     -0.7071      0.2500
      1.0000     -2.8284      4.0000
A =
      1      0      0
      1      0      0
      1      0      0
      1      0      0
```

The cascade form structure is shown in Figure 6.13(c). □

□ **EXAMPLE 6.5** For the filter in Example 6.4 what would be the structure if we desire a cascade form containing linear-phase components with real coefficients?

Solution

We are interested in cascade sections that have symmetry and real coefficients. From the properties of linear-phase FIR filters (see Chapter 7), if such a filter has an arbitrary zero at $z = r\angle\theta$, then there must be three other zeros at $(1/r)\angle\theta$, $r\angle-\theta$, and $(1/r)\angle-\theta$ to have real filter coefficients. We can now make use of this property. First we will determine the zero locations of the given eighth-order polynomial. Then we will group four zeros that satisfy the above property to obtain one (fourth-order) linear-phase section. There are two such sections, which we will connect in cascade.

```
>> b=[1,0,0,0,16+1/16,0,0,0,1];
>> broots=roots(b)
broots =
  -1.4142 + 1.4142i
  -1.4142 - 1.4142i
   1.4142 + 1.4142i
   1.4142 - 1.4142i
  -0.3536 + 0.3536i
  -0.3536 - 0.3536i
   0.3536 + 0.3536i
   0.3536 - 0.3536i
>> B1=real(poly([broots(1),broots(2),broots(5),broots(6)]))
B1 =
      1.0000      3.5355      6.2500      3.5355      1.0000
>> B2=real(poly([broots(3),broots(4),broots(7),broots(8)]))
B2 =
      1.0000     -3.5355      6.2500     -3.5355      1.0000
```

The structure is shown in Figure 6.14. □

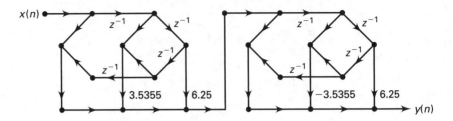

FIGURE 6.14 *Cascade of FIR linear-phase elements*

In this form we use the fact that the system function $H(z)$ of an FIR filter can be reconstructed from its samples on the unit circle. From our discussions on the DFT in Chapter 5 we recall that these samples are in fact the M-point DFT values $\{H(k),\quad 0 \leq k \leq M - 1\}$ of the M-point impulse response $h(n)$. Therefore we have

$$H(z) = \mathcal{Z}[h(n)]$$

$$= \mathcal{Z}[\text{IDFT}\{H(k)\}]$$

Using this procedure, we obtain [see (5.17) on page 127]

$$H(z) = \left(\frac{1 - z^{-M}}{M}\right) \sum_{k=0}^{M-1} \frac{H(k)}{1 - W_M^{-k}z^{-1}} \tag{6.12}$$

This shows that the DFT $H(k)$, rather than the impulse response $h(n)$ (or the difference equation), is used in this structure. It is also interesting to note that the FIR filter described by (6.12) has a recursive form similar to an IIR filter because (6.12) contains both poles and zeros. The resulting filter is an FIR filter since the poles at W_M^{-k} are canceled by the roots of

$$1 - z^{-M} = 0$$

The system function in (6.12) leads to a parallel structure as shown in Figure 6.15 for $M = 4$.

One problem with the structure in Figure 6.15 is that it requires a complex arithmetic implementation. Since an FIR filter is almost always a real-valued filter, it is possible to obtain an alternate realization in which only real arithmetic is used. This realization is derived using the symmetry properties of the DFT and the W_M^{-k} factor. Then (6.12) can be expressed

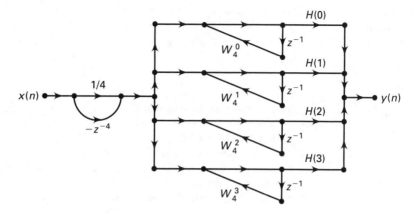

FIGURE 6.15 *Frequency sampling structure for $M = 4$*

as (see Problem 6.10)

$$H(z) = \frac{1 - z^{-M}}{M} \left\{ \sum_{k=1}^{L} 2 |H(k)| H_k(z) + \frac{H(0)}{1 - z^{-1}} + \frac{H(M/2)}{1 + z^{-1}} \right\} \quad (6.13)$$

where $L = \frac{M-1}{2}$ for M odd, $L = \frac{M}{2} - 1$ for M even, and $\{H_k(z), \quad k = 1, \ldots, L\}$ are second-order sections given by

$$H_k(z) = \frac{\cos[\angle H(k)] - z^{-1} \cos\left[\angle H(k) - \frac{2\pi k}{M}\right]}{1 - 2z^{-1} \cos\left(\frac{2\pi k}{M}\right) + z^{-2}} \quad (6.14)$$

Note that the DFT samples $H(0)$ and $H(M/2)$ are real-valued and that the third term on the right-hand side of (6.13) is absent if M is odd. Using (6.13) and (6.14), we show a frequency sampling structure in Figure 6.16 for $M = 4$ containing real coefficients.

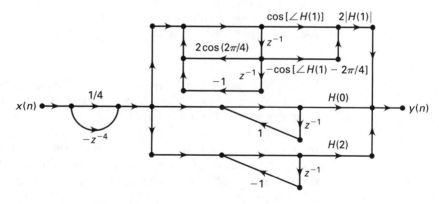

FIGURE 6.16 *Frequency sampling structure for $M = 4$ with real coefficients*

Given the impulse response $h(n)$ or the DFT $H(k)$, we have to determine the coefficients in (6.13) and (6.14). The following MATLAB function, dir2fs, converts a direct form ($h(n)$ values) to the frequency sampling form by directly implementing (6.13) and (6.14).

```
function [C,B,A] = dir2fs(h)
% Direct form to Frequency Sampling form conversion
% ---------------------------------------------------
% [C,B,A] = dir2fs(h)
% C = Row vector containing gains for parallel sections
% B = Matrix containing numerator coefficients arranged in rows
% A = Matrix containing denominator coefficients arranged in rows
% h = impulse response vector of an FIR filter
%
M = length(h);
H = fft(h,M);
magH = abs(H); phaH = angle(H)';
% check even or odd M
if (M == 2*floor(M/2))
      L = M/2-1;    % M is even
      A1 = [1,-1,0;1,1,0];
      C1 = [real(H(1)),real(H(L+2))];
else
      L = (M-1)/2; % M is odd
      A1 = [1,-1,0];
      C1 = [real(H(1))];
end
k = [1:L]';
% initialize B and A arrays
B = zeros(L,2); A = ones(L,3);
% compute denominator coefficients
A(1:L,2) = -2*cos(2*pi*k/M); A = [A;A1];
% compute numerator coefficients
B(1:L,1) = cos(phaH(2:L+1));
B(1:L,2) = -cos(phaH(2:L+1)-(2*pi*k/M));
% compute gain coefficients
C = [2*magH(2:L+1),C1]';
```

In the above function the impulse response values are supplied through the h array. After conversion, the C array contains the gain values for each parallel section. The gain values for the second-order parallel sections are given first, followed by $H(0)$ and $H(M/2)$ (if M is even). The B matrix contains the numerator coefficients, which are arranged in length-2 row vectors for each second-order section. The A matrix contains the denominator coefficients, which are arranged in length-3 row vectors for the second-order sections corresponding to those in B, followed by the coefficients for the first-order sections.

A practical problem with the structure in Figure 6.16 is that it has poles on the unit circle, which makes this filter critically unstable. If the filter is not excited by one of the pole frequencies, then the output is bounded. We can avoid this problem by sampling $H(z)$ on a circle $|z| = r$, where the radius r is very close to one but is less than one (e.g., $r = 0.99$), which results in

$$H(z) = \frac{1 - r^M z^{-M}}{M} \sum_{k=0}^{M-1} \frac{H(k)}{1 - rW_M^{-k} z^{-k}}; \quad H(k) = H\left(re^{j2\pi k/M}\right)$$

(6.15)

Now approximating $H\left(re^{j2\pi k/M}\right) \approx H\left(e^{j2\pi k/M}\right)$ for $r \approx 1$, we can obtain a stable structure similar to the one in Figure 6.16 containing real values. This is explored in Problem 6.11.

☐ **EXAMPLE 6.6** Let $h(n) = \frac{1}{9}\{1, 2, 3, 2, 1\}$. Determine and draw the frequency sampling form.
$\uparrow$

Solution

MATLAB Script _____
```
>> h = [1,2,3,2,1]/9;
>> [C,B,A] = dir2fs(h)
C =
        0.5818
        0.0849
        1.0000
B =
       -0.8090     0.8090
        0.3090    -0.3090
A =
        1.0000    -0.6180     1.0000
        1.0000     1.6180     1.0000
        1.0000    -1.0000          0
```
_____

Since $M = 5$ is odd, there is only one first-order section. Hence

$$H(z) = \frac{1 - z^{-5}}{5}\left[0.5818\frac{-0.809 + 0.809z^{-1}}{1 - 0.618z^{-1} + z^{-2}}\right.$$

$$\left. + 0.0848\frac{0.309 - 0.309z^{-1}}{1 + 1.618z^{-1} + z^{-2}} + \frac{1}{1 - z^{-1}}\right]$$

The frequency sampling form is shown in Figure 6.17. ☐

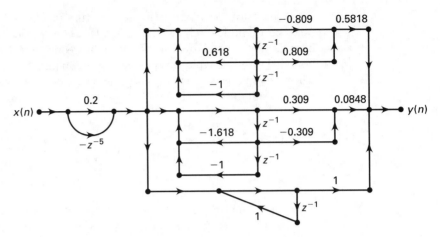

FIGURE 6.17 *Frequency sampling structure in Example 6.6*

☐   **EXAMPLE 6.7**   The frequency samples of a 32-point linear-phase FIR filter are given by

$$|H(k)| = \begin{cases} 1, & k = 0, 1, 2 \\ 0.5, & k = 3 \\ 0, & k = 4, 5, \ldots, 15 \end{cases}$$

Determine its frequency sampling form, and compare its computational complexity with the linear-phase form.

Solution   In this example since the samples of the DFT $H(k)$ are given, we could use (6.13) and (6.14) directly to determine the structure. However, we will use the `dir2fs` function for which we will have to determine the impulse response $h(n)$. Using the symmetry property and the linear-phase constraint, we assemble the DFT $H(k)$ as

$$H(k) = |H(k)| e^{j \angle H(k)}, \quad k = 0, 1, \ldots, 31$$

$$|H(k)| = |H(32 - k)|, \quad k = 1, 2, \ldots, 31; \; H(0) = 1$$

$$\angle H(k) = -\frac{31}{2} \frac{2\pi}{32} k = -\angle H(32 - k), \quad k = 0, 1, \ldots, 31$$

Now the IDFT of $H(k)$ will result in the desired impulse response.

```
>> M = 32; alpha = (M-1)/2;
>> magHk = [1,1,1,0.5,zeros(1,25),0.5,1,1];
>> k1 = 0:15; k2 = 16:M-1;
>> angHk = [-alpha*(2*pi)/M*k1, alpha*(2*pi)/M*(M-k2)];
>> H = magHk.*exp(j*angHk);
>> h = real(ifft(H,M));
>> [C,B,A] = dir2fs(h)
```

```
C =
    2.0000
    2.0000
    1.0000
    0.0000
    0.0000
    0.0000
    0.0000
         0
    0.0000
    0.0000
    0.0000
    0.0000
    0.0000
    0.0000
    0.0000
    1.0000
         0
B =
   -0.9952    0.9952
    0.9808   -0.9808
   -0.9569    0.9569
   -0.8944    0.3162
    0.9794   -0.7121
    0.8265    0.2038
   -0.6754    0.8551
    1.0000    0.0000
    0.6866   -0.5792
    0.5191    0.9883
   -0.4430    0.4993
   -0.8944   -0.3162
   -0.2766    0.3039
    0.9343    0.9996
   -0.9077   -0.8084
A =
    1.0000   -1.9616    1.0000
    1.0000   -1.8478    1.0000
    1.0000   -1.6629    1.0000
    1.0000   -1.4142    1.0000
    1.0000   -1.1111    1.0000
    1.0000   -0.7654    1.0000
    1.0000   -0.3902    1.0000
    1.0000    0.0000    1.0000
    1.0000    0.3902    1.0000
    1.0000    0.7654    1.0000
    1.0000    1.1111    1.0000
    1.0000    1.4142    1.0000
    1.0000    1.6629    1.0000
```

$$
\begin{array}{ccc}
1.0000 & 1.8478 & 1.0000 \\
1.0000 & 1.9616 & 1.0000 \\
1.0000 & -1.0000 & 0 \\
1.0000 & 1.0000 & 0
\end{array}
$$

Note that only four gain coefficients are nonzero. Hence the frequency sampling form is

$$
H(z) = \frac{1 - z^{-32}}{32} \left[2\frac{-0.9952 + 0.9952z^{-1}}{1 - 1.9616z^{-1} + z^{-2}} + 2\frac{0.9808 - 0.9808z^{-1}}{1 - 1.8478z^{-1} + z^{-2}} + \frac{-0.9569 + 0.9569z^{-1}}{1 - 1.6629z^{-1} + z^{-2}} + \frac{1}{1 - z^{-1}} \right]
$$

To determine the computational complexity, note that since $H(0) = 1$, the first-order section requires no multiplication, while the three second-order sections require three multiplications each for a total of nine multiplications per output sample. The total number of additions is 13. To implement the linear-phase structure would require 16 multiplications and 31 additions per output sample. Therefore the frequency sampling structure of this FIR filter is more efficient than the linear-phase structure. □

LATTICE FILTER STRUCTURES

■

The lattice filter is extensively used in digital speech processing and in the implementation of adaptive filters. It is a preferred form of realization over other FIR or IIR filter structures because in speech analysis and in speech synthesis the small number of coefficients allows a large number of *formants* to be modeled in real time. The all-zero lattice is the FIR filter representation of the lattice filter, while the lattice ladder is the IIR filter representation.

ALL-ZERO
LATTICE
FILTERS

An FIR filter of length M (or order $M - 1$) has a lattice structure with $M - 1$ stages as shown in Figure 6.18. Each stage of the filter has an input and output that are related by the order-recursive equations [19]:

$$
\begin{aligned}
f_m(n) &= f_{m-1}(n) + K_m g_{m-1}(n-1), & m = 1, 2, \ldots, M - 1 \\
g_m(n) &= K_m f_{m-1}(n) + g_{m-1}(n-1), & m = 1, 2, \ldots, M - 1
\end{aligned}
\tag{6.16}
$$

where the parameters K_m, $m = 1, 2, \ldots, M - 1$, called the *reflection coefficients*, are the lattice filter coefficients. If the initial values of $f_m(n)$ and $g_m(m)$ are both the scaled value (scaled by K_0) of the filter input $x(n)$, then the output of the $(M - 1)$ stage lattice filter corresponds to

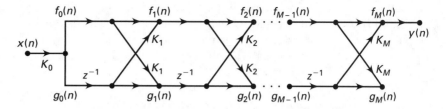

FIGURE 6.18 *All-zero lattice filter*

the output of an $(M-1)$ order FIR filter; that is,

$$f_0(n) = g_0(n) = K_0 x(n)$$
$$y(n) = f_{M-1}(n)$$

(6.17)

If the FIR filter is given by the direct form

$$H(z) = \sum_{m=0}^{M-1} b_m z^{-m} = b_0 \left(1 + \sum_{m=1}^{M-1} \frac{b_m}{b_0} z^{-m} \right) \qquad (6.18)$$

and if we denote the polynomial $A_{M-1}(z)$ by

$$A_{M-1}(z) = \left(1 + \sum_{m=1}^{M-1} \alpha_{M-1}(m) z^{-m} \right); \qquad (6.19)$$

$$\alpha_{M-1}(m) = \frac{b_m}{b_0}, \ m = 1, \ldots, M-1$$

then the lattice filter coefficients $\{K_m\}$ can be obtained by the following recursive algorithm [19]:

$$K_0 = b_0$$
$$K_{M-1} = \alpha_{M-1}(M-1)$$
$$J_m(z) = z^{-m} A_m(z^{-1}); \qquad m = M-1, \ldots, 1$$
$$A_{m-1}(z) = \frac{A_m(z) - K_m J_m(z)}{1 - K_m^2}, \qquad m = M-1, \ldots, 1$$
$$K_m = \alpha_m(m), \qquad m = M-2, \ldots, 1$$

(6.20)

Note that the above algorithm will fail if $|K_m| = 1$ for any $m = 1, \ldots, M-1$. Clearly, this condition is satisfied by linear-phase FIR filters since

$$b_0 = |b_{M-1}| \Rightarrow |K_{M-1}| = |\alpha_{M-1}(M-1)| = \left| \frac{b_{M-1}}{b_0} \right| = 1$$

Therefore linear-phase FIR filters cannot be implemented using lattice structures.

Given the coefficients $\{b_n\}$ of the direct form, we can obtain the lattice filter coefficients $\{K_m\}$ using (6.20). This is done by the MATLAB function dir2latc given below. Note that the equation to compute $J_m(z)$ implies that the polynomial $J_m(z)$ is a fliplr operation on the $A_m(z)$ polynomial.

```
function [K] = dir2latc(b)
% FIR Direct form to All-Zero Lattice form Conversion
% --------------------------------------------------------
% [K] = dir2latc(b)
%   K = Lattice filter coefficients (reflection coefficients)
%   b = FIR direct form coefficients (impulse response)
%
M = length(b);
K = zeros(1,M);
b1 = b(1);
if b1 == 0
error('b(1) is equal to zero')
end
K(1) = b1; A = b/b1;
for m=M:-1:2
K(m) = A(m);
J = fliplr(A);
A = (A-K(m)*J)/(1-K(m)*K(m));
A = A(1:m-1);
end
```

The lattice filter is implemented using (6.16) and (6.17), which is done by a latcfilt function, which is given below.

```
function [y] = latcfilt(K,x)
% LATTICE form realization of FIR filters
% ---------------------------------------
% y = latcfilt(K,x)
%   y = output sequence
%   K = LATTICE filter (reflection) coefficient array
%   x = input sequence
%
Nx = length(x)-1;
x = K(1)*x;
M = length(K)-1; K = K(2:M+1);
fg = [x; [0 x(1:Nx)]];
for m = 1:M
    fg = [1,K(m);K(m),1]*fg;
    fg(2,:) = [0 fg(2,1:Nx)];
end
y = fg(1,:);
```

The equations (6.20) can also be used to determine the direct form coefficients $\{b_m\}$ from the lattice filter coefficients $\{K_m\}$ using a recursive procedure [19]:

$$A_0(z) = J_0(z) = 1$$
$$A_m(z) = A_{m-1}(z) + K_m z^{-1} J_{m-1}(z), \quad m = 1, 2, \ldots, M-1$$
$$J_m(z) = z^{-m} A_m(z^{-1}), \quad m = 1, 2, \ldots, M-1 \quad \textbf{(6.21)}$$
$$b_m = K_0 \alpha_{M-1}(m), \quad m = 0, 1, \ldots, M-1$$

The MATLAB function `latc2dir` given below implements (6.21). Note that the product $K_m z^{-1} J_{m-1}(z)$ is obtained by convolving the two corresponding arrays, while the polynomial $J_m(z)$ is obtained by using a `fliplr` operation on the $A_m(z)$ polynomial.

```
function [b] = latc2dir(K)
% All-Zero Lattice form to FIR Direct form Conversion
% -------------------------------------------------------
% [b] = latc2dir(K)
%   b = FIR direct form coefficients (impulse response)
%   K = Lattice filter coefficients (reflection coefficients)
%
M = length(K);
J = 1; A = 1;
for m=2:1:M
A = [A,0]+conv([0,K(m)],J);
J = fliplr(A);
end
b=A*K(1);
```

☐ **EXAMPLE 6.8**   An FIR filter is given by the difference equation

$$y(n) = 2x(n) + \frac{13}{12}x(n-1) + \frac{5}{4}x(n-2) + \frac{2}{3}x(n-3)$$

Determine its lattice form.

Solution

MATLAB Script
```
>> b=[2, 13/12, 5/4, 2/3];
>> K=dir2latc(b)
K =
      2.0000    0.2500    0.5000    0.3333
```

Hence

$$K_0 = 2, \ K_1 = \frac{1}{4}, \ K_2 = \frac{1}{2}, \ K_3 = \frac{1}{3}$$

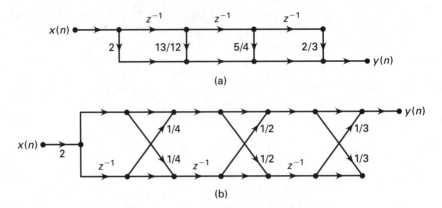

FIGURE 6.19 *FIR filter structures in Example 6.8: (a) Direct form (b) Lattice form*

The direct form and the lattice form structures are shown in Figure 6.19. To check that our lattice structure is correct, let us compute the impulse response of the filter using both forms.

```
>> [x,n] = impseq(0,0,3);
>> format long
>> hdirect=filter(b,1,delta)
hdirect =
    2.00000000000000   1.08333333333333   1.25000000000000   0.66666666666667
>> hlattice=latcfilt(K,delta)
hlattice =
    2.00000000000000   1.08333333333333   1.25000000000000   0.66666666666667
```
□

ALL-POLE LATTICE FILTERS

A lattice structure for an IIR filter is restricted to an all-pole system function. It can be developed from an FIR lattice structure. Let an all-pole system function be given by

$$H(z) = \frac{1}{1 + \sum_{m=1}^{N} a_N(m)z^{-m}} \tag{6.22}$$

which from (6.19) is equal to $H(z) = \frac{1}{A_N(z)}$. Clearly, it is an *inverse* system to the FIR lattice of Figure 6.18 (except for factor b_0). This IIR filter of order N has a lattice structure with N stages as shown in Figure 6.20. Each stage of the filter has an input and output that are related by

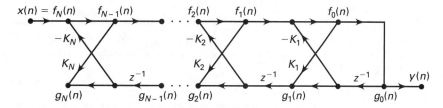

$x(n) = f_N(n)$ $f_{N-1}(n)$ $f_2(n)$ $f_1(n)$ $f_0(n)$

$-K_N$ $-K_2$ $-K_1$

K_N z^{-1} K_2 z^{-1} K_1 z^{-1} $y(n)$

$g_N(n)$ $g_{N-1}(n)$ $g_2(n)$ $g_1(n)$ $g_0(n)$

FIGURE 6.20   *All-pole lattice filter*

the order-recursive equations [19]:

$$f_N(n) = x(n)$$
$$f_{m-1}(n) = f_m(n) - K_m g_{m-1}(n-1), \qquad m = N, N-1, \ldots, 1$$
$$g_m(n) = K_m f_{m-1}(n) + g_{m-1}(n-1), \quad m = N, N-1, \ldots, 1$$
$$y(n) = f_0(n) = g_0(n)$$

(6.23)

where the parameters K_m, $\quad m = 1, 2, \ldots, M-1$, are the reflection coefficients of the all-pole lattice and are obtained from (6.20) except for K_0, which is equal to 1.

MATLAB
IMPLEMEN-
TATION

Since the IIR lattice coefficients are derived from the same (6.20) procedure used for an FIR lattice filter, we can use the `dir2latc` function in MATLAB. Care must be taken to ignore the K_0 coefficient in the K array. Similarly, the `latc2dir` function can be used to convert the lattice $\{K_m\}$ coefficients into the direct form $\{a_N(m)\}$ provided that $K_0 = 1$ is used as the first element of the K array. The implementation of an IIR lattice is given by (6.23), and we will discuss it in the next section.

□   **EXAMPLE 6.9**   Consider an all-pole IIR filter given by

$$H(z) = \frac{1}{1 + \frac{13}{24}z^{-1} + \frac{5}{8}z^{-2} + \frac{1}{3}z^{-3}}$$

Determine its lattice structure.

Solution

MATLAB Script _____

```
>> a=[1, 13/24, 5/8, 1/3];
>> K=dir2latc(b)
K =
      1.0000    0.2500    0.5000    0.3333
```

_____

Hence

$$K_1 = \frac{1}{4}, \qquad K_2 = \frac{1}{2}, \qquad \text{and} \qquad K_3 = \frac{1}{3}$$

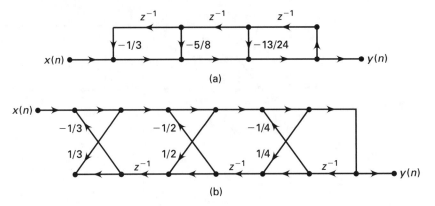

FIGURE 6.21 *IIR filter structures in Example 6.9: (a) Direct form (b) Lattice form*

The direct form and the lattice form structures of this IIR filter are shown in Figure 6.21. $\qquad\square$

LATTICE-LADDER FILTERS

A general IIR filter containing both poles and zeros can be realized as a lattice-type structure by using an all-pole lattice as the basic building block. Consider an IIR filter with system function

$$H(z) = \frac{\displaystyle\sum_{k=0}^{M} b_M(k) z^{-k}}{1 + \displaystyle\sum_{k=1}^{N} a_N(k) z^{-k}} = \frac{B_M(z)}{A_N(z)} \tag{6.24}$$

where, without loss of generality, we assume that $N \geq M$. A lattice-type structure can be constructed by first realizing an all-pole lattice with coefficients K_m, $1 \leq m \leq N$ for the denominator of (6.24), and then adding a *ladder* part by taking the output as a weighted linear combination of $\{g_m(n)\}$ as shown in Figure 6.22 for $M = N$. The result

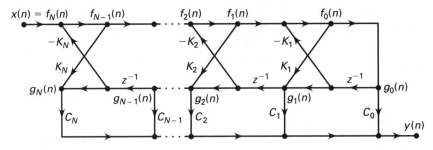

FIGURE 6.22 *Lattice-ladder structure for realizing a pole-zero IIR filter.*

is a pole-zero IIR filter that has the *lattice-ladder* structure. Its output is given by

$$y(n) = \sum_{m=0}^{M} C_m g_m(n) \tag{6.25}$$

where $\{C_m\}$ are called the *ladder coefficients* that determine the zeros of the system function $H(z)$. It can be shown [19] that $\{C_m\}$ are given by

$$B_M(z) = \sum_{m=0}^{M} C_m J_m(z) \tag{6.26}$$

where $J_m(z)$ is the polynomial in (6.20). From (6.26) one can obtain a recursive relation

$$B_m(z) = B_{m-1}(z) + C_m J_m(z); \quad m = 0, 2, \ldots, M$$

or equivalently,

$$C_m = b_m + \sum_{i=m+1}^{M} C_i \alpha_i(i-m); \quad m = M, \, M-1, \ldots, 0 \tag{6.27}$$

from the definitions of $B_m(z)$ and $A_m(z)$.

MATLAB
IMPLEMEN-
TATION

To obtain a lattice-ladder structure for a general rational IIR filter, we can first obtain the lattice coefficients $\{K_m\}$ from $A_N(z)$ using the recursion (6.20). Then we can solve (6.27) recursively for the ladder coefficients $\{C_m\}$ to realize the numerator $B_M(z)$. This is done in the MATLAB function `dir2ladr` given below. It can also be used to determine the all-pole lattice parameters when the array b is set to b=[1].

```
function [K,C] = dir2ladr(b,a)
% IIR Direct form to pole-zero Lattice/Ladder form Conversion
% -------------------------------------------------------------
% [K,C] = dir2ladr(b,a)
% K = Lattice coefficients (reflection coefficients), [K1,...,KN]
% C = Ladder Coefficients, [C0,...,CN]
% b = Numerator polynomial coefficients (deg <= Num deg)
% a = Denominator polynomial coefficients
%
a1 = a(1); a = a/a1; b = b/a1;
M = length(b); N = length(a);
if M > N
    error('    *** length of b must be <= length of a ***')
end
```

```
b = [b, zeros(1,N-M)]; K = zeros(1,N-1);
A = zeros(N-1,N-1); C = b;
for m = N-1:-1:1
      A(m,1:m) = -a(2:m+1)*C(m+1);
      K(m) = a(m+1);
      J = fliplr(a);
      a = (a-K(m)*J)/(1-K(m)*K(m));
      a = a(1:m);
      C(m) = b(m) + sum(diag(A(m:N-1,1:N-m)));
end
```

Note: To use this function, $N \geq M$. If $M > N$, then the numerator $A_N(z)$ should be divided into the denominator $B_M(z)$ using the `deconv` function to obtain a proper rational part and a polynomial part. The proper rational part can be implemented using a lattice-ladder structure, while the polynomial part is implemented using a direct structure.

To convert a lattice-ladder form into a direct form, we first use the recursive procedure in (6.21) on $\{K_m\}$ coefficients to determine $\{a_N(k)\}$ and then solve (6.27) recursively to obtain $\{b_M(k)\}$. This is done in the MATLAB function `ladr2dir` given below.

```
function [b,a] = ladr2dir(K,C)
% Lattice/Ladder form to IIR Direct form Conversion
% -------------------------------------------------
% [b,a] = ladr2dir(K,C)
%   b = numerator polynomial coefficients
%   a = denominator polymonial coefficients
%   K = Lattice coefficients (reflection coefficients)
%   C = Ladder coefficients
%
N = length(K); M = length(C);
C = [C, zeros(1,N-M+1)];
J = 1; a = 1; A = zeros(N,N);
for m=1:1:N
      a = [a,0]+conv([0,K(m)],J);
      A(m,1:m) = -a(2:m+1);
      J = fliplr(a);
end
b(N+1) = C(N+1);
for m = N:-1:1
      A(m,1:m) = A(m,1:m)*C(m+1);
      b(m) = C(m) - sum(diag(A(m:N,1:N-m+1)));
end
```

The lattice-ladder filter is implemented using (6.23) and (6.25). This is done in the MATLAB function `ladrfilt`, which is given below. It should be noted that due to the recursive nature of this implementation along with the feedback loops, this MATLAB function is neither an elegant nor an efficient method of implementation. It is not possible to exploit MATLAB's inherent parallel processing capabilities in implementing this lattice-ladder structure.

```
function [y] = ladrfilt(K,C,x)
% LATTICE/LADDER form realization of IIR filters
% ------------------------------------------------
% [y] = ladrfilt(K,C,x)
% y = output sequence
% K = LATTICE (reflection) coefficient array
% C = LADDER coefficient array
% x = input sequence
%
Nx = length(x); y = zeros(1,Nx);
N = length(C); f = zeros(N,Nx); g = zeros(N,Nx+1);
f(N,:) = x;
for n = 2:1:Nx+1
    for m = N:-1:2
        f(m-1,n-1) = f(m,n-1) - K(m-1)*g(m-1,n-1);
        g(m,n) = K(m-1)*f(m-1,n-1) + g(m-1,n-1);
    end
    g(1,n) = f(1,n-1);
end
y = C*g(:,2:Nx+1);
```

☐ **EXAMPLE 6.10**   Convert the following pole-zero IIR filter into a lattice-ladder structure.

$$H(z) = \frac{1 + 2z^{-1} + 2z^{-2} + z^{-3}}{1 + \frac{13}{24}z^{-1} + \frac{5}{8}z^{-2} + \frac{1}{3}z^{-3}}$$

Solution   **MATLAB Script** _____

```
>> b = [1,2,2,1] a = [1, 13/24, 5/8, 1/3];
>> [K,C] = dir2ladr(b)
K =
    0.2500    0.5000    0.3333
C =
   -0.2695    0.8281    1.4583    1.0000
```

Hence

$$K_1 = \frac{1}{4}, \ K_2 = \frac{1}{2}, \ K_3 = \frac{1}{3};$$

and

$$C_0 = -0.2695, \ C_1 = 0.8281, \ C_2 = 1.4583, \ C_3 = 1$$

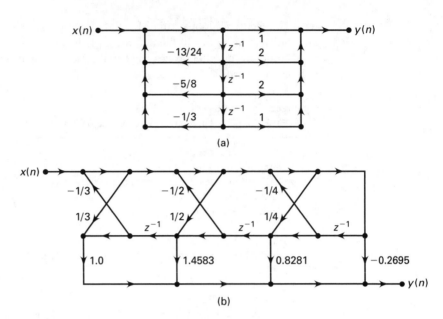

FIGURE 6.23 *IIR filter structures in Example 6.10: (a) Direct form (b) Lattice-ladder form*

The resulting direct form and the lattice-ladder form structures are shown in Figure 6.23. To check that our lattice-ladder structure is correct, let us compute the first 8 samples of its impulse response using both forms.

```
>> [x,n]=impseq(0,0,7)
>> format long
>> hdirect = filter(b,a,x)
hdirect =
  Columns 1 through 4
    1.00000000000000   1.45833333333333   0.58506944444444   -0.56170428240741
  Columns 5 through 8
   -0.54752302758488   0.45261700163162   0.28426911049255   -0.25435705167494
>> hladder = ladrfilt(K,C,x)
hladder =
  Columns 1 through 4
    1.00000000000000   1.45833333333333   0.58506944444444   -0.56170428240741
  Columns 5 through 8
   -0.54752302758488   0.45261700163162   0.28426911049255   -0.25435705167494
```

□

PROBLEMS

P6.1 A causal linear time-invariant system is described by

$$y(n) = \sum_{k=0}^{5} \left(\frac{1}{2}\right)^k x(n-k) + \sum_{\ell=1}^{5} \left(\frac{1}{3}\right)^\ell y(n-\ell)$$

Determine and draw the block diagrams of the following structures. Compute the response of the system to

$$x(n) = u(n), \quad 0 \le n \le 100$$

in each case using the corresponding structures.

a. Direct form I

b. Direct form II

c. Cascade form containing second-order direct form II sections

d. Parallel form containing second-order direct form II sections

e. Lattice-ladder form

P6.2 An IIR filter is described by the following system function:

$$H = 2 \left(\frac{1 + 0z^{-1} + z^{-2}}{1 - 0.8z^{-1} + 0.64z^{-2}}\right) \left(\frac{2 - z^{-1}}{1 - 0.75z^{-1}}\right) \left(\frac{1 + 2z^{-1} + z^{-2}}{1 + 0.81z^{-2}}\right)$$

Determine and draw the following structures.

a. Direct form I

b. Direct form II

c. Cascade form containing second-order direct form II sections

d. Parallel form containing second-order direct form II sections

e. Lattice-ladder form

P6.3 An IIR filter is described by the following system function:

$$H(z) = \left(\frac{-14.75 - 12.9z^{-1}}{1 - \frac{7}{8}z^{-1} + \frac{3}{32}z^{-2}}\right) + \left(\frac{24.5 + 26.82z^{-1}}{1 - z^{-1} + \frac{1}{2}z^{-2}}\right)$$

Determine and draw the following structures:

a. Direct form I

b. Direct form II

c. Cascade form containing second-order direct form II sections

d. Parallel form containing second-order direct form II sections

e. Lattice-ladder form

P6.4 Figure 6.24 describes a causal linear time-invariant system. Determine and draw the following structures:

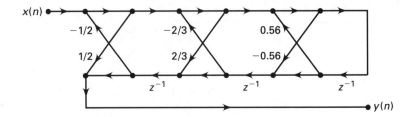

FIGURE 6.24 *Structure for Problem 6.4*

   a. Direct form I

   b. Direct form II

   c. Cascade form containing second-order direct form II sections

   d. Parallel form containing second-order direct form II sections

P6.5   A linear time-invariant system with system function

$$H(z) = \frac{0.5 \left(1 + z^{-1}\right)^6}{\left(1 - \frac{3}{2}z^{-1} + \frac{7}{8}z^{-2} - \frac{13}{16}z^{-3} - \frac{1}{8}z^{-4} - \frac{11}{32}z^{-5} + \frac{7}{16}z^{-6}\right)}$$

is to be implemented using a flowgraph of the form shown in Figure 6.25.

   a. Fill in all the coefficients in the diagram.

   b. Is your solution unique? Explain.

P6.6   A linear time-invariant system with system function

$$H(z) = \frac{5 + 11.2z^{-1} + 5.44z^{-2} - 0.384z^{-3} - 2.3552z^{-4} - 1.2288z^{-5}}{1 + 0.8z^{-1} - 0.512z^{-3} - 0.4096z^{-4}}$$

is to be implemented using a flowgraph of the form shown in Figure 6.26. Fill in all the coefficients in the diagram.

P6.7   Consider the linear time-invariant system given in Problem 6.5.

$$H(z) = \frac{0.5 \left(1 + z^{-1}\right)^6}{\left(1 - \frac{3}{2}z^{-1} + \frac{7}{8}z^{-2} - \frac{13}{16}z^{-3} - \frac{1}{8}z^{-4} - \frac{11}{32}z^{-5} + \frac{7}{16}z^{-6}\right)}$$

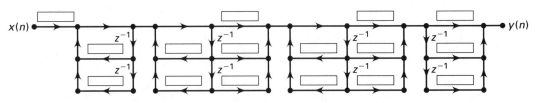

FIGURE 6.25 *Structure for Problem 6.5*

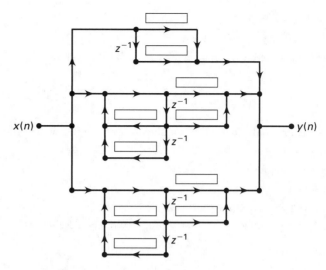

FIGURE 6.26 *Structure for Problem 6.6*

It is to be implemented using a flowgraph of the form shown in Figure 6.27.

a. Fill in all the coefficients in the diagram.

b. Is your solution unique? Explain.

P6.8 An FIR filter is described by the difference equation

$$y(n) = \sum_{k=0}^{10} \left(\frac{1}{2}\right)^{|5-k|} x(n-k)$$

Determine and draw the block diagrams of the following structures.

a. Direct form

b. Linear-phase form

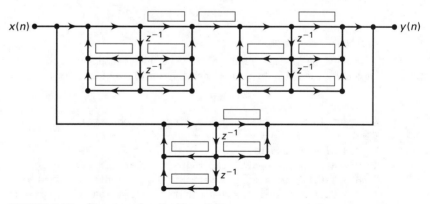

FIGURE 6.27 *Structure for Problem 6.7*

c. Cascade form

d. Frequency sampling form

P6.9   A linear time-invariant system is given by the system function

$$H(z) = \sum_{k=0}^{10} (2z)^{-k}$$

Determine and draw the block diagrams of the following structures.

a. Direct form

b. Cascade form

c. Lattice form

d. Frequency sampling form

P6.10   Using the conjugate symmetry property of the DFT

$$H(k) = \begin{cases} H(0), & k = 0 \\ H^*(M-k), & k = 1, \ldots, M-1 \end{cases}$$

and the conjugate symmetry property of the W_M^{-k} factor, show that (6.12) can be put in the form (6.13) and (6.14) for real FIR filters.

P6.11   To avoid poles on the unit circle in the frequency sampling structure, one samples $H(z)$ at $z_k = re^{j2\pi k/M}$, $k = 0, \ldots, M-1$, where $r \approx 1$(but < 1) as discussed in this chapter.

a. Using

$$H\left(re^{j2\pi k/M}\right) \approx H(k)$$

show that the frequency sampling structure is given by

$$H(z) = \frac{1-(rz)^{-M}}{M} \left\{ \sum_{k=1}^{L} 2|H(k)| H_k(z) + \frac{H(0)}{1-rz^{-1}} + \frac{H(M/2)}{1+rz^{-1}} \right\}$$

where

$$H_k(z) = \frac{\cos[\angle H(k)] - rz^{-1}\cos\left[\angle H(k) - \frac{2\pi k}{M}\right]}{1 - 2rz^{-1}\cos\left(\frac{2\pi k}{M}\right) + z^{-2}}, \quad k = 1, \ldots, L$$

and M is even.

b. Modify the MATLAB function **dir2fs** (which was developed in this chapter) to implement the above frequency sampling form. The format of this function should be

```
[C,B,A,rM] = dir2fs(h,r)
% Direct form to Frequency Sampling form conversion
% ----------------------------------------------------
% [C,B,A] = dir2fs(h)
%   C = Row vector containing gains for parallel sections
%   B = Matrix containing numerator coefficients arranged in rows
%   A = Matrix containing denominator coefficients arranged in rows
% rM = r^M factor needed in the feedforward loop
```

```
%  h = impulse response vector of an FIR filter
%  r = radius of the circle over which samples are taken (r<1)
%
```

c. Determine the frequency sampling structure for the impulse response given in Example 6.6 using the above function.

P6.12 Determine the impulse response of an FIR filter with lattice parameters

$$K_0 = 2, \ K_1 = 0.6, \ K_2 = 0.3, \ K_3 = 0.5, \ K_4 = 0.9$$

Draw the direct form and lattice form structures of the above filter.

FIR FILTER DESIGN

We now turn our attention to the inverse problem of designing systems from the given specifications. It is an important as well as a difficult problem. In digital signal processing there are two important types of systems. The first type of systems perform signal filtering in the time domain and hence are called *digital filters*. The second type of systems provide signal representation in the frequency domain and are called *spectrum analyzers*. In Chapter 5 we described signal representations using the DFT. In this and the next chapter we will study several basic design algorithms for both FIR and IIR filters. These designs are mostly of the *frequency selective* type; that is, we will design primarily multiband lowpass, highpass, bandpass, and bandstop filters. In FIR filter design we will also consider systems like differentiators or Hilbert transformers, which, although not frequency-selective filters, nevertheless follow the design techniques being considered. More sophisticated filter designs are based on arbitrary frequency-domain specifications and require tools that are beyond the scope of this book.

We first begin with some preliminary issues related to design philosophy and design specifications. These issues are applicable to both FIR and IIR filter designs. We will then study FIR filter design algorithms in the rest of this chapter. In Chapter 8 we will provide a similar treatment for IIR filters.

PRELIMINARIES

The design of a digital filter is carried out in three steps:

- **Specifications:** Before we can design a filter, we must have some specifications. These specifications are determined by the applications.
- **Approximations:** Once the specifications are defined, we use various concepts and mathematics that we studied so far to come up with a filter description that approximates the given set of specifications. This step is the topic of filter design.

- **Implementation:** The product of the above step is a filter description in the form of either a difference equation, or a system function $H(z)$, or an impulse response $h(n)$. From this description we implement the filter in hardware or through software on a computer as we discussed in Chapter 6.

In this and the next chapter we will discuss in detail only the second step, which is the conversion of specifications into a filter description.

In many applications like speech or audio signal processing, digital filters are used to implement frequency-selective operations. Therefore, specifications are required in the frequency-domain in terms of the desired magnitude and phase response of the filter. Generally a linear phase response in the passband is desirable. In the case of FIR filters, it is possible to have exact linear phase as we have seen in Chapter 6. In the case of IIR filters a linear phase in the passband is not achievable. Hence we will consider *magnitude-only* specifications.

The magnitude specifications are given in one of two ways. The first approach is called *absolute specifications*, which provide a set of requirements on the magnitude response function $\left|H(e^{j\omega})\right|$. These specifications are generally used for FIR filters. IIR filters are specified in a somewhat different way, which we will discuss in Chapter 8. The second approach is called *relative specifications*, which provide requirements in *decibels* (dB), given by

$$\text{dB scale} = -20\log_{10} \frac{\left|H(e^{j\omega})\right|}{\left|H(e^{j\omega})\right|_{\max}} \geq 0$$

This approach is the most popular one in practice and is used for both FIR and IIR filters. To illustrate these specifications, we will consider a lowpass filter design as an example.

ABSOLUTE SPECIFICATIONS

A typical absolute specification of a lowpass filter is shown in Figure 7.1a, in which

- band $[0, \omega_p]$ is called the *passband*, and δ_1 is the tolerance (or ripple) that we are willing to accept in the ideal passband response,
- band $[\omega_s, \pi]$ is called the *stopband*, and δ_2 is the corresponding tolerance (or ripple), and
- band $[\omega_p, \omega_s]$ is called the *transition band*, and there are no restrictions on the magnitude response in this band.

RELATIVE (DB) SPECIFICATIONS

A typical absolute specification of a lowpass filter is shown in Figure 7.1b, in which

- R_p is the passband ripple in dB, and
- A_s is the stopband attenuation in dB.

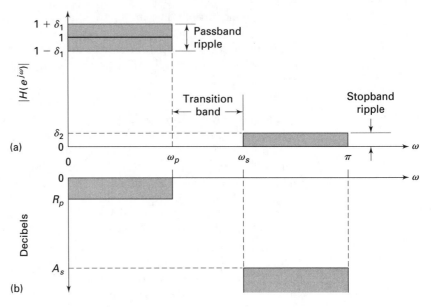

FIGURE 7.1 *FIR filter specifications: (a) Absolute (b) Relative*

The parameters given in the above two specifications are obviously related. Since $\left|H(e^{j\omega})\right|_{\max}$ in absolute specifications is equal to $(1 + \delta_1)$, we have

$$R_p = -20\log_{10}\frac{1-\delta_1}{1+\delta_1} > 0 \ (\approx 0) \qquad \textbf{(7.1)}$$

and

$$A_s = -20\log_{10}\frac{\delta_2}{1+\delta_1} > 0 \ (\gg 1) \qquad \textbf{(7.2)}$$

☐ **EXAMPLE 7.1** In a certain filter's specifications the passband ripple is 0.25 dB, and the stopband attenuation is 50 dB. Determine δ_1 and δ_2.

Solution Using (7.1), we obtain

$$R_p = 0.25 = -20\log_{10}\frac{1-\delta_1}{1+\delta_1} \Rightarrow \delta_1 = 0.0144$$

Using (7.2), we obtain

$$A_s = 50 = -20\log_{10}\frac{\delta_2}{1+\delta_1} = -20\log_{10}\frac{\delta_2}{1+0.0144} \Rightarrow \delta_2 = 0.0032 \qquad \square$$

☐ **EXAMPLE 7.2** Given the passband tolerance $\delta_1 = 0.01$ and the stopband tolerance $\delta_2 = 0.001$, determine the passband ripple R_p and the stopband attenuation A_s.

Solution   From (7.1) the passband ripple is

$$R_p = -20 \log_{10} \frac{1 - \delta_1}{1 + \delta_1} = 0.1737 \text{ dB}$$

and from (7.2) the stopband attenuation is

$$A_s = -20 \log_{10} \frac{\delta_2}{1 + \delta_1} = 60 \text{ dB} \qquad \square$$

The above specifications were given for a lowpass filter. Similar specifications can also be given for other types of frequency-selective filters, such as highpass or bandpass. However, the most important design parameters are *frequency-band tolerances* (or ripples) and *band-edge frequencies*. Whether the given band is a passband or a stopband is a relatively minor issue. Therefore in describing design techniques, we will concentrate on a lowpass filter. In the next chapter we will discuss how to transform a lowpass filter into other types of frequency-selective filters. Hence it makes more sense to develop techniques for a lowpass filter so that we can compare these techniques. However, we will also provide examples of other types of filters. In light of this discussion our design goal is the following.

Problem Statement   Design a lowpass filter (i.e., obtain its system function $H(z)$ or its difference equation) that has a passband $[0, \omega_p]$ with tolerance δ_1 (or R_p in dB) and a stopband $[\omega_s, \pi]$ with tolerance δ_2 (or A_s in dB).

In this chapter we turn our attention to the design and approximation of FIR digital filters. These filters have several design and implementational advantages:

- The phase response can be exactly linear.
- They are relatively easy to design since there are no stability problems.
- They are efficient to implement.
- The DFT can be used in their implementation.

As we discussed in Chapter 6, we are generally interested in linear-phase frequency-selective FIR filters. Advantages of a linear-phase response are:

- design problem contains only real arithmetic and not complex arithmetic;
- linear-phase filters provide no delay distortion and only a fixed amount of delay;
- for the filter of length M (or order $M - 1$) the number of operations are of the order of $M/2$ as we discussed in the linear-phase filter implementation.

We first begin with a discussion of the properties of the linear-phase FIR filters, which are required in design algorithms. Then we will discuss three design techniques, namely the window design, the frequency sampling design, and the optimal equiripple design techniques for linear-phase FIR filters.

PROPERTIES OF LINEAR-PHASE FIR FILTERS

■

In this section we discuss shapes of impulse and frequency responses and locations of system function zeros of linear-phase FIR filters. Let $h(n)$, $0 \leq n \leq M - 1$ be the impulse response of length (or duration) M. Then the system function is

$$H(z) = \sum_{n=0}^{M-1} h(n)z^{-n} = z^{-(M-1)} \sum_{n=0}^{M-1} h(n)z^{M-1-n}$$

which has $(M - 1)$ poles at the origin $z = 0$ (trivial poles) and $(M - 1)$ zeros located anywhere in the z-plane. The frequency response function is

$$H(e^{j\omega}) = \sum_{n=0}^{M-1} h(n)e^{-j\omega n}, \quad -\pi < \omega \leq \pi$$

Now we will discuss specific requirements on the forms of $h(n)$ and $H(e^{j\omega})$ as well as requirements on the specific locations of $(M - 1)$ zeros that the linear-phase constraint imposes.

IMPULSE
RESPONSE
$h(n)$

We impose a linear-phase constraint

$$\angle H(e^{j\omega}) = -\alpha\omega, \quad -\pi < \omega \leq \pi$$

where α is a *constant phase delay*. Then we know from Chapter 6 that $h(n)$ must be symmetric, that is,

$$h\,(n) = h(M - 1 - n), \quad 0 \leq n \leq (M - 1) \text{ with } \alpha = \frac{M - 1}{2} \qquad \textbf{(7.3)}$$

Hence $h(n)$ is symmetric about α, which is the index of symmetry. There are two possible types of symmetry:

- M *odd:* In this case $\alpha = (M - 1)/2$ is an integer. The impulse response is as shown below.

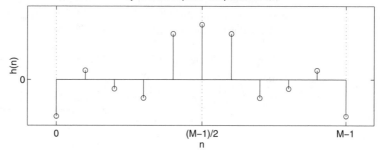

Symmetric Impulse Response: M odd

• *M even:* In this case $\alpha = (M-1)/2$ is not an integer. The impulse response is as shown below.

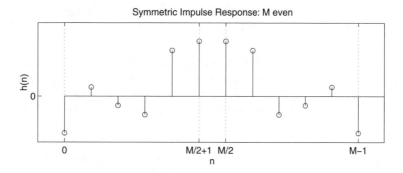

Symmetric Impulse Response: M even

We also have a second type of "linear-phase" FIR filter if we require that the phase response $\angle H(e^{j\omega})$ satisfy the condition

$$\angle H(e^{j\omega}) = \beta - \alpha\omega$$

which is a straight line but not through the origin. In this case α is not a constant phase delay, but

$$\frac{d\angle H(e^{j\omega})}{d\omega} = -\alpha$$

is constant, which is the group delay. Therefore α is called a *constant group delay.* In this case, as a group, frequencies are delayed at a constant rate. But some frequencies may get delayed more and others delayed less. For this type of linear phase one can show that

$$h(n) = -h(M-1-n), \quad 0 \le n \le (M-1); \; \alpha = \frac{M-1}{2}, \; \beta = \pm\frac{\pi}{2} \quad \text{(7.4)}$$

This means that the impulse response $h(n)$ is *antisymmetric*. The index of symmetry is still $\alpha = (M-1)/2$. Once again we have two possible types, one for M odd and one for M even.

• *M odd:* In this case $\alpha = (M-1)/2$ is an integer and the impulse response is as shown below.

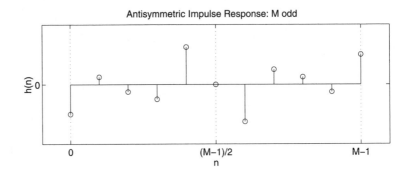

Antisymmetric Impulse Response: M odd

Note that the sample $h(\alpha)$ at $\alpha = (M-1)/2$ must necessarily be equal to zero, i.e., $h((M-1)/2) = 0$.

• *M even:* In this case $\alpha = (M-1)/2$ is not an integer and the impulse response is as shown below.

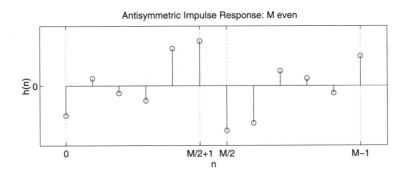

Antisymmetric Impulse Response: M even

FREQUENCY
RESPONSE
$H(e^{j\omega})$

When the cases of symmetry and antisymmetry are combined with odd and even M, we obtain four types of linear-phase FIR filters. Frequency response functions for each of these types have some peculiar expressions and shapes. To study these responses, we write $H(e^{j\omega})$ as

$$H(e^{j\omega}) = H_r(\omega)e^{j(\beta - \alpha\omega)}; \quad \beta = \pm\frac{\pi}{2}, \ \alpha = \frac{M-1}{2} \qquad (7.5)$$

Chapter 7 ■ FIR FILTER DESIGN

where $H_r(\omega)$ is an *amplitude response* function and not a magnitude response function. The amplitude response is a real function, but unlike the magnitude response, which is always positive, the amplitude response may be both positive and negative. The phase response associated with the magnitude response is a *discontinuous* function, while that associated with the amplitude response is a *continuous linear* function. To illustrate the difference between these two types of responses, consider the following example.

□ **EXAMPLE 7.3** Let the impulse response be $h(n) = \{1, 1, 1\}$. Determine and draw frequency responses.
$\uparrow$

Solution The frequency response function is

$$H(e^{j\omega}) = \sum_{0}^{2} h(n)e^{j\omega n} = 1 + 1e^{-j\omega} + e^{-j2\omega} = \left\{ e^{j\omega} + 1 + e^{-j\omega} \right\} e^{-j\omega}$$

$$= \left\{ 1 + 2\cos\omega \right\} e^{-j\omega}$$

From this the magnitude and the phase responses are

$$\left| H(e^{j\omega}) \right| = |1 + 2\cos\omega|, \quad 0 < \omega \le \pi$$

$$\angle H\left(e^{j\omega}\right) = \begin{cases} -\omega, & 0 < \omega < 2\pi/3 \\ \pi - \omega, & 2\pi/3 < \omega < \pi \end{cases}$$

since $\cos\omega$ can be both positive and negative. In this case the phase response is *piecewise linear*. On the other hand, the amplitude and the corresponding phase responses are

$$H_r(\omega) = 1 + 2\cos\omega,$$
$$\angle H\left(e^{j\omega}\right) = -\omega, \qquad -\pi < \omega \le \pi$$

In this case the phase response is *truly linear*. These responses are shown in Figure 7.2. From this example the difference between the magnitude and the amplitude (or between the piecewise linear and the linear-phase) responses should be clear. □

Type-1 linear-phase FIR filter: Symmetrical impulse response, M odd In this case $\beta = 0$, $\alpha = (M-1)/2$ is an integer, and $h(n) = h(M-1-n)$, $0 \le n \le M-1$. Then we can show (see Problem 7.1) that

$$H(e^{j\omega}) = \left[\sum_{n=0}^{(M-1)/2} a(n)\cos\omega n \right] e^{-j\omega(M-1)/2} \qquad (7.6)$$

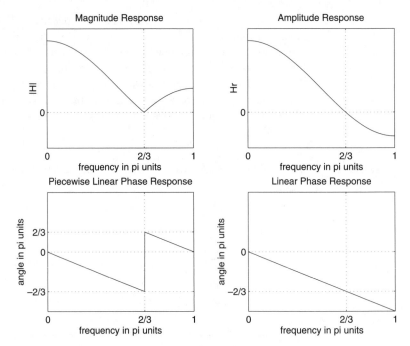

FIGURE 7.2 *Frequency responses in Example 7.3*

where sequence $a(n)$ is obtained from $h(n)$ as

$$a(0) = h\left(\frac{M-1}{2}\right) \quad : \quad \text{the middle sample}$$

$$a(n) = 2h\left(\frac{M-1}{2} - n\right), \quad 1 \le n \le \frac{M-3}{2} \tag{7.7}$$

Comparing (7.5) with (7.6), we have

$$H_r(\omega) = \sum_{n=0}^{(M-1)/2} a(n) \cos \omega n \tag{7.8}$$

Type-2 linear-phase FIR filter: Symmetrical impulse response, M even In this case again $\beta = 0$, $h(n) = h(M-1-n)$, $0 \le n \le M-1$, but $\alpha = (M-1)/2$ is not an integer. Then we can show (see Problem 7.2) that

$$H(e^{j\omega}) = \left[\sum_{n=1}^{M/2} b(n) \cos\left\{\omega\left(n - \frac{1}{2}\right)\right\}\right] e^{-j\omega(M-1)/2} \tag{7.9}$$

where

$$b(n) = 2h\left(\frac{M}{2} - n\right), \quad n = 1, 2, \ldots, \frac{M}{2} \tag{7.10}$$

Hence

$$H_r(\omega) = \sum_{n=1}^{M/2} b(n) \cos\left\{\omega\left(n - \frac{1}{2}\right)\right\} \tag{7.11}$$

Note: At $\omega = \pi$ we get

$$H_r(\pi) = \sum_{n=1}^{M/2} b(n) \cos\left\{\pi\left(n - \frac{1}{2}\right)\right\} = 0$$

regardless of $b(n)$ or $h(n)$. Hence we cannot use this type (i.e., symmetric $h(n)$, M even) for highpass or bandstop filters.

Type-3 linear-phase FIR filter: Antisymmetric impulse response, M ***odd***   In this case $\beta = \pi/2$, $\alpha = (M-1)/2$ is an integer, $h(n) = -h(M-1-n)$, $0 \leq n \leq M-1$, and $h((M-1)/2) = 0$. Then we can show (see Problem 7.3) that

$$H(e^{j\omega}) = \left[\sum_{n=1}^{(M-1)/2} c(n) \sin \omega n\right] e^{j\left[\frac{\pi}{2} - \left(\frac{M-1}{2}\right)\omega\right]} \tag{7.12}$$

where

$$c(n) = 2h\left(\frac{M-1}{2} - n\right), \quad n = 1, 2, \ldots, \frac{M-1}{2} \tag{7.13}$$

and

$$H_r(\omega) = \sum_{n=1}^{(M-1)/2} c(n) \sin \omega n \tag{7.14}$$

Note: At $\omega = 0$ and $\omega = \pi$ we have $H_r(\omega) = 0$, regardless of $c(n)$ or $h(n)$. Furthermore, $e^{j\pi/2} = j$, which means that $jH_r(\omega)$ is purely imaginary. Hence this type of filter is not suitable for designing a lowpass filter or a highpass filter. However, this behavior is suitable for approximating ideal digital Hilbert transformers and differentiators. An ideal Hilbert transformer [19] is an all-pass filter that imparts a 90° phase shift on the input signal. It is frequently used in communication systems for modulation purposes. Differentiators are used in many analog and digital systems to take the derivative of a signal.

Type-4 linear-phase FIR filter: Antisymmetric impulse response, M even   This case is similar to Type-2. We have (see Problem 7.4)

$$H(e^{j\omega}) = \left[\sum_{n=1}^{M/2} d(n) \sin \left\{ \omega \left(n - \frac{1}{2} \right) \right\} \right] e^{j[\frac{\pi}{2} - \omega(M-1)/2]} \tag{7.15}$$

where

$$d(n) = 2h \left(\frac{M}{2} - n \right), \quad n = 1, 2, \ldots, \frac{M}{2} \tag{7.16}$$

and

$$H_r(\omega) = \sum_{n=1}^{M/2} d(n) \sin \left\{ \omega \left(n - \frac{1}{2} \right) \right\} \tag{7.17}$$

Note: At $\omega = 0$, $H_r(0) = 0$ and $e^{j\pi/2} = j$. Hence this type is also suitable for designing digital Hilbert transformers and differentiators.

MATLAB
IMPLEMEN-
TATION

The MATLAB routine **freqz** computes the frequency response but we cannot determine the amplitude response from it because there is no function in MATLAB comparable to the **abs** function that can find amplitude. However, it easy to write simple routines to compute amplitude responses for each of the four types. We provide four functions to do this.

1. **Hr_type1**:

```
function [Hr,w,a,L] = Hr_Type1(h);
% Computes Amplitude response Hr(w) of a Type-1 LP FIR filter
% ------------------------------------------------------------
% [Hr,w,a,L] = Hr_Type1(h)
% Hr = Amplitude Response
%  w = 500 frequencies between [0 pi] over which Hr is computed
%  a = Type-1 LP filter coefficients
%  L = Order of Hr
%  h = Type-1 LP filter impulse response
%
 M = length(h);
 L = (M-1)/2;
 a = [h(L+1) 2*h(L:-1:1)]; % 1x(L+1) row vector
 n = [0:1:L];              % (L+1)x1 column vector
 w = [0:1:500]'*pi/500;
Hr = cos(w*n)*a';
```

2. Hr_type2:

```
function [Hr,w,b,L] = Hr_Type2(h);
% Computes Amplitude response of a Type-2 LP FIR filter
% -------------------------------------------------------
% [Hr,w,b,L] = Hr_Type2(h)
% Hr = Amplitude Response
%  w = frequencies between [0 pi] over which Hr is computed
%  b = Type-2 LP filter coefficients
%  L = Order of Hr
%  h = Type-2 LP impulse response
%
 M = length(h);
 L = M/2;
 b = 2*[h(L:-1:1)];
 n = [1:1:L]; n = n-0.5;
 w = [0:1:500]'*pi/500;
Hr = cos(w*n)*b';
```

3. Hr_type3:

```
function [Hr,w,c,L] = Hr_Type3(h);
% Computes Amplitude response Hr(w) of a Type-3 LP FIR filter
% ------------------------------------------------------------
% [Hr,w,c,L] = Hr_Type3(h)
% Hr = Amplitude Response
%  w = frequencies between [0 pi] over which Hr is computed
%  c = Type-3 LP filter coefficients
%  L = Order of Hr
%  h = Type-3 LP impulse response
%
 M = length(h);
 L = (M-1)/2;
 c = [2*h(L+1:-1:1)];
 n = [0:1:L];
 w = [0:1:500]'*pi/500;
Hr = sin(w*n)*c';
```

4. Hr_type4:

```
function [Hr,w,d,L] = Hr_Type4(h);
% Computes Amplitude response of a Type-4 LP FIR filter
% -------------------------------------------------------
% [Hr,w,d,L] = Hr_Type4(h)
% Hr = Amplitude Response
%  w = frequencies between [0 pi] over which Hr is computed
%  d = Type-4 LP filter coefficients
%  L = Order of d
%  h = Type-4 LP impulse response
```

```
%
M = length(h);
L = M/2;
d = 2*[h(L:-1:1)];
n = [1:1:L]; n = n-0.5;
w = [0:1:500]'*pi/500;
Hr = sin(w*n)*d';
```

These four functions can be combined into one function. This function can be written to determine the type of the linear-phase filter and to implement the appropriate amplitude response expression. This is explored in Problem 7.5. The use of these functions is described in Examples 7.4 through 7.7.

ZERO
LOCATIONS

Recall that for an FIR filter there are $(M-1)$ (trivial) poles at the origin and $(M-1)$ zeros located somewhere in the z-plane. For linear-phase FIR filters, these zeros possess certain symmetries that are due to the symmetry constraints on $h(n)$. It can be shown (see [19] and Problem 7.6) that if $H(z)$ has a zero at

$$z = z_1 = re^{j\theta}$$

then for linear phase there must be a zero at

$$z = \frac{1}{z_1} = \frac{1}{r}e^{-j\theta}$$

For a real-valued filter we also know that if z_1 is complex, then there must be a conjugate zero at $z_1^* = re^{-j\theta}$, which implies that there must be a zero at $1/z_1^* = (1/r)e^{j\theta}$. Thus a general *zero constellation* is a quadruplet

$$re^{j\theta}, \qquad \frac{1}{r}e^{j\theta}, \qquad re^{-j\theta}, \qquad \text{and} \qquad \frac{1}{r}e^{-j\theta}$$

as shown in Figure 7.3. Clearly, if $r = 1$, then $1/r = 1$, and hence the zeros are on the unit circle and occur in pairs

$$e^{j\theta} \qquad \text{and} \qquad e^{-j\theta}$$

If $\theta = 0$ or $\theta = \pi$, then the zeros are on the real line and occur in pairs

$$r \qquad \text{and} \qquad \frac{1}{r}$$

Finally, if $r = 1$ and $\theta = 0$ or $\theta = \pi$, the zeros are either at $z = 1$ or $z = -1$. These symmetries can be used to implement cascade forms with linear-phase sections.

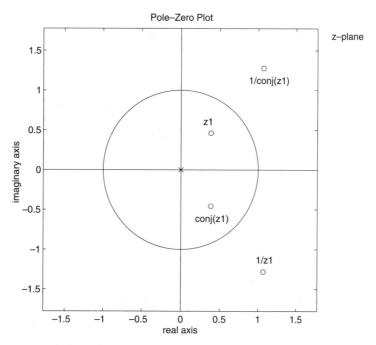

Pole–Zero Plot

z–plane

FIGURE 7.3   *A general zero constellation*

In the following examples we illustrate the above described properties of linear-phase FIR filters.

☐   **EXAMPLE 7.4**   Let $h(n) = \{-4, 1, -1, -2, 5, 6, 5, -2, -1, 1, -4\}$. Determine the amplitude response $H_r(\omega)$ and the locations of the zeros of $H(z)$.

Solution   Since $M = 11$, which is odd, and since $h(n)$ is symmetric about $\alpha = (11-1)/2 = 5$, this is a Type-1 linear-phase FIR filter. From (7.7) we have

$$a(0) = h(\alpha) = h(5) = 6, \quad a(1) = 2h(5-1) = 10, \quad a(2) = 2h(5-2) = -4$$
$$a(3) = 2h(5-3) = -2, \quad a(4) = 2h(5-4) = 2, \quad a(5) = 2h(5-5) = -8$$

From (7.8), we obtain

$$H_r(\omega) = a(0) + a(1)\cos\omega + a(2)\cos 2\omega + a(3)\cos 3\omega + a(4)\cos 4\omega + a(5)\cos 5\omega$$
$$= 6 + 10\cos\omega - 4\cos 2\omega - 2\cos 3\omega + 2\cos 4\omega - 8\cos 5\omega$$

MATLAB Script _____

```
>> h = [-4,1,-1,-2,5,6,5,-2,-1,1,-4];
>> M = length(h); n = 0:M-1;
>> [Hr,w,a,L] = Hr_Type1(h);
```

```
>> a,L
a = 6      10     -4     -2      2     -8
L = 5
>> amax = max(a)+1; amin = min(a)-1;
>> subplot(2,2,1); stem(n,h); axis([-1 2*L+1 amin amax])
>> xlabel('n'); ylabel('h(n)'); title('Impulse Response')
>> subplot(2,2,3); stem(0:L,a); axis([-1 2*L+1 amin amax])
>> xlabel('n'); ylabel('a(n)'); title('a(n) coefficients')
>> subplot(2,2,2); plot(w/pi,Hr);grid
>> xlabel('frequency in pi units'); ylabel('Hr')
>> title('Type-1 Amplitude Response')
>> subplot(2,2,4); zplane(h,1)
```

The plots and the zero locations are shown in Figure 7.4. From these plots we observe that there are no restrictions on $H_r(\omega)$ either at $\omega = 0$ or at $\omega = \pi$. There is one zero-quadruplet constellation and three zero pairs. □

□   **EXAMPLE 7.5**   Let $h(n) = \{-4, 1, -1, -2, 5, 6, 6, 5, -2, -1, 1, -4\}$. Determine the amplitude response $H_r(\omega)$ and the locations of the zeros of $H(z)$.

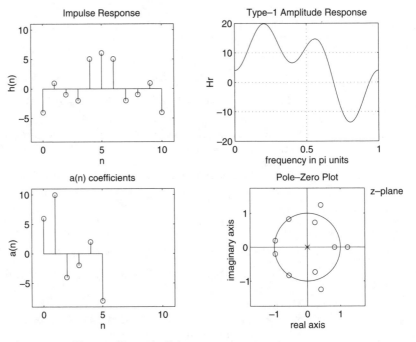

FIGURE 7.4   *Plots in Example 7.4*

Solution

This is a Type-2 linear-phase FIR filter since $M = 12$ and since $h(n)$ is symmetric with respect to $\alpha = (12 - 1)/2 = 5.5$. From (7.10) we have

$$b(1) = 2h\left(\tfrac{12}{2} - 1\right) = 12, \quad b(2) = 2h\left(\tfrac{12}{2} - 2\right) = 10, \quad b(3) = 2h\left(\tfrac{12}{2} - 3\right) = -4$$

$$b(4) = 2h\left(\tfrac{12}{2} - 4\right) = -2, \quad b(5) = 2h\left(\tfrac{12}{2} - 5\right) = 2, \quad b(6) = 2h\left(\tfrac{12}{2} - 6\right) = -8$$

Hence from (7.11) we obtain

$$H_r(\omega) = b(1)\cos\left[\omega\left(1 - \tfrac{1}{2}\right)\right] + b(2)\cos\left[\omega\left(2 - \tfrac{1}{2}\right)\right] + b(3)\cos\left[\omega\left(3 - \tfrac{1}{2}\right)\right]$$

$$+ b(4)\cos\left[\omega\left(4 - \tfrac{1}{2}\right)\right] + b(5)\cos\left[\omega\left(5 - \tfrac{1}{2}\right)\right] + b(6)\cos\left[\omega\left(6 - \tfrac{1}{2}\right)\right]$$

$$= 12\cos\left(\frac{\omega}{2}\right) + 10\cos\left(\frac{3\omega}{2}\right) - 4\cos\left(\frac{5\omega}{2}\right) - 2\cos\left(\frac{7\omega}{2}\right)$$

$$+ 2\cos\left(\frac{9\omega}{2}\right) - 8\cos\left(\frac{11\omega}{2}\right)$$

MATLAB Script

```
>> h = [-4,1,-1,-2,5,6,6,5,-2,-1,1,-4];
>> M = length(h); n = 0:M-1;
>> [Hr,w,a,L] = Hr_Type2(h);
>> b,L
b = 12    10    -4    -2    2    -8
L = 6
>> bmax = max(b)+1; bmin = min(b)-1;
>> subplot(2,2,1); stem(n,h); axis([-1 2*L+1 bmin bmax])
>> xlabel('n'); ylabel('h(n)'); title('Impulse Response')
>> subplot(2,2,3); stem(1:L,b); axis([-1 2*L+1 bmin bmax])
>> xlabel('n'); ylabel('b(n)'); title('b(n) coefficients')
>> subplot(2,2,2); plot(w/pi,Hr);grid
>> xlabel('frequency in pi units'); ylabel('Hr')
>> title('Type-1 Amplitude Response')
>> subplot(2,2,4); zplane(h,1)
```

The plots and the zero locations are shown in Figure 7.5. From these plots we observe that $H_r(\omega)$ is zero at $\omega = \pi$. There is one zero-quadruplet constellation, three zero pairs, and one zero at $\omega = \pi$ as expected. □

□ **EXAMPLE 7.6** Let $h(n) = \{-4, 1, -1, -2, 5, 0, -5, 2, 1, -1, 4\}$. Determine the amplitude response $H_r(\omega)$ and the locations of the zeros of $H(z)$.

Solution

Since $M = 11$, which is odd, and since $h(n)$ is antisymmetric about $\alpha = (11 - 1)/2 = 5$, this is a Type-3 linear-phase FIR filter. From (7.13) we have

$$c(0) = h(\alpha) = h(5) = 0, \quad c(1) = 2h(5 - 1) = 10, \quad c(2) = 2h(2 - 2) = -4$$

$$c(3) = 2h(5 - 3) = -2, \quad c(4) = 2h(5 - 4) = 2, \quad c(5) = 2h(5 - 5) = -8$$

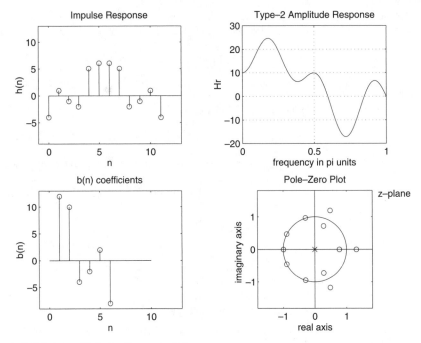

FIGURE 7.5 *Plots in Example 7.5*

From (7.14) we obtain

$$H_r(\omega) = c(0) + c(1)\sin\omega + c(2)\sin 2\omega + c(3)\sin 3\omega + c(4)\sin 4\omega + c(5)\sin 5\omega$$

$$= 0 + 10\sin\omega - 4\sin 2\omega - 2\sin 3\omega + 2\sin 4\omega - 8\sin 5\omega$$

MATLAB Script _____

```
>> h = [-4,1,-1,-2,5,0,-5,2,1,-1,4];
>> M = length(h); n = 0:M-1;
>> [Hr,w,c,L] = Hr_Type3(h);
>> c,L
a = 0      10     -4     -2      2     -8
L = 5
>> cmax = max(c)+1; cmin = min(c)-1;
>> subplot(2,2,1); stem(n,h); axis([-1 2*L+1 cmin cmax])
>> xlabel('n'); ylabel('h(n)'); title('Impulse Response')
>> subplot(2,2,3); stem(0:L,c); axis([-1 2*L+1 cmin cmax])
>> xlabel('n'); ylabel('c(n)'); title('c(n) coefficients')
>> subplot(2,2,2); plot(w/pi,Hr);grid
>> xlabel('frequency in pi units'); ylabel('Hr')
>> title('Type-1 Amplitude Response')
>> subplot(2,2,4); zplane(h,1)
```

The plots and the zero locations are shown in Figure 7.6. From these plots we observe that $H_r(\omega) = 0$ at $\omega = 0$ and at $\omega = \pi$. There is one zero-quadruplet constellation, two zero pairs, and zeros at $\omega = 0$ and $\omega = \pi$ as expected. $\square$

☐ **EXAMPLE 7.7** Let $h(n) = \{-4, 1, -1, -2, 5, 6, -6, -5, 2, 1, -1, 4\}$. Determine the amplitude
$\qquad\qquad\qquad\quad\uparrow$
response $H_r(\omega)$ and the locations of the zeros of $H(z)$.

Solution This is a Type-4 linear-phase FIR filter since $M = 12$ and since $h(n)$ is anti-symmetric with respect to $\alpha = (12 - 1)/2 = 5.5$. From (7.16) we have

$$d(1) = 2h\left(\tfrac{12}{2} - 1\right) = 12, \quad d(2) = 2h\left(\tfrac{12}{2} - 2\right) = 10, \quad d(3) = 2h\left(\tfrac{12}{2} - 3\right) = -4$$

$$d(4) = 2h\left(\tfrac{12}{2} - 4\right) = -2, \quad d(5) = 2h\left(\tfrac{12}{2} - 5\right) = 2, \quad d(6) = 2h\left(\tfrac{12}{2} - 6\right) = -8$$

Hence from (7.17) we obtain

$$H_r(\omega) = d(1)\sin\left[\omega\left(1 - \tfrac{1}{2}\right)\right] + d(2)\sin\left[\omega\left(2 - \tfrac{1}{2}\right)\right] + d(3)\sin\left[\omega\left(3 - \tfrac{1}{2}\right)\right]$$

$$+ d(4)\sin\left[\omega\left(4 - \tfrac{1}{2}\right)\right] + d(5)\sin\left[\omega\left(5 - \tfrac{1}{2}\right)\right] + d(6)\sin\left[\omega\left(6 - \tfrac{1}{2}\right)\right]$$

$$= 12\sin\left(\tfrac{\omega}{2}\right) + 10\sin\left(\tfrac{3\omega}{2}\right) - 4\sin\left(\tfrac{5\omega}{2}\right) - 2\sin\left(\tfrac{7\omega}{2}\right)$$

$$+ 2\sin\left(\tfrac{9\omega}{2}\right) - 8\sin\left(\tfrac{11\omega}{2}\right)$$

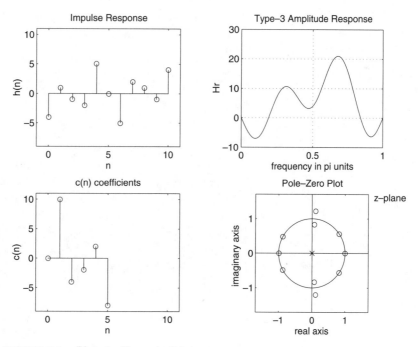

FIGURE 7.6 *Plots in Example 7.6*

```
>> h = [-4,1,-1,-2,5,6,-6,-5,2,1,-1,4];
>> M = length(h); n = 0:M-1;
>> [Hr,w,d,L] = Hr_Type4(h);
>> b,L
d = 12    10    -4    -2    2    -8
L = 6
>> dmax = max(d)+1; dmin = min(d)-1;
>> subplot(2,2,1); stem(n,h); axis([-1 2*L+1 dmin dmax])
>> xlabel('n'); ylabel('h(n)'); title('Impulse Response')
>> subplot(2,2,3); stem(1:L,d); axis([-1 2*L+1 dmin dmax])
>> xlabel('n'); ylabel('d(n)'); title('d(n) coefficients')
>> subplot(2,2,2); plot(w/pi,Hr);grid
>> xlabel('frequency in pi units'); ylabel('Hr')
>> title('Type-1 Amplitude Response')
>> subplot(2,2,4); zplane(h,1)
```

The plots and the zero locations are shown in Figure 7.7. From these plots we observe that $H_r(\omega)$ is zero at $\omega = 0$. There is one zero-quadruplet constellation, three zero pairs, and one zero at $\omega = 0$ as expected. □

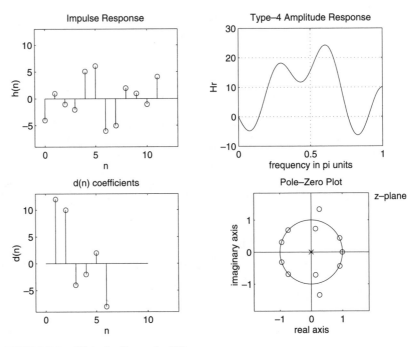

FIGURE 7.7 *Plots in Example 7.7*

WINDOW DESIGN TECHNIQUES

The basic idea behind the window design is to choose a proper ideal frequency-selective filter (which always has a noncausal, infinite-duration impulse response) and then truncate (or window) its impulse response to obtain a linear-phase and causal FIR filter. Therefore the emphasis in this method is on selecting an appropriate *windowing* function and an appropriate *ideal* filter. We will denote an ideal frequency-selective filter by $H_d(e^{j\omega})$, which has a unity magnitude gain and linear-phase characteristics over its passband, and zero response over its stopband. An ideal LPF of bandwidth $\omega_c < \pi$ is given by

$$H_d(e^{j\omega}) = \begin{cases} 1 \cdot e^{-j\alpha\omega}, & |\omega| \leq \omega_c \\ 0, & \omega_c < |\omega| \leq \pi \end{cases} \tag{7.18}$$

where ω_c is also called the *cutoff* frequency, and α is called the sample *delay* (note that from the DTFT properties, $e^{-j\alpha\omega}$ implies shift in the positive n direction or delay). The impulse response of this filter is of infinite duration and is given by

$$h_d(n) = \mathcal{F}^{-1}\left[H_d(e^{j\omega})\right] = \frac{1}{2\pi} \int_{-\pi}^{\pi} H_d(e^{j\omega})e^{j\omega n}d\omega \tag{7.19}$$

$$= \frac{1}{2\pi} \int_{-\omega_c}^{\omega_c} 1 \cdot e^{-j\alpha\omega}e^{j\omega n}d\omega$$

$$= \frac{\sin\left[\omega_c(n-\alpha)\right]}{\pi(n-\alpha)}$$

Note that $h_d(n)$ is symmetric with respect to α, a fact useful for linear-phase FIR filters.

To obtain an FIR filter from $h_d(n)$, one has to truncate $h_d(n)$ on both sides. To obtain a causal and linear-phase FIR filter $h(n)$ of length M, we must have

$$h(n) = \begin{cases} h_d(n), & 0 \leq n \leq M-1 \\ 0, & \text{elsewhere} \end{cases} \quad \text{and} \quad \alpha = \frac{M-1}{2} \tag{7.20}$$

This operation is called "windowing." In general, $h(n)$ can be thought of as being formed by the product of $h_d(n)$ and a window function $w(n)$ as follows:

$$h(n) = h_d(n)w(n) \tag{7.21}$$

where

$$w(n) = \begin{cases} \text{some symmetric function with respect to} \\ \alpha \text{ over } 0 \leq n \leq M - 1 \\ 0, \text{ otherwise} \end{cases}$$

Depending on how we define $w(n)$ above, we obtain different window designs. For example, in (7.20) above

$$w(n) = \begin{cases} 1, & 0 \leq n \leq M - 1 \\ 0, & \text{otherwise} \end{cases} = \mathcal{R}_M(n)$$

which is the *rectangular window* defined earlier.

In the frequency domain the causal FIR filter response $H(e^{j\omega})$ is given by the periodic convolution of $H_d(e^{j\omega})$ and the window response $W(e^{j\omega})$; that is,

$$H(e^{j\omega}) = H_d(e^{j\omega}) \circledast W(e^{j\omega}) = \frac{1}{2\pi} \int_{-\pi}^{\pi} W\left(e^{j\lambda}\right) H_d\left(e^{j(\omega-\lambda)}\right) d\lambda$$

$$\tag{7.22}$$

This is shown pictorially in Figure 7.8 for a typical window response, from which we have the following observations:

1. Since the window $w(n)$ has a finite length equal to M, its response has a *peaky main lobe* whose width is proportional to $1/M$, and has side lobes of smaller heights.

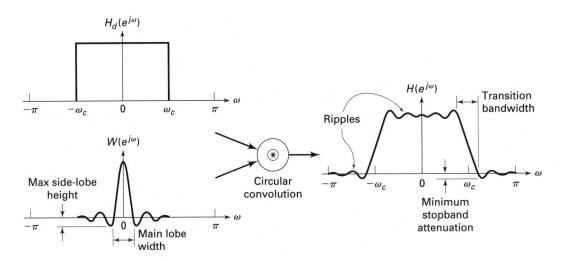

FIGURE 7.8   *Windowing operation in the frequency domain*

2. The periodic convolution (7.22) produces a smeared version of the ideal response $H_d(e^{j\omega})$.

3. The main lobe produces a transition band in $H(e^{j\omega})$ whose width is responsible for the transition width. This width is then proportional to $1/M$. The wider the main lobe, the wider will be the transition width.

4. The side lobes produce ripples that have similar shapes in both the passband and stopband.

Basic Window Design Idea For the given filter specifications choose the filter length M and a window function $w(n)$ for the narrowest main lobe width and the smallest side lobe attenuation possible.

From observation 4 above we note that the passband tolerance δ_1 and the stopband tolerance δ_2 cannot be specified independently. We generally take care of δ_2 alone, which results in $\delta_2 = \delta_1$. We now briefly describe various well-known window functions. We will use the rectangular window as an example to study their performances in the frequency domain.

RECTANGULAR WINDOW

This is the simplest window function but provides the worst performance from the viewpoint of stopband attenuation. It was defined earlier by

$$w(n) = \begin{cases} 1, & 0 \leq n \leq M - 1 \\ 0, & \text{otherwise} \end{cases} \tag{7.23}$$

Its frequency response function is

$$W(e^{j\omega}) = \left[\frac{\sin\left(\frac{\omega M}{2}\right)}{\sin\left(\frac{\omega}{2}\right)} \right] e^{-j\omega \frac{M-1}{2}} \Rightarrow W_r(\omega) = \frac{\sin\left(\frac{\omega M}{2}\right)}{\sin\left(\frac{\omega}{2}\right)}$$

which is the amplitude response. From (7.22) the actual amplitude response $H_r(\omega)$ is given by

$$H_r(\omega) \simeq \frac{1}{2\pi} \int\limits_{-\pi}^{\omega+\omega_c} W_r(\lambda)\, d\lambda = \frac{1}{2\pi} \int\limits_{-\pi}^{\omega+\omega_c} \frac{\sin\left(\frac{\omega M}{2}\right)}{\sin\left(\frac{\omega}{2}\right)} d\lambda, \quad M \gg 1 \tag{7.24}$$

This implies that the running integral of the window amplitude response (or *accumulated* amplitude response) is necessary in the accurate analysis of the transition bandwidth and the stopband attenuation. Figure 7.9 shows the rectangular window function $w(n)$, its amplitude response $W(\omega)$, the amplitude response in dB, and the accumulated amplitude response (7.24) in dB. From the observation of plots in Figure 7.9 we can make several observations.

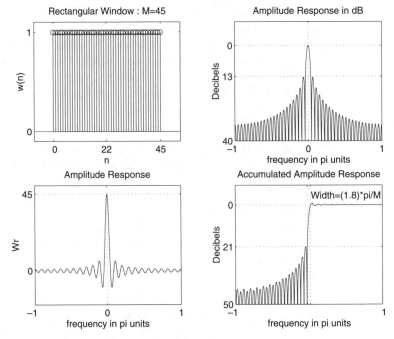

FIGURE 7.9 *Rectangular window:* $M = 45$

1. The amplitude response $W_r(\omega)$ has the first zero at $\omega = \omega_1$, where

$$\frac{\omega_1 M}{2} = \pi \qquad \text{or} \qquad \omega_1 = \frac{2\pi}{M}$$

Hence the width of the main lobe is $2\omega_1 = 4\pi/M$. Therefore the *approximate transition bandwidth* is $4\pi/M$.

2. The magnitude of the first side lobe (which is also the peak side lobe magnitude) is approximately at $\omega = 3\pi/M$ and is given by

$$\left| W_r\left(\omega = \frac{3\pi}{M}\right) \right| = \left| \frac{\sin\left(\frac{3\pi}{2}\right)}{\sin\left(\frac{3\pi}{2M}\right)} \right| \simeq \frac{2M}{3\pi} \qquad \text{for } M \gg 1$$

Comparing this with the main lobe amplitude, which is equal to M, the *peak side lobe magnitude* is

$$\frac{2}{3\pi} = 21.22\% \equiv 13 \text{ dB}$$

of the main lobe amplitude.

3. The accumulated amplitude response has the first side lobe magnitude at 21 dB. This results in the *minimum stopband attenuation* of 21 dB irrespective of the window length M.

4. Using the minimum stopband attenuation, the transition bandwidth can be accurately computed. It is shown in the accumulated amplitude response plot in Figure 7.9. This computed *exact transition bandwidth* is

$$\omega_s - \omega_p = \frac{1.8\pi}{M}$$

which is about half the approximate bandwidth of $4\pi/M$.

Clearly, this is a simple window operation in the time domain and an easy function to analyze in the frequency domain. However, there are two main problems. First, the minimum stopband attenuation of 21 dB is insufficient in practical applications. Second, the rectangular windowing being a direct truncation of the infinite length $h_d(n)$, it suffers from the *Gibbs phenomenon*. If we increase M, the width of each side lobe will decrease, but the area under each lobe will remain constant. Therefore the *relative amplitudes* of side lobes will remain constant, and the minimum stopband attenuation will remain at 21 dB. This implies that all ripples will bunch up near the band edges. It is shown in Figure 7.10.

Since the rectangular window is impractical in many applications, we consider other window functions, many of which bear the names of the people who first proposed them. Although these window functions can

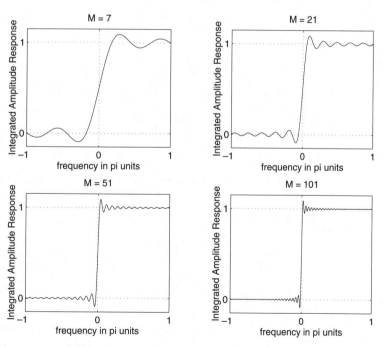

FIGURE 7.10 *Gibbs phenomenon*

also be analyzed similar to the rectangular window, we present only their results.

BARTLETT WINDOW

Since the Gibbs phenomenon results from the fact that the rectangular window has a sudden transition from 0 to 1 (or 1 to 0), Bartlett suggested a more gradual transition in the form of a triangular window, which is given by

$$
w(n) = \begin{cases}
\dfrac{2n}{M-1}, & 0 \le n \le \dfrac{M-1}{2} \\[2ex]
2 - \dfrac{2n}{M-1}, & \dfrac{M-1}{2} \le n \le M-1 \\[2ex]
0, & \text{otherwise}
\end{cases}
\tag{7.25}
$$

This window and its frequency-domain responses are shown in Figure 7.11.

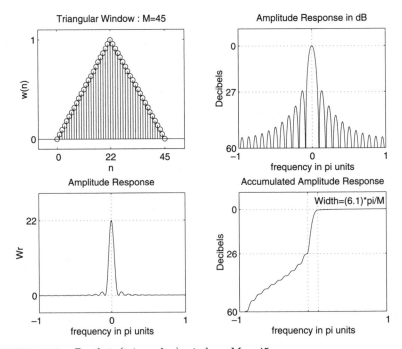

FIGURE 7.11 *Bartlett (triangular) window:* $M = 45$

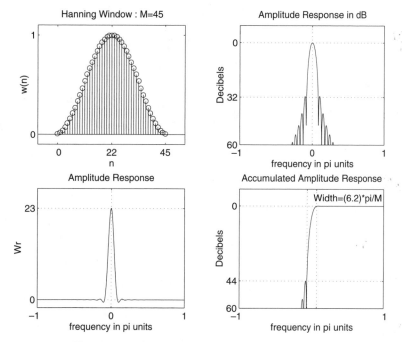

FIGURE 7.12 *Hanning window:* $M = 45$

HANNING
WINDOW

This is a raised cosine window function given by

$$w(n) = \begin{cases} 0.5 \left[1 - \cos\left(\frac{2\pi n}{M-1}\right)\right], & 0 \leq n \leq M - 1 \\ 0, & \text{otherwise} \end{cases} \qquad (7.26)$$

This window and its frequency-domain responses are shown in Figure 7.12.

HAMMING
WINDOW

This window is similar to the Hanning window except that it has a small amount of discontinuity and is given by

$$w(n) = \begin{cases} 0.54 - 0.46 \cos\left(\frac{2\pi n}{M-1}\right), & 0 \leq n \leq M - 1 \\ 0, & \text{otherwise} \end{cases} \qquad (7.27)$$

This window and its frequency-domain responses are shown in Figure 7.13.

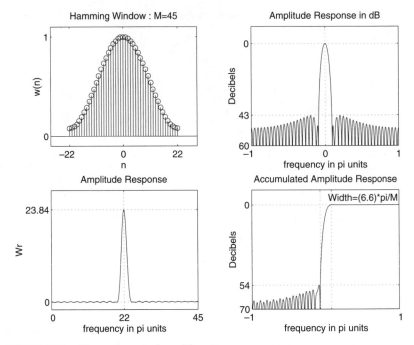

FIGURE 7.13   *Hamming window:* $M = 45$

BLACKMAN WINDOW

This window is also similar to the previous two but contains a second harmonic term and is given by

$$w(n) = \begin{cases} 0.42 - 0.5\cos\left(\frac{2\pi n}{M-1}\right) + 0.08\cos\left(\frac{4\pi n}{M-1}\right), & 0 \le n \le M - 1 \\ 0, & \text{otherwise} \end{cases}$$

(7.28)

This window and its frequency-domain responses are shown in Figure 7.14.

In Table 7.1 we provide a summary of window function characteristics in terms of their transition widths (as a function of M) and their minimum stopband attenuations in dB. Both the approximate as well as the exact transition bandwidths are given. Note that the transition widths and the stopband attenuations increase as we go down the table. The Hamming window appears to be the best choice for many applications.

KAISER WINDOW

This is one of the most useful and optimum windows. It is optimum in the sense of providing a large main lobe width for the given stopband attenuation, which implies the sharpest transition width. The window

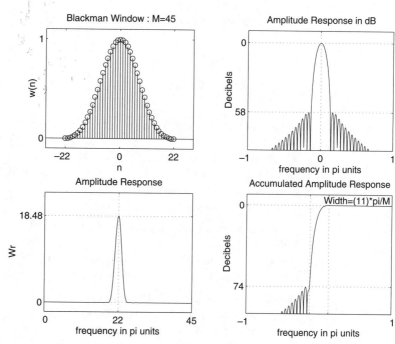

FIGURE 7.14 *Blackman window:* $M = 45$

TABLE 7.1 *Summary of commonly used window function characteristics*

Window Name	Transition Width $\Delta\omega$ Approximate	Exact Values	Min. Stopband Attenuation
Rectangular	$\dfrac{4\pi}{M}$	$\dfrac{1.8\pi}{M}$	21 dB
Bartlett	$\dfrac{8\pi}{M}$	$\dfrac{6.1\pi}{M}$	25 dB
Hanning	$\dfrac{8\pi}{M}$	$\dfrac{6.2\pi}{M}$	44 dB
Hamming	$\dfrac{8\pi}{M}$	$\dfrac{6.6\pi}{M}$	53 dB
Blackman	$\dfrac{12\pi}{M}$	$\dfrac{11\pi}{M}$	74 dB

function is due to J. F. Kaiser and is given by

$$w(n) = \frac{I_0\left[\beta\sqrt{1 - \left(1 - \frac{2n}{M-1}\right)^2}\right]}{I_0[\beta]}, \quad 0 \le n \le M-1 \qquad \textbf{(7.29)}$$

where $I_0[\cdot]$ is the *modified zero-order Bessel function*, and β is a parameter that depends on M and that can be chosen to yield various transition widths and near-optimum stopband attenuation. This window can provide different transition widths for the same M, which is something other windows lack. For example,

- if $\beta = 5.658$, then the transition width is equal to $7.8\pi/M$, and the minimum stopband attenuation is equal to 60 dB. This is shown in Figure 7.15.
- if $\beta = 4.538$, then the transition width is equal to $5.8\pi/M$, and the minimum stopband attenuation is equal to 50 dB.

Hence the performance of this window is comparable to that of the Hamming window. In addition, the Kaiser window provides flexible transition bandwidths. Due to the complexity involved in the Bessel functions, the design equations for this window are not easy to derive. Fortunately,

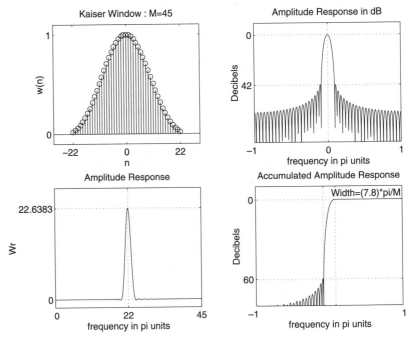

FIGURE 7.15   *Kaiser window:* $M = 45$, $\beta = 5.658$

Kaiser has developed *empirical* design equations, which we provide below without proof.

<table>
<tr><td>DESIGN
EQUATIONS</td><td>

Given ω_p, ω_s, R_p, and A_s

$$\text{Norm. transition width} = \Delta f \triangleq \frac{\omega_s - \omega_p}{2\pi}$$

$$\text{Filter order } M \simeq \frac{A_s - 7.95}{14.36\Delta f} + 1 \qquad \textbf{(7.30)}$$

$$\text{Parameter } \beta = \begin{cases} 0.1102\,(A_s - 8.7), & A_s \geq 50 \\[2mm] 0.5842\,(A_s - 21)^{0.4} \\ \quad + 0.07886\,(A_s - 21), & 21 < A_s < 50 \end{cases}$$

</td></tr>
</table>

DESIGN
EQUATIONS

Given ω_p, ω_s, R_p, and A_s

$$\text{Norm. transition width} = \Delta f \triangleq \frac{\omega_s - \omega_p}{2\pi}$$

$$\text{Filter order } M \simeq \frac{A_s - 7.95}{14.36\Delta f} + 1 \qquad \textbf{(7.30)}$$

$$\text{Parameter } \beta = \begin{cases} 0.1102\,(A_s - 8.7), & A_s \geq 50 \\[2mm] 0.5842\,(A_s - 21)^{0.4} \\ \quad + 0.07886\,(A_s - 21), & 21 < A_s < 50 \end{cases}$$

MATLAB
IMPLEMEN-
TATION

MATLAB provides several routines to implement window functions discussed in this section. A brief description of these routines is given below.

- w=boxcar(M) returns the M-point rectangular window function in array w.
- w=triang(M) returns the M-point Bartlett (triangular) window function in array w.
- w=hanning(M) returns the M-point Hanning window function in array w.
- w=hamming(M) returns the M-point Hamming window function in array w.
- w=blackman(M) returns the M-point Blackman window function in array w.
- w=kaiser(M,beta) returns the beta-valued M-point rectangular window function in array w.

Using these routines, we can use MATLAB to design FIR filters based on the window technique, which also requires an ideal lowpass impulse response $h_d(n)$. Therefore it is convenient to have a simple routine that creates $h_d(n)$ as shown below.

```
function hd = ideal_lp(wc,M);
% Ideal LowPass filter computation
% --------------------------------
% [hd] = ideal_lp(wc,M)
%  hd = ideal impulse response between 0 to M-1
%  wc = cutoff frequency in radians
%   M = length of the ideal filter
%
```

```
alpha = (M-1)/2;
n = [0:1:(M-1)];
m = n - alpha + eps;              % add smallest number to avoid divide by zero
hd = sin(wc*m) ./ (pi*m);
```

In the Signal Processing toolbox MATLAB provides a routine called fir1, which designs FIR filters using windows. However, this routine is not available in the Student Edition. To display the frequency-domain plots of digital filters, MATLAB provides the freqz routine, which we used in earlier chapters. Using this routine, we have developed a modified version, called freqz_m, which returns the magnitude response in absolute as well as in relative dB scale, the phase response, and the group delay response. We will need the group delay response in the next chapter.

```
function [db,mag,pha,grd,w] = freqz_m(b,a);
% Modified version of freqz subroutine
% ------------------------------------
% [db,mag,pha,grd,w] = freqz_m(b,a);
%   db = Relative magnitude in dB computed over 0 to pi radians
%  mag = absolute magnitude computed over 0 to pi radians
%  pha = Phase response in radians over 0 to pi radians
%  grd = Group delay over 0 to pi radians
%    w = 501 frequency samples between 0 to pi radians
%    b = numerator polynomial of H(z)    (for FIR: b=h)
%    a = denominator polynomial of H(z) (for FIR: a=[1])
%
[H,w] = freqz(b,a,1000,'whole');
    H = (H(1:1:501))'; w = (w(1:1:501))';
  mag = abs(H);
   db = 20*log10((mag+eps)/max(mag));
  pha = angle(H);
  grd = grpdelay(b,a,w);
```

DESIGN EXAMPLES

We now provide several examples of FIR filter design using window techniques and MATLAB routines.

□ **EXAMPLE 7.8** Design a digital FIR lowpass filter with the following specifications:

$$\omega_p = 0.2\pi, \qquad R_p = 0.25 \text{ dB}$$

$$\omega_s = 0.3\pi, \qquad A_s = 50 \text{ dB}$$

Choose an appropriate window function from Table 7.1. Determine the impulse response and provide a plot of the frequency response of the designed filter.

Solution Both the Hamming and Blackman windows can provide attenuation of more than 50 dB. Let us choose the Hamming window, which provides the smaller transition band and hence has the smaller order. Although we do not use the passband ripple value of $R_p = 0.25$ dB in the design, we will have to check the actual ripple from the design and verify that it is indeed within the given tolerance. The design steps are given in the following MATLAB script.

```
>> wp = 0.2*pi; ws = 0.3*pi;
>> tr_width = ws - wp;
>> M = ceil(6.6*pi/tr_width) + 1
M = 67
>> n=[0:1:M-1];
>> wc = (ws+wp)/2, % Ideal LPF cutoff frequency
>> hd = ideal_lp(wc,M);
>> w_ham = (hamming(M))';
>> h = hd .* w_ham;
>> [db,mag,pha,grd,w] = freqz_m(h,[1]);
>> delta_w = 2*pi/1000;
>> Rp = -(min(db(1:1:wp/delta_w+1)));      % Actual Passband Ripple
Rp = 0.0394
>> As = -round(max(db(ws/delta_w+1:1:501)))  % Min Stopband attenuation
As = 52
% plots
>> subplot(1,1,1)
>> subplot(2,2,1); stem(n,hd); title('Ideal Impulse Response')
>> axis([0 M-1 -0.1 0.3]); xlabel('n'); ylabel('hd(n)')
>> subplot(2,2,2); stem(n,w_ham);title('Hamming Window')
>> axis([0 M-1 0 1.1]); xlabel('n'); ylabel('w(n)')
>> subplot(2,2,3); stem(n,h);title('Actual Impulse Response')
>> axis([0 M-1 -0.1 0.3]); xlabel('n'); ylabel('h(n)')
>> subplot(2,2,4); plot(w/pi,db);title('Magnitude Response in dB');grid
>> axis([0 1 -100 10]); xlabel('frequency in pi units'); ylabel('Decibels')
```

Note that the filter length is $M = 67$, the actual stopband attenuation is 52 dB, and the actual passband ripple is 0.0394 dB. Clearly, the passband ripple is satisfied by this design. This practice of verifying the passband ripple is strongly recommended. The time- and the frequency-domain plots are shown in Figure 7.16. □

□ **EXAMPLE 7.9** For the design specifications given in Example 7.8, choose the Kaiser window and design the necessary lowpass filter.

Solution The design steps are given in the following MATLAB script.

```
>> wp = 0.2*pi; ws = 0.3*pi; As = 50;
>> tr_width = ws - wp;
>> M = ceil((As-7.95)/(14.36*tr_width/(2*pi))+1) + 1
M = 61
>> n=[0:1:M-1];
>> beta = 0.1102*(As-8.7)
```

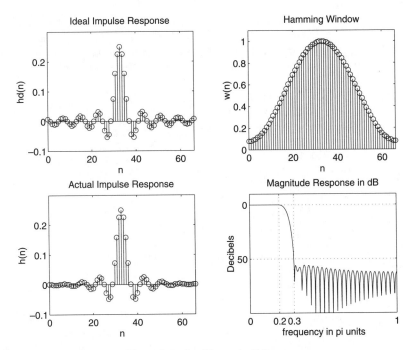

FIGURE 7.16 *Lowpass filter plots for Example 7.8*

```
beta = 4.5513
>> wc = (ws+wp)/2;
>> hd = ideal_lp(wc,M);
>> w_kai = (kaiser(M,beta))';
>> h = hd .* w_kai;
>> [db,mag,pha,grd,w] = freqz_m(h,[1]);
>> delta_w = 2*pi/1000;
>> As = -round(max(db(ws/delta_w+1:1:501))) % Min Stopband Attenuation
As = 52
% Plots
>> subplot(1,1,1)
>> subplot(2,2,1); stem(n,hd); title('Ideal Impulse Response')
>> axis([0 M-1 -0.1 0.3]); xlabel('n'); ylabel('hd(n)')
>> subplot(2,2,2); stem(n,w_kai);title('Kaiser Window')
>> axis([0 M-1 0 1.1]);  xlabel('n'); ylabel('w(n)')
>> subplot(2,2,3); stem(n,h);title('Actual Impulse Response')
>> axis([0 M-1 -0.1 0.3]); xlabel('n'); ylabel('h(n)')
>> subplot(2,2,4);plot(w/pi,db);title('Magnitude Response in dB');grid
>> axis([0 1 -100 10]); xlabel('frequency in pi units'); ylabel('Decibels')
```

Note that the Kaiser window parameters are $M = 61$ and $\beta = 4.5513$ and that the actual stopband attenuation is 52 dB. The time- and the frequency-domain plots are shown in Figure 7.17. □

Chapter 7 ■ FIR FILTER DESIGN

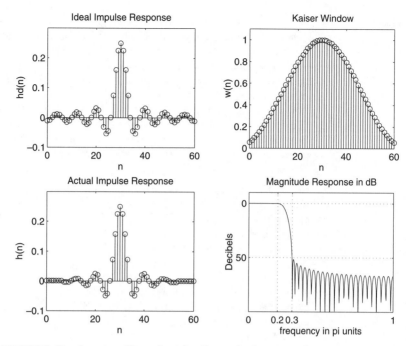

FIGURE 7.17 *Lowpass filter plots for Example 7.9*

☐   **EXAMPLE 7.10**   Let us design the following digital bandpass filter.

$$\begin{aligned}
\text{lower stopband edge:} \quad & \omega_{1s} = 0.2\pi, \quad && A_s = 60 \text{ dB} \\
\text{lower passband edge:} \quad & \omega_{1p} = 0.35\pi, \quad && R_p = 1 \text{ dB} \\
\text{upper passband edge:} \quad & \omega_{2p} = 0.65\pi \quad && R_p = 1 \text{ dB} \\
\text{upper stopband edge:} \quad & \omega_{2s} = 0.8\pi \quad && A_s = 60 \text{ dB}
\end{aligned}$$

These quantities are shown in Figure 7.18.

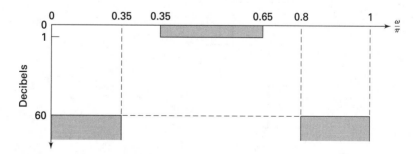

FIGURE 7.18 *Bandpass filter specifications in Example 7.10*

Solution

There are two transition bands, namely, $\Delta\omega_1 \triangleq \omega_{1p} - \omega_{1s}$ and $\Delta\omega_2 \triangleq \omega_{2s} - \omega_{2p}$. These two bandwidths must be the same in the window design; that is, there is no independent control over $\Delta\omega_1$ and $\Delta\omega_2$. Hence $\Delta\omega_1 = \Delta\omega_2 = \Delta\omega$. For this design we can use either the Kaiser window or the Blackman window. Let us use the Blackman Window. We will also need the ideal bandpass filter impulse response $h_d(n)$. Note that this impulse response can be obtained from two ideal lowpass magnitude responses, provided they have the same phase response. This is shown in Figure 7.19. Therefore the MATLAB routine `ideal-lp(wc,M)` is sufficient to determine the impulse response of an ideal bandpass filter. The design steps are given in the following MATLAB script.

```
>> ws1 = 0.2*pi; wp1 = 0.35*pi;
>> wp2 = 0.65*pi; ws2 = 0.8*pi;
>> As = 60;
>> tr_width = min((wp1-ws1),(ws2-wp2));
>> M = ceil(11*pi/tr_width) + 1
M = 75
>> n=[0:1:M-1];
>> wc1 = (ws1+wp1)/2; wc2 = (wp2+ws2)/2;
>> hd = ideal_lp(wc2,M) - ideal_lp(wc1,M);
>> w_bla = (blackman(M))';
>> h = hd .* w_bla;
>> [db,mag,pha,grd,w] = freqz_m(h,[1]);
>> delta_w = 2*pi/1000;
>> Rp = -min(db(wp1/delta_w+1:1:wp2/delta_w)) % Actua; Passband Ripple
Rp = 0.0030
>> As = -round(max(db(ws2/delta_w+1:1:501))) % Min Stopband Attenuation
As = 75
%Plots
>> subplot(2,2,1); stem(n,hd); title('Ideal Impulse Response')
>> axis([0 M-1 -0.4 0.5]); xlabel('n'); ylabel('hd(n)')
>> subplot(2,2,2); stem(n,w_bla);title('Blackman Window')
>> axis([0 M-1 0 1.1]); xlabel('n'); ylabel('w(n)')
```

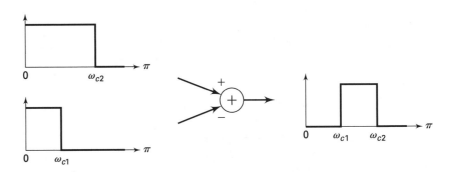

FIGURE 7.19 *Ideal bandpass filter from two lowpass filters*

Chapter 7 ■ FIR FILTER DESIGN

```
>> subplot(2,2,3); stem(n,h);title('Actual Impulse Response')
>> axis([0 M-1 -0.4 0.5]); xlabel('n'); ylabel('h(n)')
>> subplot(2,2,4);plot(w/pi,db);axis([0 1 -150 10]);
>> title('Magnitude Response in dB');grid;
>> xlabel('frequency in pi units'); ylabel('Decibels')
```

Note that the Blackman window length is $M = 61$ and that the actual stopband attenuation is 75 dB. The time- and the frequency-domain plots are shown in Figure 7.20. □

□ **EXAMPLE 7.11**   The frequency response of an ideal bandstop filter is given by

$$H_e\left(e^{j\omega}\right) = \begin{cases} 1, & 0 \le |\omega| < \pi/3 \\ 0, & \pi/3 \le |\omega| \le 2\pi/3 \\ 1, & 2\pi/3 < |\omega| \le \pi \end{cases}$$

Using a Kaiser window, design a bandstop filter of length 45 with stopband attenuation of 60 dB.

Solution   Note that in these design specifications, the transition bandwidth is not given. It will be determined by the length $M = 45$ and the parameter β of the Kaiser

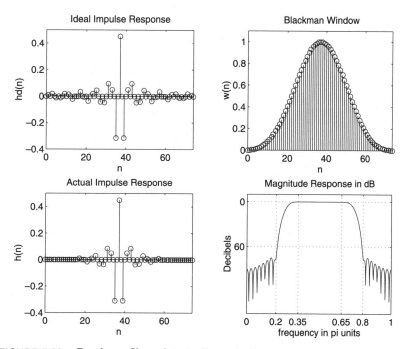

FIGURE 7.20   *Bandpass filter plots in Example 7.10*

window. From the design equations (7.30) we can determine β from A_s; that is,

$$\beta = 0.1102 \times (A_s - 8.7)$$

The ideal bandstop impulse response can also be determined from the ideal lowpass impulse response using a method similar to Figure 7.19. We can now implement the Kaiser window design and check for the minimum stopband attenuation. This is shown in the following MATLAB script.

```
>> M = 45; As = 60; n=[0:1:M-1];
>> beta = 0.1102*(As-8.7)
beta = 5.6533
>> w_kai = (kaiser(M,beta))';
>> wc1 = pi/3; wc2 = 2*pi/3;
>> hd = ideal_lp(wc1,M) + ideal_lp(pi,M) - ideal_lp(wc2,M);
>> h = hd .* w_kai;
>> [db,mag,pha,grd,w] = freqz_m(h,[1]);
>> subplot(1,1,1);
>> subplot(2,2,1); stem(n,hd); title('Ideal Impulse Response')
>> axis([-1 M -0.2 0.8]); xlabel('n'); ylabel('hd(n)')
>> subplot(2,2,2); stem(n,w_kai);title('Kaiser Window')
>> axis([-1 M 0 1.1]); xlabel('n'); ylabel('w(n)')
>> subplot(2,2,3); stem(n,h);title('Actual Impulse Response')
>> axis([-1 M -0.2 0.8]); xlabel('n'); ylabel('h(n)')
>> subplot(2,2,4);plot(w/pi,db); axis([0 1 -80 10]);
>> title('Magnitude Response in dB');grid;
>> xlabel('frequency in pi units'); ylabel('Decibels')
```

The β parameter is equal to 5.6533, and from the magnitude plot in Figure 7.21 we observe that the minimum stopband attenuation is smaller than 60 dB. Clearly, we have to increase β to increase the attenuation to 60 dB. The required value was found to be $\beta = 5.9533$.

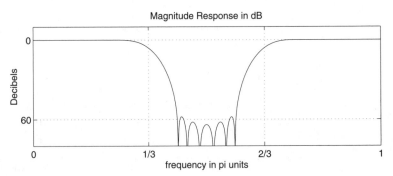

FIGURE 7.21 *Bandstop filter magnitude response in Example 7.11 for* $\beta = 5.6533$

Chapter 7 ■ FIR FILTER DESIGN

```
>> M = 45; As = 60; n=[0:1:M-1];
>> beta = 0.1102*(As-8.7)+0.3
beta = 5.9533
>> w_kai = (kaiser(M,beta))';
>> wc1 = pi/3; wc2 = 2*pi/3;
>> hd = ideal_lp(wc1,M) + ideal_lp(pi,M) - ideal_lp(wc2,M);
>> h = hd .* w_kai;
>> [db,mag,pha,grd,w] = freqz_m(h,[1]);
>> subplot(1,1,1);
>> subplot(2,2,1); stem(n,hd); title('Ideal Impulse Response')
>> axis([-1 M -0.2 0.8]); xlabel('n'); ylabel('hd(n)')
>> subplot(2,2,2); stem(n,w_kai);title('Kaiser Window')
>> axis([-1 M 0 1.1]); xlabel('n'); ylabel('w(n)')
>> subplot(2,2,3); stem(n,h);title('Actual Impulse Response')
>> axis([-1 M -0.2 0.8]); xlabel('n'); ylabel('h(n)')
>> subplot(2,2,4);plot(w/pi,db); axis([0 1 -80 10]);
>> title('Magnitude Response in dB');grid;
>> xlabel('frequency in pi units'); ylabel('Decibels')
```

The time- and the frequency-domain plots are shown in Figure 7.22, in which
the designed filter satisfies the necessary requirements. □

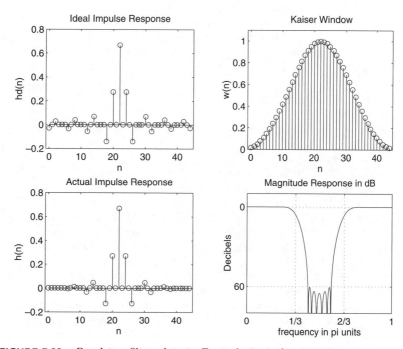

FIGURE 7.22 *Bandstop filter plots in Example 7.11: $\beta = 5.9533$*

The frequency response of an ideal digital differentiator is given by

$$H_d\left(e^{j\omega}\right) = \begin{cases} j\omega, & 0 < \omega \le \pi \\ -j\omega, & -\pi < \omega < 0 \end{cases} \qquad (7.31)$$

Using a Hamming window of length 21, design a digital FIR differentiator. Plot the time- and the frequency-domain responses.

Solution   The ideal impulse response of a digital differentiator with linear phase is given by

$$h_d\left(n\right) = \mathcal{F}\left[H_d\left(e^{j\omega}\right)e^{-j\alpha\omega}\right] = \frac{1}{2\pi}\int\limits_{-\pi}^{\pi} H_d\left(e^{j\omega}\right)e^{-j\alpha\omega}e^{j\omega n}d\omega$$

$$= \frac{1}{2\pi}\int\limits_{-\pi}^{0}\left(-j\omega\right)e^{-j\alpha\omega}e^{j\omega n}d\omega + \frac{1}{2\pi}\int\limits_{0}^{\pi}\left(j\omega\right)e^{-j\alpha\omega}e^{j\omega n}d\omega$$

$$= \begin{cases} \dfrac{\cos\pi\left(n-\alpha\right)}{\left(n-\alpha\right)}, & n \ne \alpha \\ 0, & n = \alpha \end{cases}$$

The above impulse response can be implemented in MATLAB along with the Hamming window to design the required differentiator. Note that if M is an even number, then $\alpha = (M-1)/2$ is not an integer and $h_d\left(n\right)$ will be zero for all n. Hence M must be an odd number, and this will be a Type-3 linear-phase FIR filter. However, the filter will not be a full-band differentiator since $H_r\left(\pi\right) = 0$ for Type-3 filters.

```
>> M = 21; alpha = (M-1)/2;
>> n = 0:M-1;
>> hd = (cos(pi*(n-alpha)))./(n-alpha); hd(alpha+1)=0;
>> w_ham = (hamming(M))';
>> h = hd .* w_ham;
>> [Hr,w,P,L] = Hr_Type3(h);
% plots
>> subplot(1,1,1);
>> subplot(2,2,1); stem(n,hd); title('Ideal Impulse Response')
>> axis([-1 M -1.2 1.2]); xlabel('n'); ylabel('hd(n)')
>> subplot(2,2,2); stem(n,w_ham);title('Hamming Window')
>> axis([-1 M 0 1.2]); xlabel('n'); ylabel('w(n)')
>> subplot(2,2,3); stem(n,h);title('Actual Impulse Response')
>> axis([-1 M -1.2 1.2]); xlabel('n'); ylabel('h(n)')
>> subplot(2,2,4);plot(w/pi,Hr/pi); title('Amplitude Response');grid;
>> xlabel('frequency in pi units'); ylabel('slope in pi units'); axis([0 1 0 1]);
```

The plots are shown in Figure 7.23.   □

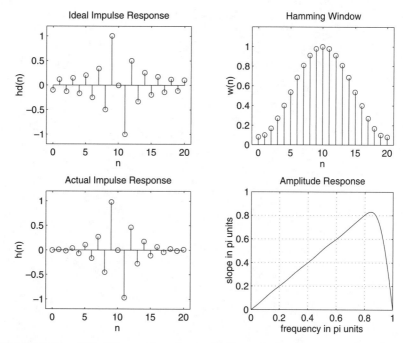

FIGURE 7.23 *FIR differentiator design in Example 7.12*

☐ **EXAMPLE 7.13** Design a length-25 digital Hilbert transformer using a Hanning window.

Solution The ideal frequency response of a linear-phase Hilbert transformer is given by

$$H_d\left(e^{j\omega}\right) = \begin{cases} -je^{-j\alpha\omega}, & 0 < \omega < \pi \\ +je^{-j\alpha\omega}, & -\pi < \omega < 0 \end{cases} \tag{7.32}$$

After inverse transformation the ideal impulse response is given by

$$h_d\left(n\right) = \begin{cases} \dfrac{2}{\pi} \dfrac{\sin^2 \pi\left(n - \alpha\right)/2}{n - \alpha}, & n \neq \alpha \\ 0, & n = \alpha \end{cases}$$

which can be easily implemented in MATLAB. Note that since $M = 25$, the designed filter is of Type-3.

```
>> M = 25; alpha = (M-1)/2;
>> n = 0:M-1;
>> hd = (2/pi)*((sin((pi/2)*(n-alpha)).^2)./(n-alpha)); hd(alpha+1)=0;
>> w_han = (hanning(M))';
>> h = hd .* w_han;
>> [Hr,w,P,L] = Hr_Type3(h);
% plots
>> subplot(1,1,1);
```

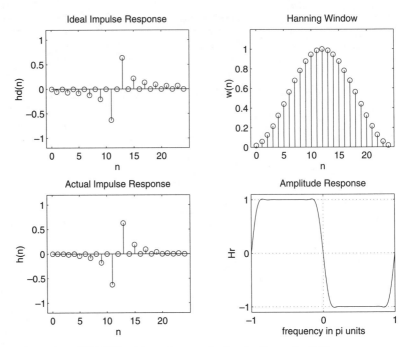

FIGURE 7.24 *FIR Hilbert transformer design in Example 7.13*

```
>> subplot(2,2,1); stem(n,hd); title('Ideal Impulse Response')
>> axis([-1 M -1.2 1.2]); xlabel('n'); ylabel('hd(n)')
>> subplot(2,2,2); stem(n,w_han);title('Hanning Window')
>> axis([-1 M 0 1.2]); xlabel('n'); ylabel('w(n)')
>> subplot(2,2,3); stem(n,h);title('Actual Impulse Response')
>> axis([-1 M -1.2 1.2]); xlabel('n'); ylabel('h(n)')
>> w = w'; Hr = Hr';
>> w = [-fliplr(w), w(2:501)]; Hr = [-fliplr(Hr), Hr(2:501)];
>> subplot(2,2,4);plot(w/pi,Hr); title('Amplitude Response');grid;
>> xlabel('frequency in pi units'); ylabel('Hr'); axis([-1 1 -1.1 1.1]);
```

The plots are shown in Figure 7.24. Observe that the amplitude response is plotted over $-\pi \le \omega \le \pi$. □

FREQUENCY SAMPLING DESIGN TECHNIQUES

In this design approach we use the fact that the system function $H(z)$ can be obtained from the samples $H(k)$ of the frequency response $H(e^{j\omega})$. Furthermore, this design technique fits nicely with the frequency sampling structure that we discussed in Chapter 6. Let $h(n)$ be the impulse response of an M-point FIR filter, $H(k)$ be its M-point DFT, and $H(z)$ be its system function. Then from (6.12) we have

$$H(z) = \sum_{n=0}^{M-1} h(n) z^{-n} = \frac{1-z^{-M}}{M} \sum_{k=0}^{M-1} \frac{H(k)}{1-z^{-1}e^{j2\pi k/M}} \qquad (7.33)$$

and

$$H\left(e^{j\omega}\right) = \frac{1-e^{-j\omega M}}{M} \sum_{k=0}^{M-1} \frac{H(k)}{1-e^{-j\omega}e^{j2\pi k/M}} \qquad (7.34)$$

with

$$H(k) = H\left(e^{j2\pi k/M}\right) = \begin{cases} H(0), & k=0 \\ H^*(M-k), & k=1,\ldots,M-1 \end{cases}$$

For a linear-phase FIR filter we have

$$h(n) = \pm h(M-1-n), \quad n=0,1,\ldots,M-1$$

where the positive sign is for the Type-1 and Type-2 linear-phase filters, while the negative sign is for the Type-3 and Type-4 linear-phase filters. Then $H(k)$ is given by

$$H(k) = H_r\left(\frac{2\pi k}{M}\right) e^{j\angle H(k)} \qquad (7.35)$$

where

$$H_r\left(\frac{2\pi k}{M}\right) = \begin{cases} H_r(0), & k=0 \\ H_r\left(\frac{2\pi(M-k)}{M}\right), & k=1,\ldots,M-1 \end{cases} \qquad (7.36)$$

and

$$\angle H(k) = \begin{cases} -\left(\dfrac{M-1}{2}\right)\left(\dfrac{2\pi k}{M}\right), & k=0,\ldots,\left\lfloor\dfrac{M-1}{2}\right\rfloor \\ +\left(\dfrac{M-1}{2}\right)\dfrac{2\pi}{M}(M-k), & k=\left\lfloor\dfrac{M-1}{2}\right\rfloor+1,\ldots,M-1 \end{cases} \text{, (Type-1 \& 2)}$$

$$(7.37)$$

or

$$\angle H(k) = \begin{cases} \left(\pm\dfrac{\pi}{2}\right)-\left(\dfrac{M-1}{2}\right)\left(\dfrac{2\pi k}{M}\right), & k=0,\ldots,\left\lfloor\dfrac{M-1}{2}\right\rfloor \\ -\left(\pm\dfrac{\pi}{2}\right)+\left(\dfrac{M-1}{2}\right)\dfrac{2\pi}{M}(M-k), & \\ & k=\left\lfloor\dfrac{M-1}{2}\right\rfloor+1,\ldots,M-1 \end{cases} \text{, (Type-3 \& 4)}$$

$$(7.38)$$

Finally, we have

$$h(n) = \text{IDFT}\,[H(k)] \tag{7.39}$$

Note that several textbooks (e.g., [19, 20, 16]) provide explicit formulas to compute $h(n)$, given $H(k)$. We will use MATLAB's `ifft` routine to compute $h(n)$ from (7.39).

Basic Idea   Given the ideal lowpass filter $H_d(e^{j\omega})$, choose the filter length M and then sample $H_d(e^{j\omega})$ at M equispaced frequencies between 0 and 2π. The actual response $H(e^{j\omega})$ is the interpolation of the samples $H(k)$ given by (7.34). This is shown in Figure 7.25. The impulse response is given by (7.39). Similar steps apply to other frequency-selective filters. Furthermore, this idea can also be extended for approximating arbitrary frequency-domain specifications.

From Figure 7.25 we observe the following:

    1. The approximation error—that is, the difference between the ideal and the actual response—is zero at the sampled frequencies.

    2. The approximation error at all other frequencies depends on the shape of the ideal response; that is, the sharper the ideal response, the larger the approximation error.

    3. The error is larger near the band edges and smaller within the band.

There are two design approaches. In the first approach we use the basic idea literally and provide no constraints on the approximation error; that is, we accept whatever error we get from the design. This approach is called a *naive design* method. In the second approach we try to minimize error in the stopband by varying values of the transition band samples. It results in a much better design called an *optimum design* method.

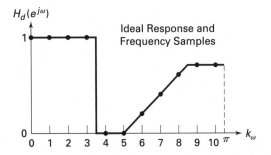 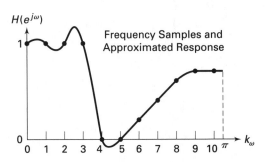

FIGURE 7.25   *Pictorial description of frequency sampling technique*

NAIVE DESIGN
METHOD
In this method we set $H(k) = H_d(e^{j2\pi k/M})$, $k = 0, \ldots, M-1$ and use (7.35) through (7.39) to obtain the impulse response $h(n)$.

☐ **EXAMPLE 7.14** Consider the lowpass filter specifications from Example 7.8.

$$\omega_p = 0.2\pi, \quad R_p = 0.25 \text{ dB}$$

$$\omega_s = 0.3\pi, \quad A_s = 50 \text{ dB}$$

Design an FIR filter using the frequency sampling approach.

Solution   Let us choose $M = 20$ so that we have a frequency sample at ω_p, that is, at $k = 2$:

$$\omega_p = 0.2\pi = \frac{2\pi}{20}2$$

and the next sample at ω_s, that is, at $k = 3$:

$$\omega_s = 0.3\pi = \frac{2\pi}{20}3$$

Thus we have 3 samples in the passband $[0 \leq \omega \leq \omega_p]$ and 7 samples in the stopband $[\omega_s \leq \omega \leq \pi]$. From (7.36) we have

$$H_r(k) = [1, 1, 1, \underbrace{0, \ldots, 0}_{15 \text{ zeros}}, 1, 1]$$

Since $M = 20$, $\alpha = \frac{20-1}{2} = 9.5$ and since this is a Type-2 linear-phase filter, from (7.37) we have

$$\angle H(k) = \begin{cases} -9.5\dfrac{2\pi}{20}k = -0.95\pi k, & 0 \leq k \leq 9 \\ +0.95\pi(20-k), & 10 \leq k \leq 19 \end{cases}$$

Now from (7.35) we assemble $H(k)$ and from (7.39) determine the impulse response $h(n)$. The MATLAB script follows:

```
>> M = 20; alpha = (M-1)/2; l = 0:M-1; wl = (2*pi/M)*l;
>> Hrs = [1,1,1,zeros(1,15),1,1]; %Ideal Amp Res sampled
>> Hdr = [1,1,0,0]; wdl = [0,0.25,0.25,1]; %Ideal Amp Res for plotting
>> k1 = 0:floor((M-1)/2); k2 = floor((M-1)/2)+1:M-1;
>> angH = [-alpha*(2*pi)/M*k1, alpha*(2*pi)/M*(M-k2)];
>> H = Hrs.*exp(j*angH);
>> h = real(ifft(H,M));
>> [db,mag,pha,grd,w] = freqz_m(h,1);
>> [Hr,ww,a,L] = Hr_Type2(h);
>> subplot(1,1,1)
>> subplot(2,2,1);plot(wl(1:11)/pi,Hrs(1:11),'o',wdl,Hdr);
>> axis([0,1,-0.1,1.1]); title('Frequency Samples: M=20')
>> xlabel('frequency in pi units'); ylabel('Hr(k)')
```

```
>> subplot(2,2,2); stem(l,h); axis([-1,M,-0.1,0.3])
>> title('Impulse Response'); xlabel('n'); ylabel('h(n)');
>> subplot(2,2,3); plot(ww/pi,Hr,wl(1:11)/pi,Hrs(1:11),'o');
>> axis([0,1,-0.2,1.2]); title('Amplitude Response')
>> xlabel('frequency in pi units'); ylabel('Hr(w)')
>> subplot(2,2,4);plot(w/pi,db); axis([0,1,-60,10]); grid
>> title('Magnitude Response'); xlabel('frequency in pi units'); ylabel('Decibels');
```

The time- and the frequency-domain plots are shown in Figure 7.26. Observe that the minimum stopband attenuation is about 16 dB, which is clearly unacceptable. If we increase M, then there will be samples in the transition band, for which we do not precisely know the frequency response. Therefore the naive design method is seldom used in practice. □

OPTIMUM
DESIGN
METHOD

To obtain more attenuation, we will have to increase M and make the transition band samples free samples—that is, we vary their values to obtain the largest attenuation for the given M and the transition width. This problem is known as an optimization problem, and it is solved using linear programming techniques. We demonstrate the effect of transition band sample variation on the design using the following example.

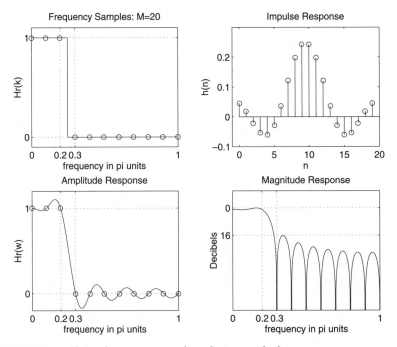

FIGURE 7.26 *Naive frequency sampling design method*

□ **EXAMPLE 7.15**   Using the optimum design method, design a better lowpass filter of Example 7.14.

Solution   Let us choose $M = 40$ so that we have one sample in the transition band $0.2\pi < \omega < 0.3\pi$. Since $\omega_1 \triangleq 2\pi/40$, the transition band samples are at $k = 5$ and at $k = 40 - 5 = 35$. Let us denote the value of these samples by T_1, $0 < T_1 < 1$; then the sampled amplitude response is

$$H_r(k) = [1, 1, 1, 1, 1, T_1, \underbrace{0, \ldots, 0}_{29 \text{ zeros}}, T_1, 1, 1, 1, 1]$$

Since $\alpha = \frac{40-1}{2} = 19.5$, the samples of the phase response are

$$\angle H(k) = \begin{cases} -19.5\dfrac{2\pi}{40}k = -0.975\pi k, & 0 \le k \le 19 \\ +0.975\pi(40 - k), & 20 \le k \le 39 \end{cases}$$

Now we can vary T_1 to get the best minimum stopband attenuation. This will result in the widening of the transition width. We first see what happens when $T_1 = 0.5$.

```
% T1 = 0.5
>> M = 40; alpha = (M-1)/2;
>> Hrs = [ones(1,5),0.5,zeros(1,29),0.5,ones(1,4)];
>> k1 = 0:floor((M-1)/2); k2 = floor((M-1)/2)+1:M-1;
>> angH = [-alpha*(2*pi)/M*k1, alpha*(2*pi)/M*(M-k2)];
>> H = Hrs.*exp(j*angH);
>> h = real(ifft(H,M));
```

From the plots of this design in Figure 7.27 we observe that the minimum stopband attenuation is now 30 dB, which is better than the naive design attenuation but is still not at the acceptable level of 50 dB. The best value for T_1 was obtained by varying it manually (although more efficient linear programming techniques are available, these were not used in this case), and the near optimum solution was found at $T_1 = 0.39$.

```
% T1 = 0.39
>> M = 40; alpha = (M-1)/2;
>> Hrs = [ones(1,5),0.39,zeros(1,29),0.39,ones(1,4)];
>> k1 = 0:floor((M-1)/2); k2 = floor((M-1)/2)+1:M-1;
>> angH = [-alpha*(2*pi)/M*k1, alpha*(2*pi)/M*(M-k2)];
>> H = Hrs.*exp(j*angH);
>> h = real(ifft(H,M));
```

From the plots in Figure 7.28 we observe that the optimum stopband attenuation is 43 dB. It is obvious that to further increase the attenuation, we will have to vary more than one sample in the transition band.   □

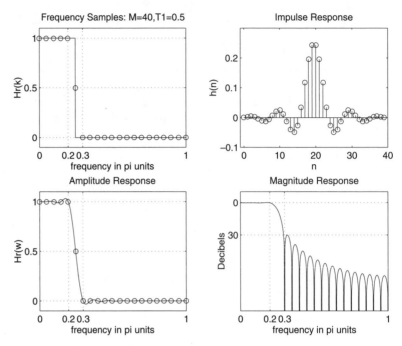

FIGURE 7.27 *Optimum frequency design method:* $T_1 = 0.5$

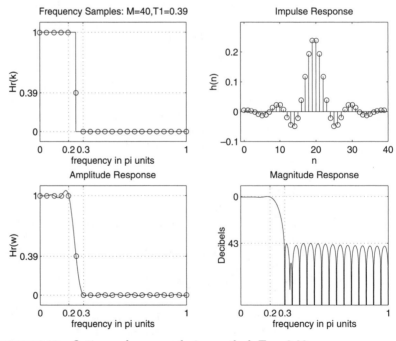

FIGURE 7.28 *Optimum frequency design method:* $T_1 = 0.39$

Clearly, this method is superior in that by varying one sample we can get a much better design. In practice the transition bandwidth is generally small, containing either one or two samples. Hence we need to optimize at most two samples to obtain the largest minimum stopband attenuation. This is also equivalent to minimizing the maximum side lobe magnitudes in the absolute sense. Hence this optimization problem is also called a *minimax* problem. This problem is solved by Rabiner et al. [20], and the solution is available in the form of tables of transition values. A selected number of tables are also available [19, Appendix B]. This problem can also be solved in MATLAB, but it would require the use of the Optimization toolbox. We will consider a more general version of this problem in the next section. We now illustrate the use of these tables in the following examples.

☐ **EXAMPLE 7.16**   Let us revisit our lowpass filter design in Example 7.14. We will solve it using two samples in the transition band so that we can get a better stopband attenuation.

Solution   Let us choose $M = 60$ so that there are two samples in the transition band. Let the values of these transition band samples be T_1 and T_2. Then $H_r(\omega)$ is given by

$$H(\omega) = [\underbrace{1, \ldots, 1}_{7 \text{ ones}}, T_1, T_2, \underbrace{0, \ldots, 0}_{43 \text{ zeros}}, T_2, T_1, \underbrace{1, \ldots, 1}_{6 \text{ ones}}]$$

From tables [19, Appendix B] $T_1 = 0.5925$ and $T_2 = 0.1099$. Using these values, we use MATLAB to compute $h(n)$.

```
>> M = 60; alpha = (M-1)/2; l = 0:M-1; wl = (2*pi/M)*l;
>> Hrs = [ones(1,7),0.5925,0.1099,zeros(1,43),0.1099,0.5925,ones(1,6)];
>> Hdr = [1,1,0,0]; wdl = [0,0.2,0.3,1];
>> k1 = 0:floor((M-1)/2); k2 = floor((M-1)/2)+1:M-1;
>> angH = [-alpha*(2*pi)/M*k1, alpha*(2*pi)/M*(M-k2)];
>> H = Hrs.*exp(j*angH);
>> h = real(ifft(H,M));
>> [db,mag,pha,grd,w] = freqz_m(h,1);
>> [Hr,ww,a,L] = Hr_Type2(h);
```

The time- and the frequency-domain plots are shown in Figure 7.29. The minimum stopband attenuation is now at 63 dB, which is acceptable.   ☐

☐ **EXAMPLE 7.17**   Design the bandpass filter of Example 7.10 using the frequency sampling technique. The design specifications are these:

$$\text{lower stopband edge:} \quad \omega_{1s} = 0.2\pi, \quad A_s = 60 \text{ dB}$$

$$\text{lower passband edge:} \quad \omega_{1p} = 0.35\pi, \quad R_p = 1 \text{ dB}$$

$$\text{upper passband edge:} \quad \omega_{2p} = 0.65\pi \quad R_p = 1 \text{ dB}$$

$$\text{upper stopband edge:} \quad \omega_{2s} = 0.8\pi \quad A_s = 60 \text{ dB}$$

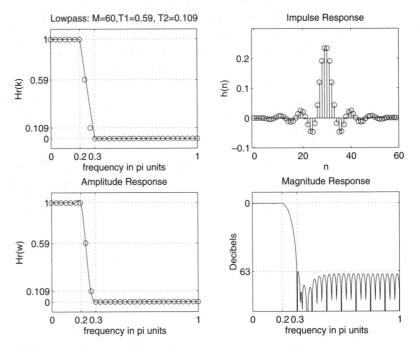

FIGURE 7.29 *Lowpass filter design plots in Example 7.16*

Solution

Let us choose $M = 40$ so that we have two samples in the transition band. Let the frequency samples in the lower transition band be T_1 and T_2. Then the samples of the amplitude response are

$$H_r(\omega) = [\underbrace{0, \ldots, 0}_{5}, T_1, T_2, \underbrace{1, \ldots, 1}_{7}, T_2, T_1, \underbrace{0, \ldots, 0}_{9}, T_1, T_2, \underbrace{1, \ldots, 1}_{7}, T_2, T_1, \underbrace{0, \ldots, 0}_{4}]$$

The optimum values of T_1 and T_2 for $M = 40$ and seven samples in the passband [19, Appendix B] are

$$T_1 = 0.109021, \ T_2 = 0.59417456$$

The MATLAB script is

```
>> M = 40; alpha = (M-1)/2; l = 0:M-1; wl = (2*pi/M)*l;
>> T1 = 0.109021; T2 = 0.59417456;
>> Hrs = [zeros(1,5),T1,T2,ones(1,7),T2,T1,zeros(1,9),T1,T2,ones(1,7),T2,T1,zeros(1,4)];
>> Hdr = [0,0,1,1,0,0]; wdl = [0,0.2,0.35,0.65,0.8,1];
>> k1 = 0:floor((M-1)/2); k2 = floor((M-1)/2)+1:M-1;
>> angH = [-alpha*(2*pi)/M*k1, alpha*(2*pi)/M*(M-k2)];
>> H = Hrs.*exp(j*angH);
```

```
>> h = real(ifft(H,M));
>> [db,mag,pha,grd,w] = freqz_m(h,1);
>> [Hr,ww,a,L] = Hr_Type2(h);
```

The plots in Figure 7.30 show an acceptable bandpass filter design.   □

□ **EXAMPLE 7.18**   Design the following highpass filter:

$$\text{Stopband edge:} \quad \omega_s = 0.6\pi \quad A_s = 50 \text{ dB}$$

$$\text{Passband edge:} \quad \omega_p = 0.8\pi \quad R_p = 1 \text{ dB}$$

Solution   Recall that for a highpass filter M must be odd (or Type-1 filter). Hence we will choose $M = 33$ to get two samples in the transition band. With this choice of M it is not possible to have frequency samples at ω_s and ω_p. The samples of the amplitude response are

$$H_r(k) = [\underbrace{0,\ldots,0}_{11}, T_1, T_2, \underbrace{1,\ldots,1}_{8}, T_2, T_1, \underbrace{0,\ldots,0}_{10}]$$

while the phase response samples are

$$\angle H(k) = \begin{cases} -\dfrac{33-1}{2}\dfrac{2\pi}{33}k = -\dfrac{32}{33}\pi k, & 0 \le k \le 16 \\ +\dfrac{32}{33}\pi(33-k), & 17 \le k \le 32 \end{cases}$$

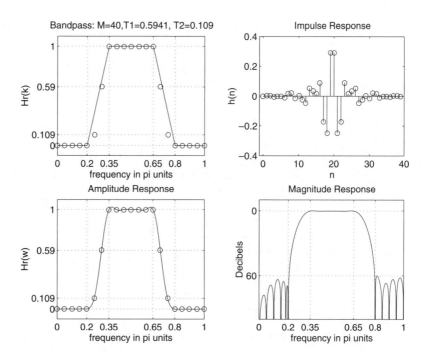

FIGURE 7.30 *Bandpass filter design plots in Example 7.17*

The optimum values of transition samples are $T_1 = 0.1095$ and $T_2 = 0.598$. Using these values, the MATLAB design is

```
>> M = 33; alpha = (M-1)/2; l = 0:M-1; wl = (2*pi/M)*l;
>> T1 = 0.1095; T2 = 0.598;
>> Hrs = [zeros(1,11),T1,T2,ones(1,8),T2,T1,zeros(1,10)];
>> Hdr = [0,0,1,1]; wdl = [0,0.6,0.8,1];
>> k1 = 0:floor((M-1)/2); k2 = floor((M-1)/2)+1:M-1;
>> angH = [-alpha*(2*pi)/M*k1, alpha*(2*pi)/M*(M-k2)];
>> H = Hrs.*exp(j*angH);
>> h = real(ifft(H,M));
>> [db,mag,pha,grd,w] = freqz_m(h,1);
>> [Hr,ww,a,L] = Hr_Type1(h);
```

The time- and the frequency-domain plots of the design are shown in Figure 7.31. □

☐   **EXAMPLE 7.19**   Design a 33-point digital differentiator based on the ideal differentiator of (7.31) given in Example 7.12.

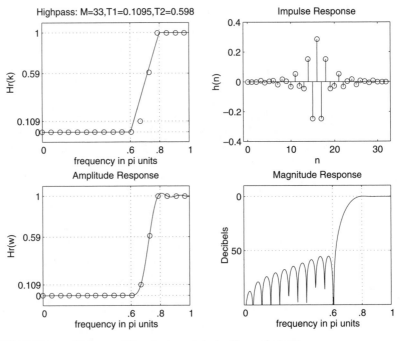

FIGURE 7.31   *Highpass filter design plots in Example 7.18*

From (7.31) the samples of the (imaginary-valued) amplitude response are given by

$$jH_r(k) = \begin{cases} +j\dfrac{2\pi}{M}k, & k = 0,\ldots,\left\lfloor\dfrac{M-1}{2}\right\rfloor \\[3mm] -j\dfrac{2\pi}{M}(M-k), & k = \left\lfloor\dfrac{M-1}{2}\right\rfloor + 1,\ldots,M-1 \end{cases}$$

and for linear phase the phase samples are

$$\angle H(k) = \begin{cases} -\dfrac{M-1}{2}\dfrac{2\pi}{M}k = -\dfrac{M-1}{M}\pi k, & k = 0,\ldots,\left\lfloor\dfrac{M-1}{2}\right\rfloor \\[3mm] +\dfrac{M-1}{M}\pi(M-k), & k = \left\lfloor\dfrac{M-1}{2}\right\rfloor + 1,\ldots,M-1 \end{cases}$$

Therefore

$$H(k) = jH_r(k)\,e^{j\angle H(k)}, \quad 0 \le k \le M-1 \quad \text{and} \quad h(n) = \text{IDFT}\left[H(k)\right]$$

```
>> M = 33; alpha = (M-1)/2; Dw = 2*pi/M;
>> l = 0:M-1; wl = Dw*l;
>> k1 = 0:floor((M-1)/2); k2 = floor((M-1)/2)+1:M-1;
>> Hrs = [j*Dw*k1,-j*Dw*(M-k2)];
>> angH = [-alpha*Dw*k1, alpha*Dw*(M-k2)];
>> H = Hrs.*exp(j*angH);
>> h = real(ifft(H,M));
>> [Hr,ww,a,P]=Hr_Type3(h);
```

The time- and the frequency-domain plots are shown in Figure 7.32. We observe that the differentiator is not a full-band differentiator. □

☐ **EXAMPLE 7.20** Design a 51-point digital Hilbert transformer based on the ideal Hilbert transformer of (7.32).

From (7.32) the samples of the (imaginary-valued) amplitude response are given by

$$jH_r(k) = \begin{cases} -j, & k = 1,\ldots,\left\lfloor\dfrac{M-1}{2}\right\rfloor \\[3mm] 0, & k = 0 \\[3mm] +j, & k = \left\lfloor\dfrac{M-1}{2}\right\rfloor + 1,\ldots,M-1 \end{cases}$$

Since this is a Type-3 linear-phase filter, the amplitude response will be zero at $\omega = \pi$. Hence to reduce the ripples, we should choose the two samples (in transition bands) near $\omega = \pi$ optimally between 0 and j. Using our previous experience, we could select this value as $0.39j$. The samples of the phase response are selected similar to those in Example 7.19.

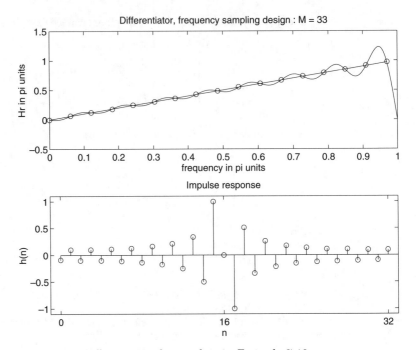

FIGURE 7.32 *Differentiator design plots in Example 7.19*

```
>> M = 51; alpha = (M-1)/2; Dw = 2*pi/M;
>> l = 0:M-1; wl = Dw*l;
>> k1 = 0:floor((M-1)/2); k2 = floor((M-1)/2)+1:M-1;
>> Hrs = [0,-j*ones(1,(M-3)/2),-0.39j,0.39j,j*ones(1,(M-3)/2)];
>> angH = [-alpha*Dw*k1, alpha*Dw*(M-k2)];
>> H = Hrs.*exp(j*angH);
>> h = real(ifft(H,M));
>> [Hr,ww,a,P]=Hr_Type3(h);
```

The plots in Figure 7.33 show the effect of the transition band samples.   □

The type of frequency sampling filter that we considered is called a Type-A filter, in which the sampled frequencies are

$$\omega_k = \frac{2\pi}{M}k, \quad 0 \leq k \leq M - 1$$

There is a second set of uniformly spaced samples given by

$$\omega_k = \frac{2\pi \left(k + \frac{1}{2}\right)}{M}, \quad 0 \leq k \leq M - 1$$

This is called a Type-B filter, for which a frequency sampling structure is also available. The expressions for the magnitude response $H(e^{j\omega})$ and the

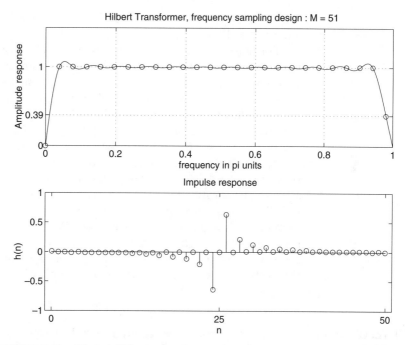

FIGURE 7.33 *Digital Hilbert transformer design plots in Example 7.20*

impulse response $h(n)$ are somewhat more complicated and are available in [19]. Their design can also be done in MATLAB using the approach discussed in this section.

OPTIMAL EQUIRIPPLE DESIGN TECHNIQUE

The last two techniques—namely, the window design and the frequency sampling design—were easy to understand and implement. However, they have some disadvantages. First, we cannot specify the band frequencies ω_p and ω_s precisely in the design; that is, we have to accept whatever values we obtain after the design. Second, we cannot specify both δ_1 and δ_2 ripple factors simultaneously. Either we have $\delta_1 = \delta_2$ in the window design method, or we can optimize only δ_2 in the frequency sampling method. Finally, the approximation error—that is, the difference between the ideal response and the actual response—is not uniformly distributed over the band intervals. It is higher near the band edges and smaller in the regions away from band edges. By distributing the error uniformly, we can obtain a lower-order filter satisfying the same specifications. Fortunately, a technique exists that can eliminate the above three problems.

This technique is somewhat difficult to understand and requires a computer for its implementation.

For linear-phase FIR filters it is possible to derive a set of conditions for which it can be proved that the design solution is optimal in the sense of *minimizing the maximum approximation error* (sometimes called the *minimax* or the *Chebyshev* error). Filters that have this property are called *equiripple* filters because the approximation error is uniformly distributed in both the passband and the stopband. This results in lower-order filters.

In the following we first formulate a minimax optimal FIR design problem and discuss the total number of maxima and minima (collectively called *extrema*) that one can obtain in the amplitude response of a linear-phase FIR filter. Using this, we then discuss a general equiripple FIR filter design algorithm, which uses polynomial interpolation for its solution. This algorithm is known as the Parks-McClellan algorithm, and it incorporates the Remez exchange routine for polynomial solution. This algorithm is available as a subroutine on many computing platforms. In this section we will use MATLAB to design equiripple FIR filters.

DEVELOPMENT OF THE MINIMAX PROBLEM

Earlier in this chapter we showed that the frequency response of the four cases of linear-phase FIR filters can be written in the form

$$H(e^{j\omega}) = e^{j\beta}e^{-j\frac{M-1}{2}\omega}H_r(w)$$

where the values for β and the expressions for $H_r(\omega)$ are given in Table 7.2.

TABLE 7.2 *Amplitude response and β-values for linear-phase FIR filters*

Linear-phase FIR Filter Type	β	$H_r(e^{j\omega})$
Type-1: M odd, symmetric $h(n)$	0	$\displaystyle\sum_{0}^{(M-1)/2} a(n)\cos\omega n$
Type-2: M even, symmetric $h(n)$	0	$\displaystyle\sum_{1}^{M/2} b(n)\cos\left[\omega(n-1/2)\right]$
Type-3: M odd, antisymmetric $h(n)$	$\dfrac{\pi}{2}$	$\displaystyle\sum_{1}^{(M-1)/2} c(n)\sin\omega n$
Type-4: M even, antisymmetric $h(n)$	$\dfrac{\pi}{2}$	$\displaystyle\sum_{1}^{M/2} d(n)\sin\left[\omega(n-1/2)\right]$

TABLE 7.3 $Q(\omega)$, L, and $P(\omega)$ for linear-phase FIR filters

LP FIR Filter Type	$Q(\omega)$	L	$P(\omega)$
Type-1	1	$\dfrac{M-1}{2}$	$\sum\limits_{0}^{L} a(n)\cos\omega n$
Type-2	$\cos\dfrac{\omega}{2}$	$\dfrac{M}{2}-1$	$\sum\limits_{0}^{L} \tilde{b}(n)\cos\omega n$
Type-3	$\sin\omega$	$\dfrac{M-3}{2}$	$\sum\limits_{0}^{L} \tilde{c}(n)\cos\omega n$
Type-4	$\sin\dfrac{\omega}{2}$	$\dfrac{M}{2}-1$	$\sum\limits_{0}^{L} \tilde{d}(n)\cos\omega n$

Using simple trigonometric identities, each expression for $H_r(\omega)$ above can be written as a product of a fixed function of ω (call this $Q(\omega)$) and a function that is a sum of cosines (call this $P(\omega)$). For details see [19] and Problems 7.1–7.4. Thus

$$H_r(\omega) = Q(\omega)P(\omega) \tag{7.40}$$

where $P(\omega)$ is of the form

$$P(\omega) = \sum_{n=0}^{L} \alpha(n)\cos\omega n \tag{7.41}$$

and $Q(\omega)$, L, $P(\omega)$ for the four cases are given in Table 7.3.

The purpose of this analysis is to have a *common form* for $H_r(\omega)$ across all four cases. It makes the problem formulation much easier. To formulate our problem as a Chebyshev approximation problem, we have to define the desired amplitude response $H_{dr}(\omega)$ and a weighting function $W(\omega)$, both defined over passbands and stopbands. The weighting function is necessary so that we can have an independent control over δ_1 and δ_2. The weighted error is defined as

$$E(\omega) \triangleq W(\omega)\left[H_{dr}(\omega) - H_r(\omega)\right], \quad \omega \in \mathcal{S} \triangleq [0, \omega_p] \cup [\omega_s, \pi] \tag{7.42}$$

These concepts are made clear in the following set of figures. It shows a typical equiripple filter response along with its ideal response.

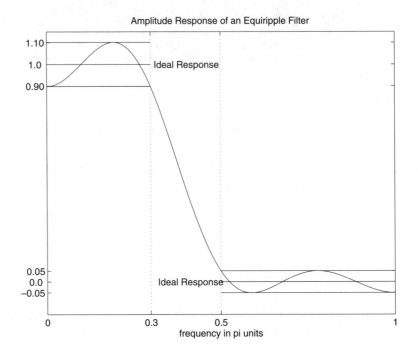

Amplitude Response of an Equiripple Filter

The error $[H_{dr}(\omega) - H_r(\omega)]$ response is shown below.

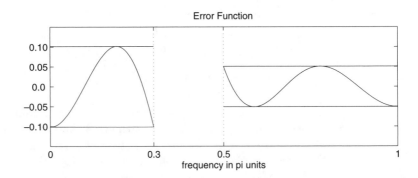

Error Function

Now if we choose

$$W(\omega) = \begin{cases} \dfrac{\delta_2}{\delta_1}, & \text{in the passband} \\ 1, & \text{in the stopband} \end{cases} \qquad \textbf{(7.43)}$$

Then the weighted error $E(\omega)$ response is

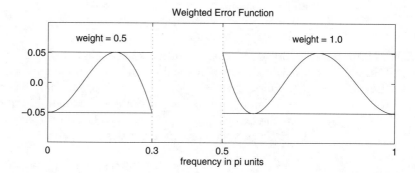

Thus the maximum error in both the passband and stopband is δ_2. Therefore, if we succeed in minimizing the maximum weighted error to δ_2, we automatically also satisfy the specification in the passband to δ_1. Substituting $H_r(\omega)$ from (7.40) into (7.42), we obtain

$$E(\omega) = W(\omega)\left[H_{dr}(\omega) - Q(\omega)P(\omega)\right]$$
$$= W(\omega)Q(\omega)\left[\frac{H_{dr}(\omega)}{Q(\omega)} - P(\omega)\right], \quad \omega \in \mathcal{S}$$

If we define

$$\hat{W}(\omega) \triangleq W(\omega)Q(w) \qquad \text{and} \qquad \hat{H}_{dr}(\omega) \triangleq \frac{H_{dr}(\omega)}{Q(\omega)}$$

then we obtain

$$E(\omega) = \hat{W}(\omega)\left[\hat{H}_{dr}(\omega) - P(\omega)\right], \quad \omega \in \mathcal{S} \tag{7.44}$$

Thus we have a common form of $E(\omega)$ for all four cases.

Problem Statement The Chebyshev approximation problem can now be defined as:

Determine the set of coefficients $a(n)$ or $\tilde{b}(n)$ or $\tilde{c}(n)$ or $\tilde{d}(n)$ [or equivalently $a(n)$ or $b(n)$ or $c(n)$ or $d(n)$] to minimize the maximum absolute value of $E(\omega)$ over the passband and stopband, i.e.,

$$\min_{\substack{\text{over coeff.}}}\left[\max_{\omega \in \mathcal{S}}|E(\omega)|\right] \tag{7.45}$$

Now we have succeeded in specifying the exact ω_p, ω_s, δ_1, and δ_2. In addition the error can now be distributed uniformly in both the passband and stopband.

<table>
<tr><td>

CONSTRAINT ON THE NUMBER OF EXTREMA

</td><td>

Before we give the solution to the above problem, we will first discuss the issue: how many local maxima and minima exist in the error function $E(\omega)$ for a given M-point filter? This information is used by the Parks-McClellan algorithm to obtain the polynomial interpolation. The answer is in the expression $P(\omega)$. From (7.41) $P(\omega)$ is a trigonometric function in ω. Using trigonometric identities of the form

</td></tr>
</table>

$$\cos(2\omega) = 2\cos^2(\omega) - 1$$

$$\cos(3\omega) = 4\cos^3(\omega) - 3\cos(\omega)$$

$$\vdots \quad = \quad \vdots$$

$P(\omega)$ can be converted to a trigonometric polynomial in $\cos(\omega)$, which we can write (7.41) as

$$P(\omega) = \sum_{n=0}^{L} \beta(n) \cos^n \omega \qquad \textbf{(7.46)}$$

□ **EXAMPLE 7.21**   Let $h(n) = \frac{1}{15}[1, 2, 3, 4, 3, 2, 1]$. Then $M = 7$ and $h(n)$ is symmetric, which means that we have a Type-1 linear-phase filter. Hence $L = (M-1)/2 = 3$. Now from (7.7)

$$\alpha(n) = a(n) = 2h(3-n), \quad 1 \le n \le 2; \quad \text{and} \quad \alpha(0) = a(0) = h(3)$$

or $\alpha(n) = \frac{1}{15}[4, 6, 4, 2]$. Hence

$$P(\omega) = \sum_{0}^{3} \alpha(n) \cos \omega n = \frac{1}{15}(4 + 6\cos\omega + 4\cos 2\omega + 2\cos 3\omega)$$

$$= \frac{1}{15}\left\{4 + 6\cos\omega + 4(2\cos^2\omega - 1) + 2(4\cos^3\omega - 3\cos\omega)\right\}$$

$$= 0 + 0 + \frac{8}{15}\cos^2\omega + \frac{8}{15}\cos^3\omega = \sum_{0}^{3} \beta(n)\cos^n\omega$$

or $\beta(n) = \left[0, 0, \dfrac{8}{15}, \dfrac{8}{15}\right]$.

     From (7.46) we note that $P(\omega)$ is an Lth-order polynomial in $\cos(\omega)$. Since $\cos(\omega)$ is a *monotone* function in the *open* interval $0 < \omega < \pi$, then it follows that the Lth-order polynomial $P(\omega)$ in $\cos(\omega)$ should behave like an ordinary Lth-order polynomial $P(x)$ in x. Therefore $P(\omega)$ has *at most* (i.e., no more than) $(L-1)$ local extrema in the open interval $0 < \omega < \pi$. For example,

$$\cos^2(\omega) = \frac{1 + \cos 2\omega}{2}$$

has only one minimum at $\omega = \pi/2$. However, it has three extrema in the closed interval $0 \leq \omega \leq \pi$ (i.e., a maximum at $\omega = 0$, a minimum at $\omega = \pi/2$, and a maximum at $\omega = \pi$). Now if we include the end points $\omega = 0$ and $\omega = \pi$, then $P(\omega)$ has at most $(L+1)$ local extrema in the closed interval $0 \leq \omega \leq \pi$. Finally, we would like the filter specifications to be met exactly at band edges ω_p and ω_s. Then the specifications can be met at no more than $(L+3)$ extremal frequencies in the $0 \leq \omega \leq \pi$ interval.

Conclusion   The error function $E(\omega)$ has at most $(L+3)$ extrema in S.

□   **EXAMPLE 7.22**   Let us plot the amplitude response of the filter given in Example 7.21 and count the total number of extrema in the corresponding error function.

Solution   The impulse response is

$$h(n) = \frac{1}{15}[1, 2, 3, 4, 3, 2, 1], \quad M = 7 \quad \text{or} \quad L = 3$$

and $\alpha(n) = \frac{1}{15}[4, 6, 4, 2]$ and $\beta(n) = \left[0, 0, \frac{8}{15}, \frac{8}{15}\right]$ from Example 7.21. Hence

$$P(\omega) = \frac{8}{15}\cos^2\omega + \frac{8}{15}\cos^3\omega$$

which is shown in Figure 7.34. Clearly, $P(\omega)$ has $(L-1) = 2$ extrema in the open interval $0 < \omega < \pi$. Also shown in Figure 7.34 is the error function, which has $(L+3) = 6$ extrema.   □

Let us now turn our attention to the problem statement and equation (7.45). It is a well-known problem in *approximation theory*, and the solution is given by the following important theorem.

■   **THEOREM 1**   *Alternation Theorem*
        Let S be any closed subset of the closed interval $[0, \pi]$. In order that $P(\omega)$ be the unique minimax approximation to $H_{dr}(\omega)$ on S, it is necessary and sufficient that the error function $E(\omega)$ exhibit at least $(L+2)$ "alternations" or extremal frequencies in S; that is, there must exist $(L+2)$ frequencies ω_i in S such that

$$E(\omega_i) = -E(\omega_{i-1}) = \pm\max_{S}|E(\omega)| \tag{7.47}$$

$$\overset{\triangle}{=} \pm\delta, \ \forall \ \omega_0 < \omega_1 < \cdots < \omega_{L+1} \in S$$

Combining this theorem with our earlier conclusion, we infer that the optimal equiripple filter has either $(L+2)$ or $(L+3)$ alternations

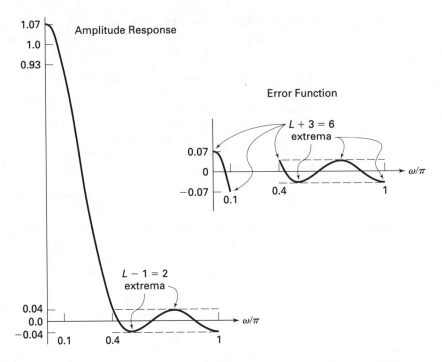

FIGURE 7.34 *Amplitude response and the error function in Example 7.22*

in its error function over $\mathcal{S}$. Most of the equiripple filters have $(L+2)$ alternations. However, for some combinations of ω_p and ω_s, we can get filters with $(L+3)$ alternations. These filters have one extra ripple in their response and hence are called *Extra-ripple* filters.

PARKS-McCLELLAN ALGORITHM

The alternation theorem ensures that the solution to our minimax approximation problem exists and is unique, but it does not tell us how to obtain this solution. We know neither the length M (or equivalently, L), nor the extremal frequencies ω_i, nor the parameters $\{\alpha(n)\}$, nor the maximum error δ. Parks and McClellan [17] provided an iterative solution using the Remez exchange algorithm. It assumes that the filter length M (or L) and the ratio δ_2/δ_1 are known. If we choose the weighting function as in (7.43), and if we choose the order M correctly, then $\delta = \delta_2$ when the solution is obtained. Clearly, δ and M are related; the larger the M, the smaller the δ. In the filter specifications δ_1, δ_2, ω_p, and ω_s are given. Therefore M has to be assumed. Fortunately, a simple formula, due to Kaiser, exists for approximating M. It is given by

$$\hat{M} = \frac{-20\log_{10}\sqrt{\delta_1\delta_2} - 13}{14.6\Delta f} + 1; \quad \Delta f = \frac{\omega_s - \omega_p}{2\pi} \quad \textbf{(7.48)}$$

The Parks-McClellan algorithm begins by guessing $(L+2)$ extremal frequencies $\{\omega_i\}$ and estimating the maximum error δ at these frequencies. It then fits an Lth-order polynomial (7.46) through points given in (7.47). Local maximum errors are determined over a finer grid, and the extremal frequencies $\{\omega_i\}$ are adjusted at these new extremal values. A new Lth-order polynomial is fit through these new frequencies, and the procedure is repeated. This iteration continues until the optimum set $\{\omega_i\}$ and the global maximum error δ are found. The iterative procedure is guaranteed to converge, yielding the polynomial $P(\omega)$. From (7.46) coefficients $\beta(n)$ are determined. Finally, the coefficients $a(n)$ as well as the impulse response $h(n)$ are computed. This algorithm is available in MATLAB as the `remez` function, which is described below.

Since we approximated M, the maximum error δ may not be equal to δ_2. If this is the case, then we have to increase M (if $\delta > \delta_2$) or decrease M (if $\delta < \delta_2$) and use the `remez` algorithm again to determine a new δ. We repeat this procedure until $\delta \leq \delta_2$. The optimal equiripple FIR filter, which satisfies all the three requirements discussed earlier is now determined.

MATLAB
IMPLEMEN-
TATION

The Parks-McClellan algorithm is available in MATLAB as a function called `remez`, the most general syntax of which is

```
[h] = remez(N,f,m,weights,ftype)
```

There are several versions of this syntax:[1]

• `[h] = remez(N,f,m)` designs an Nth-order (note that the length of the filter is $M = N + 1$) FIR digital filter whose frequency response is specified by the arrays `f` and `m`. The filter coefficients (or the impulse response) are returned in array `h` of length M. The array `f` contains band-edge frequencies in units of π, that is, $0.0 \leq \mathtt{f} \leq 1.0$. These frequencies must be in increasing order, starting with 0.0 and ending with 1.0. The array `m` contains the desired magnitude response at frequencies specified in `f`. The lengths of `f` and `m` arrays must be same and must be an even number. The weighting function used in each band is equal to unity, which means that the tolerances (δ_i's) in every band are the same.

• `[h] = remez(N,f,m,weights)` is similar to the above case except that the array `weights` specifies the weighting function in each band.

• `[h] = remez(N,f,m,ftype)` is similar to the first case except when `ftype` is the string 'differentiator' or 'hilbert', it designs digital dif-

---

[1]It should be noted that the `remez` function underwent a small change from the old Student Edition to the new Student Edition of MATLAB (or from the Signal Processing Toolbox version 2.0b to version 3.0). The description given here applies to the new version.

---

ferentiators or digital Hilbert transformers, respectively. For the digital Hilbert transformer the lowest frequency in the f array should not be 0, and the highest frequency should not be 1. For the digital differentiator, the m vector does not specify the desired slope in each band but the desired magnitude.

- [h] = remez(N,f,m,weights,ftype) is similar to the above case except that the array weights specifies the weighting function in each band.

As explained during the description of the Parks-McClellan algorithm, we have to first guess the order of the filter using (7.48) to use the routine remez. After we obtain the filter coefficients in array h, we have to check the minimum stopband attenuation and compare it with the given A_s and then increase (or decrease) the filter order. We have to repeat this procedure until we obtain the desired A_s. We illustrate this procedure in the following several MATLAB examples.

☐   **EXAMPLE 7.23**   Let us design the lowpass filter described in Example 7.8 using the Parks-McClellan algorithm. The design parameters are

$$\omega_p = 0.2\pi \;, \quad R_p = 0.25 \text{ dB}$$

$$\omega_s = 0.3\pi \;, \quad A_s = 50 \text{ dB}$$

We provide a MATLAB script to design this filter.

```
>> wp = 0.2*pi; ws = 0.3*pi; Rp = 0.25; As = 50;
>> wsi = ws/delta_w+1;
>> delta1 = (10^(Rp/20)-1)/(10^(Rp/20)+1);
>> delta2 = (1+delta1)*(10^(-As/20));
>> deltaH = max(delta1,delta2); deltaL = min(delta1,delta2);
>> weights = [delta2/delta1 1];
>> deltaf = (ws-wp)/(2*pi);
>> M = ceil((-20*log10(sqrt(delta1*delta2))-13)/(14.6*deltaf)+1)
M = 43
>> f = [0 wp/pi ws/pi 1];
>> m = [1 1 0 0];
>> h = remez(M-1,f,m,weights);
>> [db,mag,pha,grd,w] = freqz_m(h,[1]);
>> delta_w = 2*pi/1000;
>> wsi = ws/delta_w+1;
>> Asd = -max(db(wsi:1:501))
Asd = 47.8562
>> M = M+1;
>> h = remez(M-1,f,m,weights);
>> [db,mag,pha,grd,w] = freqz_m(h,[1]);
>> Asd = -max(db(wsi:1:501))
Asd = 48.2155
>> M = M+1;
>> h = remez(M-1,f,m,weights);
>> [db,mag,pha,grd,w] = freqz_m(h,[1]);
```

```
>> Asd = -max(db(wsi:1:501))
Asd = 48.8632
>> M = M+1;
>> h = remez(M-1,f,m,weights);
>> [db,mag,pha,grd,w] = freqz_m(h,[1]);
>> Asd = -max(db(wsi:1:501))
Asd = 49.8342
>> M = M+1;
>> h = remez(M-1,f,m,weights);
>> [db,mag,pha,grd,w] = freqz_m(h,[1]);
>> Asd = -max(db(wsi:1:501))
Asd = 51.0896
>> M
M = 47
```

Note that we stopped the above iterative procedure when the computed stopband attenuation exceeded the given stopband attenuation A_s, and the optimal value of M was found to be 47. This value is considerably lower than the window design techniques ($M = 61$ for a Kaiser window) or the frequency sampling technique ($M = 60$). In Figure 7.35 we show the time- and the frequency-domain plots of the designed filter along with the error function in both the passband and the stopband to illustrate the equiripple behavior.

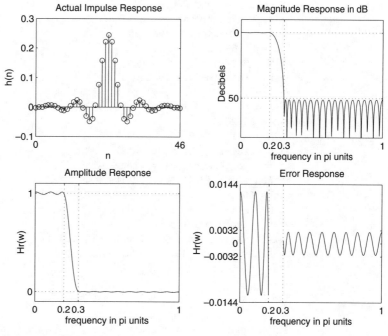

FIGURE 7.35 *Plots for equiripple lowpass FIR filter in Example 7.23*

☐ **EXAMPLE 7.24** Let us design the bandpass filter described in Example 7.10 using the Parks-McClellan algorithm. The design parameters are:

$$\begin{matrix} \omega_{1s} = 0.2\pi \\ \omega_{1p} = 0.35\pi \end{matrix} \quad ; \quad R_p = 1 \text{ dB}$$

$$\begin{matrix} \omega_{2p} = 0.65\pi \\ \omega_{2s} = 0.8\pi \end{matrix} \quad ; \quad A_s = 60 \text{ db}$$

Solution

The following MATLAB script shows how to design this filter.

```
>> ws1 = 0.2*pi; wp1 = 0.35*pi; wp2 = 0.65*pi; ws2 = 0.8*pi;
>> ws1i = floor(ws1/delta_w)+1;
>> Rp = 1.0; As = 60;
>> delta1 = (10^(Rp/20)-1)/(10^(Rp/20)+1);
>> delta2 = (1+delta1)*(10^(-As/20));
>> deltaH = max(delta1,delta2); deltaL = min(delta1,delta2);
>> weights = [1 delta2/delta1 1];
>> delta_f =min((ws2-wp2)/(2*pi), (wp1-ws1)/(2*pi));
>> M = ceil((-20*log10(sqrt(delta1*delta2))-13)/(14.6*delta_f)+1)
M = 28
>> f = [0 ws1/pi wp1/pi wp2/pi ws2/pi 1];
>> m = [0 0 1 1 0 0];
>> h = remez(M-1,f,m,weights);
>> [db,mag,pha,grd,w] = freqz_m(h,[1]);
>> delta_w=2*pi/1000;
>> ws1i = floor(ws1/delta_w)+1;
>> Asd = -max(db(1:1:ws1i))
Asd = 56.5923
>> M = M+1;
>> h = remez(M-1,f,m,weights);
>> [db,mag,pha,grd,w] = freqz_m(h,[1]);
>> Asd = -max(db(1:1:ws1/delta_w))
Asd = 61.2818
>> M = M+1;
>> h = remez(M-1,f,m,weights);
>> [db,mag,pha,grd,w] = freqz_m(h,[1]);
>> Asd = -max(db(1:1:ws1/delta_w))
Asd = 60.3820
>> M = M+1;
>> h = remez(M-1,f,m,weights);
>> [db,mag,pha,grd,w] = freqz_m(h,[1]);
>> Asd = -max(db(1:1:ws1/delta_w))
Asd = 61.3111
>> M
M = 31
```

The optimal value of M was found to be 31. The time- and the frequency-domain plots of the designed filter are shown in Figure 7.36. ☐

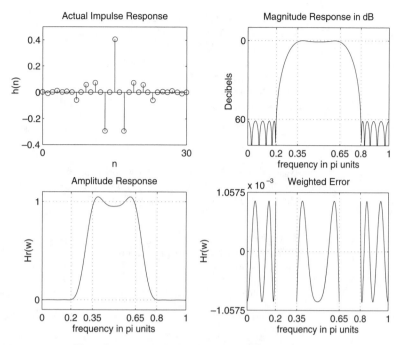

FIGURE 7.36 *Plots for equiripple bandpass FIR filter in Example 7.24*

☐ **EXAMPLE 7.25** Design a highpass filter that has the following specifications:

$$\omega_s = 0.6\pi, \quad A_s = 50 \text{ dB}$$

$$\omega_p = 0.75\pi, \quad R_p = 0.5 \text{ dB}$$

Solution Since this is a highpass filter, we must ensure that the length M is an odd number. This is shown in the following MATLAB script.

```
>> ws = 0.6*pi; wp = 0.75*pi; Rp = 0.5; As = 50;
>> delta1 = (10^(Rp/20)-1)/(10^(Rp/20)+1);
>> delta2 = (1+delta1)*(10^(-As/20));
>> deltaH = max(delta1,delta2); deltaL = min(delta1,delta2);
>> weights = [1 delta2/delta1];
>> deltaf = (wp-ws)/(2*pi);
>> M = ceil((-20*log10(sqrt(delta1*delta2))-13)/(14.6*deltaf)+1);
% M must be odd
>> M = 2*floor(M/2)+1
M =   27
>> f = [0 ws/pi wp/pi 1];
>> m = [0 0 1 1];
>> h = remez(M-1,f,m,weights);
>> [db,mag,pha,grd,w] = freqz_m(h,[1]);
>> delta_w = 2*pi/1000; wsi=ws/delta_w; wpi = wp/delta_w;
```

```
>> Asd = -max(db(1:1:wsi))
Asd =   49.5918
>> M = M+2;   % M must be odd
>> h = remez(M-1,f,m,weights);
>> [db,mag,pha,grd,w] = freqz_m(h,[1]);
>> Asd = -max(db(1:1:wsi))
Asd =   50.2253
>> M
M =   29
```

Note also that we increased the value of M to maintain its odd value. The optimum M was found to be 37. The time- and the frequency-domain plots of the designed filter are shown in Figure 7.37. □

□   **EXAMPLE 7.26**   In this example we will design a "staircase" filter, which has three bands with different ideal responses and different tolerances in each band. The design specifications are

Band-1: $0 \leq \omega \leq 0.3\pi$,   Ideal gain $= 1$,   Tolerance $\delta_1 = 0.01$

Band-2: $0.4\pi \leq \omega \leq 0.7\pi$,   Ideal gain $= 0.5$,   Tolerance $\delta_2 = 0.005$

Band-3: $0.8\pi \leq \omega \leq \pi$,   Ideal gain $= 0$,   Tolerance $\delta_3 = 0.001$

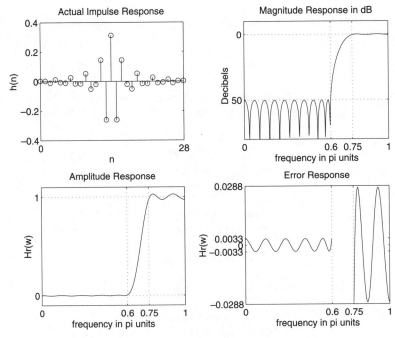

FIGURE 7.37   *Plots for equiripple highpass FIR filter in Example 7.25*

Solution The following MATLAB script describes the design procedure.

```
> w1=0; w2=0.3*pi; delta1=0.01;
> w3=0.4*pi; w4=0.7*pi; delta2=0.005;
> w5=0.8*pi; w6=pi; delta3=0.001;
> weights=[delta3/delta1 delta3/delta2 1];
> delta_f=min((w3-w2)/(2*pi), (w5-w3)/(2*pi));
> M=ceil(((-20*log10(sqrt(delta1*delta2))-13)/(14.6*delta_f)+1)
M = 43
> f=[0 w2/pi w3/pi w4/pi w5/pi 1];
> m=[1 1 0.5 0.5 0 0];
> h=remez(M-1,f,m,weights);
> [db,mag,pha,grd,w]=freqz_m(h,[1]);
> delta_w=2*pi/680;
> Asd=-max(db(w5/delta_w+10:1:341))
Asd = 56.2181
```

The optimum value of M was found at $M = 49$.

```
> M = 49;
> h=remez(M-1,f,m,weights);
> [db,mag,pha,grd,w]=freqz_m(h,[1]);
> Asd=-max(db(w5/delta_w+10:1:341))
Asd = 60.6073
```

The time- and the frequency-domain plots of the designed filter are shown in Figure 7.38. □

□ **EXAMPLE 7.27** In this example we will design a digital differentiator with different slopes in each band. The specifications are

$$\text{Band-1:} \quad 0 \leq \omega \leq 0.2\pi, \quad \text{Slope} = 1 \text{ sam/cycle}$$

$$\text{Band-2:} \quad 0.4\pi \leq \omega \leq 0.6\pi, \quad \text{Slope} = 2 \text{ sam/cycle}$$

$$\text{Band-3:} \quad 0.8\pi \leq \omega \leq \pi, \quad \text{Slope} = 3 \text{ sam/cycle}$$

Solution We need desired magnitude response values in each band. These can be obtained by multiplying band-edge frequencies in cycles/sam by the slope values in sam/cycle

$$\text{Band-1:} \quad 0 \leq f \leq 0.1, \quad \text{Slope} = 1 \text{ sam/cycle} \Rightarrow 0.0 \leq |H| \leq 0.1$$

$$\text{Band-2:} \quad 0.2 \leq f \leq 0.3, \quad \text{Slope} = 2 \text{ sam/cycle} \Rightarrow 0.4 \leq |H| \leq 0.6$$

$$\text{Band-3:} \quad 0.4 \leq f \leq 0.5, \quad \text{Slope} = 3 \text{ sam/cycle} \Rightarrow 1.2 \leq |H| \leq 1.5$$

Let the weights be equal in all bands. The MATLAB script is:

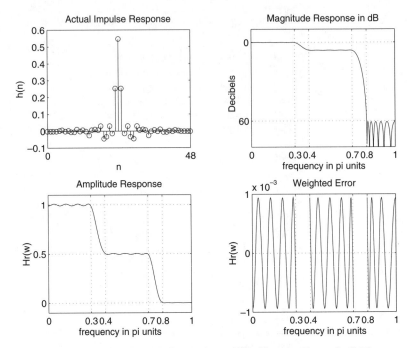

FIGURE 7.38 *Plots for equiripple staircase FIR filter in Example 7.26*

```
>> f = [0 0.2 0.4 0.6 0.8 1];        % in w/pi unis
>> m = [0,0.1,0.4,0.6,1.2,1.5];      % magnitude values
>> h = remez(25,f,m,'differentiator');
>> [db,mag,pha,grd,w] = freqz_m(h,[1]);
>> figure(1); subplot(1,1,1)
>> subplot(2,1,1); stem([0:25],h); title('Impulse Response');
>> xlabel('n'); ylabel('h(n)'); axis([0,25,-0.6,0.6])
>> set(gca,'XTickMode','manual','XTick',[0,25])
>> set(gca,'YTickMode','manual','YTick',[-0.6:0.2:0.6]);
>> subplot(2,1,2); plot(w/(2*pi),mag); title('Magnitude Response')
>> xlabel('Normalized frequency f'); ylabel('|H|')
>> set(gca,'XTickMode','manual','XTick',f/2)
>> set(gca,'YTickMode','manual','YTick',[0,0.1,0.4,0.6,1.2,1.5]); grid
```

The frequency-domain response is shown in Figure 7.39.   □

□   **EXAMPLE 7.28**   Finally, we design a Hilbert transformer over the band $0.05\pi \leq \omega \leq 0.95\pi$.

Solution   Since this is a wideband Hilbert transformer, we will choose an odd length for our filter (i.e., a Type-3 filter). Let us choose $M = 51$.

```
>> f = [0.05,0.95]; m = [1 1];
>> h = remez(50,f,m,'hilbert');
```

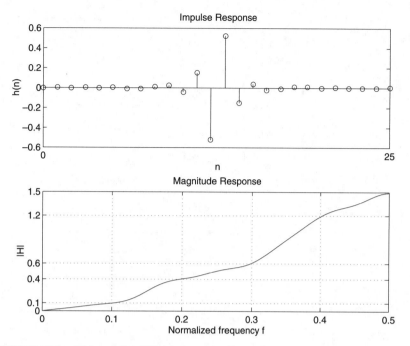

FIGURE 7.39 *Plots of the differentiator in Example 7.27*

```
>> [db,mag,pha,grd,w] = freqz_m(h,[1]);
>> figure(1); subplot(1,1,1)
>> subplot(2,1,1); stem([0:50],h); title('Impulse Response');
>> xlabel('n'); ylabel('h(n)'); axis([0,50,-0.8,0.8])
>> set(gca,'XTickMode','manual','XTick',[0,50])
>> set(gca,'YTickMode','manual','YTick',[-0.8:0.2:0.8]);
>> subplot(2,1,2); plot(w/pi,mag); title('Magnitude Response')
>> xlabel('frequency in pi units'); ylabel('|H|')
>> set(gca,'XTickMode','manual','XTick',[0,f,1])
>> set(gca,'YTickMode','manual','YTick',[0,1]);grid
```

The plots of this Hilbert transformer are shown in Figure 7.40. □

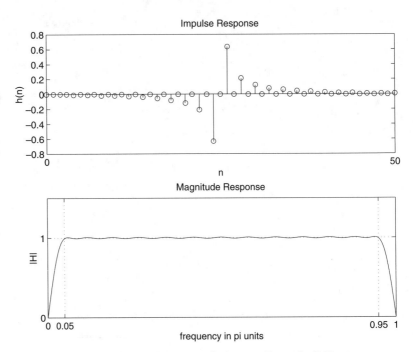

Impulse Response

Magnitude Response

FIGURE 7.40   *Plots of the Hilbert transformer in Example 7.28*

PROBLEMS

P7.1   The Type-1 linear-phase FIR filter is characterized by

$$h(n) = h(M - 1 - n), \quad 0 \le n \le M - 1, \ M \text{ odd}$$

Show that its amplitude response $H_r(\omega)$ is given by

$$H_r(\omega) = \sum_{n=0}^{L} a(n) \cos \omega n, \quad L = \frac{M-1}{2}$$

where coefficients $\{a(n)\}$ are obtained from $h(n)$.

P7.2   The Type-2 linear-phase FIR filter is characterized by

$$h(n) = h(M - 1 - n), \ 0 \le n \le M - 1, \ M \text{ even}$$

a. Show that its amplitude response $H_r(\omega)$ is given by

$$H_r(\omega) = \sum_{n=1}^{M/2} b(n) \cos \left\{ \omega \left(n - \frac{1}{2} \right) \right\}$$

where coefficients $\{b(n)\}$ are obtained from $h(n)$.

b. Show that the above $H_r(\omega)$ can be further expressed as

$$H_r(\omega) = \cos\frac{\omega}{2} \sum_{n=0}^{L} \tilde{b}(n) \cos\omega n, \quad L = \frac{M}{2} - 1$$

where $\tilde{b}(n)$ is derived from $b(n)$.

P7.3 The Type-3 linear-phase FIR filter is characterized by

$$h(n) = -h(M-1-n), \quad 0 \le n \le M-1, \ M \text{ odd}$$

a. Show that its amplitude response $H_r(\omega)$ is given by

$$H_r(\omega) = \sum_{n=1}^{(M-1)/2} c(n) \sin\omega n$$

where coefficients $\{c(n)\}$ are obtained from $h(n)$.

b. Show that the above $H_r(\omega)$ can be further expressed as

$$H_r(\omega) = \sin\omega \sum_{n=0}^{L} \tilde{c}(n) \cos\omega n, \ L = \frac{M-3}{2}$$

where $\tilde{c}(n)$ is derived from $c(n)$.

P7.4 The Type-4 linear-phase FIR filter is characterized by

$$h(n) = -h(M-1-n), \quad 0 \le n \le M-1, \ M \text{ even}$$

a. Show that its amplitude response $H_r(\omega)$ is given by

$$H_r(\omega) = \sum_{n=1}^{M/2} d(n) \sin\left\{\omega\left(n - \frac{1}{2}\right)\right\}$$

where coefficients $\{d(n)\}$ are obtained from $h(n)$.

b. Show that the above $H_r(\omega)$ can be further expressed as

$$H_r(\omega) = \sin\frac{\omega}{2} \sum_{n=0}^{L} \tilde{d}(n) \cos\omega n, \ L = \frac{M}{2} - 1$$

where $\tilde{d}(n)$ is derived from $d(n)$.

P7.5 Write a MATLAB function to compute the amplitude response $H_r(\omega)$, given a linear-phase impulse response $h(n)$. The format of this function should be

```
function [Hr,w,P,L] = Ampl_Res(h);
%
% function [Hr,w,P,L] = Ampl_Res(h)
% Computes Amplitude response Hr(w) and its polynomial P of order L,
% given a linear-phase FIR filter impulse response h.
% The type of filter is determined automatically by the subroutine.
```

```
%
% Hr = Amplitude Response
% w = frequencies between [0 pi] over which Hr is computed
% P = Polynomial coefficients
% L = Order of P
% h = Linear Phase filter impulse response
```

The subroutine should first determine the type of the linear-phase FIR filter and then use the appropriate **Hr_Type#** function discussed in the chapter. It should also check if the given $h(n)$ is of a linear-phase type. Check your subroutine on sequences given in Examples 7.4 through 7.7.

P7.6 If $H(z)$ has zeros at

$$z_1 = re^{j\theta}, \quad z_2 = \frac{1}{r}e^{-j\theta}, \quad z_3 = re^{-j\theta}, \quad z_4 = \frac{1}{r}e^{-j\theta}$$

show that $H(z)$ represents a linear-phase FIR filter.

P7.7 Design a bandstop filter using the Hanning window design technique. The specifications are

$$\begin{array}{ll} \text{lower stopband edge:} & 0.4\pi \\ & \qquad\qquad A_s = 40 \text{ dB} \\ \text{upper stopband edge:} & 0.6\pi \end{array}$$

$$\begin{array}{ll} \text{lower passband edge:} & 0.3\pi \\ & \qquad\qquad R_p = 0.5 \text{ dB} \\ \text{upper passband edge:} & 0.7\pi \end{array}$$

Plot the impulse response and the magnitude response (in dB) of the designed filter.

P7.8 Design a bandpass filter using the Hamming window design technique. The specifications are

$$\begin{array}{ll} \text{lower stopband edge:} & 0.3\pi \\ & \qquad\qquad A_s = 50 \text{ dB} \\ \text{upper stopband edge:} & 0.6\pi \end{array}$$

$$\begin{array}{ll} \text{lower passband edge:} & 0.4\pi \\ & \qquad\qquad R_p = 0.5 \text{ dB} \\ \text{upper passband edge:} & 0.5\pi \end{array}$$

Plot the impulse response and the magnitude response (in dB) of the designed filter.

P7.9 Design a highpass filter using the Kaiser window design technique. The specifications are

$$\text{stopband edge:} \quad 0.4\pi, \quad A_s = 60 \text{ dB}$$

$$\text{passband edge:} \quad 0.6\pi, \quad R_p = 0.5 \text{ dB}$$

Plot the impulse response and the magnitude response (in dB) of the designed filter.

P7.10 We wish to use the Kaiser window method to design a linear-phase FIR digital filter that meets the following specifications:

$$0 \le \left| H\left(e^{j\omega}\right) \right| \le 0.01, \qquad 0 \le \omega \le 0.25\pi$$

$$0.95 \le \left| H\left(e^{j\omega}\right) \right| \le 1.05, \quad 0.35\pi \le \omega \le 0.65\pi$$

$$0 \le \left| H\left(e^{j\omega}\right) \right| \le 0.01, \quad 0.75\pi \le \omega \le \pi$$

Determine the minimum-length impulse response $h(n)$ of such a filter. Provide a plot containing subplots of the amplitude response and the magnitude response in dB.

P7.11 Following the procedure used in this chapter, develop the following MATLAB functions to design FIR filters via the Kaiser window technique. These functions should check for the valid band-edge frequencies and restrict the filter length to 255.

a. Lowpass filter: The format should be

```
function [h,M] = kai_lpf(wp,ws,As);
% [h,M] = kai_lpf(wp,ws,As);
% Low-Pass FIR filter design using Kaiser window
%
%  h = Impulse response of length M of the designed filter
%  M = Length of h which is an odd number
% wp = Pass-band edge in radians (0 < wp < ws < pi)
% ws = Stop-band edge in radians (0 < wp < ws < pi)
% As = Stop-band attenuation in dB (As > 0)
```

b. Highpass filter: The format should be

```
function [h,M] = kai_hpf(ws,wp,As);
% [h,M] = kai_hpf(ws,wp,As);
% HighPass FIR filter design using Kaiser window
%
%  h = Impulse response of length M of the designed filter
%  M = Length of h which is an odd number
% ws = Stop-band edge in radians (0 < ws < wp < pi)
% wp = Pass-band edge in radians (0 < ws < wp < pi)
% As = Stop-band attenuation in dB (As > 0)
```

c. Bandpass filter: The format should be

```
function [h,M] = kai_bpf(ws1,wp1,wp2,ws2,As);
% [h,M] = kai_bpf(ws1,wp1,wp2,ws2,As);
% Band-Pass FIR filter design using Kaiser window
%
%  h = Impulse response of length M of the designed filter
%  M = Length of h which is an odd number
% ws1 = Lower stop-band edge in radians
% wp1 = Lower pass-band edge in radians
% wp2 = Upper pass-band edge in radians
% ws2 = Upper stop-band edge in radians
%      0 < ws1 < wp1 < wp2 < ws2< pi
% As = Stop-band attenuation in dB (As > 0)
```

d. Bandstop filter: The format should be

```
function [h,M] = kai_bsf(wp1,ws1,ws2,wp2,As);
% [h,M] = kai_bsf(wp1,ws1,ws2,wp2,As);
% Band-Pass FIR filter design using Kaiser window
%
```

```
%   h = Impulse response of length M of the designed filter
%   M = Length of h which is an odd number
%   wp1 = Lower pass-band edge in radians
%   ws1 = Lower stop-band edge in radians
%   ws2 = Upper stop-band edge in radians
%   wp2 = Upper pass-band edge in radians
%         0 < wp1 < ws1 < ws2 < wp2 < pi
%   As = Stop-band attenuation in dB (As > 0)
```

You can now develop similar functions for other windows discussed in this chapter.

P7.12 Design the staircase filter of Example 7.26 using the Blackman window approach. The specifications are

$$\text{Band-1:} \quad 0 \leq \omega \leq 0.3\pi, \quad \text{Ideal gain} = 1, \quad \delta_1 = 0.01$$

$$\text{Band-2:} \quad 0.4\pi \leq \omega \leq 0.7\pi, \quad \text{Ideal gain} = 0.5, \quad \delta_2 = 0.005$$

$$\text{Band-3:} \quad 0.8\pi \leq \omega \leq \pi, \quad \text{Ideal gain} = 0, \quad \delta_3 = 0.001$$

Compare the filter length of this design with that of Example 7.26. Provide a plot of the magnitude response in dB.

P7.13 Consider an ideal lowpass filter with the cutoff frequency $\omega_c = 0.3\pi$. We want to approximate this filter using a frequency sampling design in which we choose 40 samples.

a. Choose the sample at ω_c equal to 0.5 and use the naive design method to compute $h(n)$. Determine the minimum stopband attenuation.

b. Now vary the sample at ω_c and determine the optimum value to obtain the largest minimum stopband attenuation.

c. Plot the magnitude responses in dB of the above two designs in one plot and comment on the results.

P7.14 Design the bandstop filter of Problem 7.7 using the frequency sampling method. Choose the order of the filter appropriately so that there is one sample in the transition band. Use optimum value for this sample.

P7.15 Design the bandpass filter of Problem 7.8 using the frequency sampling method. Choose the order of the filter appropriately so that there are two samples in the transition band. Use optimum values for these samples.

P7.16 Design the highpass filter of Problem 7.9 using the frequency sampling method. Choose the order of the filter appropriately so that there are two samples in the transition band. Use optimum values.

P7.17 We want to design a narrow bandpass filter to pass the center frequency at $\omega_0 = 0.5\pi$. The bandwidth should be no more than 0.1π.

a. Use the frequency sampling technique and choose M so that there is one sample in the transition band. Use the optimum value for transition band samples and draw the frequency sampling structure.

b. Use the Kaiser window technique so that the stopband attenuation is the same as that of the above frequency sampling design. Determine the impulse response $h(n)$ and draw the linear-phase structure.

c. Compare the above two filter designs in terms of their implementation and their filtering effectiveness.

P7.18 The frequency response of an ideal bandpass filter is given by

$$H_d\left(e^{j\omega}\right) = \begin{cases} 0, & 0 \le |\omega| < \pi/3 \\ 1, & \pi/3 \le |\omega| \le 2\pi/3 \\ 0, & 2\pi/3 < |\omega| \le \pi \end{cases}$$

a. Determine the coefficients of a 25-tap filter based on the Parks-McClellan algorithm with stopband attenuation of 50 dB. The designed filter should have the smallest possible transition width.

b. Plot the amplitude response of the filter using the function developed in Problem 7.5.

P7.19 Consider the bandstop filter given in Problem 7.7.

a. Design a linear-phase bandstop FIR filter using the Parks-McClellan algorithm. Note that the length of the filter must be odd. Provide a plot of the impulse response and the magnitude response in dB of the designed filter.

b. Plot the amplitude response of the designed filter and count the total number of extrema in the stopband and passbands. Verify this number with the theoretical estimate of the total number of extrema.

c. Compare the order of this filter with those of the filters in Problems 7.7 and 7.14.

d. Verify the operation of the designed filter on the following signal:

$$x(n) = 5 - 5\cos\left(\frac{\pi n}{2}\right); \quad 0 \le n \le 300$$

P7.20 Using the Parks-McClellan algorithm, design a 25-tap FIR differentiator with slope equal to 1 sample/cycle.

a. Choose the frequency band of interest between 0.1π and 0.9π. Plot the impulse response and the amplitude response.

b. Generate 100 samples of the sinusoid

$$x(n) = 3\sin(0.25\pi n), \quad n = 0, \ldots, 100$$

and process through the above FIR differentiator. Compare the result with the theoretical "derivative" of $x(n)$. Note: Don't forget to take the 12-sample delay of the FIR filter into account.

P7.21 Design a lowest-order equiripple linear-phase FIR filter to satisfy the specifications given in Figure 7.41. Provide a plot of the amplitude response and a plot of the impulse response.

P7.22 A digital signal $x(n)$ contains a sinusoid of frequency $\pi/2$ and a Gaussian noise $w(n)$ of zero mean and unit variance; that is,

$$x(n) = 2\cos\frac{\pi n}{2} + w(n)$$

We want to filter out the noise component using a 50th-order causal and linear-phase FIR filter.

a. Using the Parks-McClellan algorithm, design a narrow bandpass filter with passband width of no more than 0.02π and stopband attenuation of at least 30 dB. Note that no other parameters are given, and you have to choose the remaining parameters for the `remez`

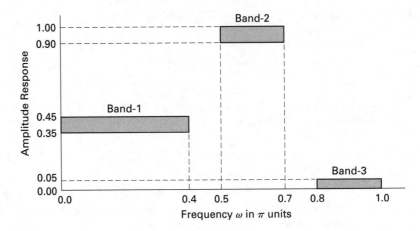

FIGURE 7.41 *Filter specification for Problem 7.21*

function to satisfy the requirements. Provide a plot of the log-magnitude response in dB of the designed filter.

b. Generate 200 samples of the sequence $x(n)$ and process through the above filter to obtain the output $y(n)$. Provide subplots of $x(n)$ and $y(n)$ for $100 \leq n \leq 200$ on one plot and comment on your results.

P7.23 Design an equiripple digital Hilbert transformer for the following specifications:

$$\text{passband:} \quad 0.1\pi \leq |\omega| \leq 0.5\pi \quad \text{ripple } \delta_1 = 0.01$$
$$\text{stopband:} \quad 0.5\pi \leq |\omega| \leq \pi \quad \text{ripple } \delta_2 = 0.01$$

Plot the amplitude response over $-\pi \leq \omega \leq \pi$.

IIR FILTER DESIGN

IIR filters have infinite-duration impulse responses, hence they can be matched to analog filters, all of which generally have infinitely long impulse responses. Therefore the basic technique of IIR filter design transforms well-known analog filters into digital filters using *complex-valued* mappings. The advantage of this technique lies in the fact that both analog filter design (AFD) tables and the mappings are available extensively in the literature. This basic technique is called the A/D (analog-to-digital) filter transformation. However, the AFD tables are available only for lowpass filters. We also want to design other frequency-selective filters (highpass, bandpass, bandstop, etc.). To do this, we need to apply frequency-band transformations to lowpass filters. These transformations are also complex-valued mappings, and they are also available in the literature. There are two approaches to this basic technique of IIR filter design:

Approach 1:

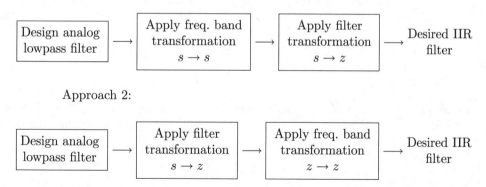

Approach 2:

The first approach is used in MATLAB to design IIR filters. A straightforward use of these MATLAB functions does not provide any insight into the design methodology. Therefore we will study the second approach because it involves the frequency-band transformation in the digital domain. Hence in this IIR filter design technique we will follow the following steps:

- Design analog lowpass filters.
- Study and apply filter transformations to obtain digital lowpass filters.
- Study and apply frequency-band transformations to obtain other digital filters from digital lowpass filters.

The main problem with these approaches is that we have no control over the phase characteristics of the IIR filter. Hence IIR filter designs will be treated as *magnitude-only* designs. More sophisticated techniques, which can simultaneously approximate both the magnitude and the phase responses, require advanced optimization tools and hence will not be covered in this book.

We begin with a discussion on the analog filter specifications and the properties of the magnitude-squared response used in specifying analog filters. This will lead us into the characteristics of three widely used analog filters, namely, *Butterworth*, *Chebyshev*, and *Elliptic* filters. We will then study transformations to convert these prototype analog filters into different frequency-selective digital filters. Finally, we will conclude this chapter with a discussion on the merits and comparisons of FIR and IIR digital filters.

SOME PRELIMINARIES

━━━━━━━━━━■━━━━━━━━

We discuss two preliminary issues in this section. First, we consider the magnitude-squared response specifications, which are more typical of analog (and hence of IIR) filters. These specifications are given on the *relative linear scale*. Second, we study the properties of the magnitude-squared response.

RELATIVE LINEAR SCALE

Let $H_a(j\Omega)$ be the frequency response of an analog filter. Then the lowpass filter specifications on the magnitude-squared response are given by

$$
\begin{aligned}
\frac{1}{1+\epsilon^2} \le |H_a(j\Omega)|^2 &\le 1, \quad |\Omega| \le \Omega_p \\
0 \le |H_a(j\Omega)|^2 &\le \frac{1}{A^2}, \quad \Omega_s \le |\Omega|
\end{aligned}
\tag{8.1}
$$

where ϵ is a passband *ripple parameter*, Ω_p is the passband cutoff frequency in rad/sec, A is a stopband *attenuation parameter*, and Ω_s is the stopband cutoff in rad/sec. These specifications are shown in Figure 8.1,

---

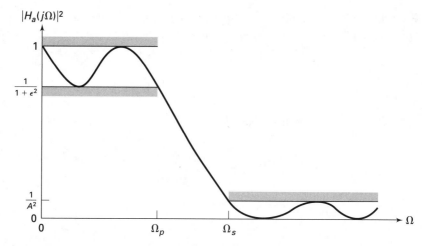

FIGURE 8.1 *Analog lowpass filter specifications*

from which we observe that $|H_a(j\Omega)|^2$ must satisfy

$$
\begin{aligned}
|H_a(j\Omega_p)|^2 &= \frac{1}{1+\epsilon^2} \quad \text{at } \Omega = \Omega_p \\
|H_a(j\Omega)|^2 &= \frac{1}{A^2} \quad\;\; \text{at } \Omega = \Omega_s
\end{aligned}
\tag{8.2}
$$

The parameters ϵ and A are related to parameters R_p and A_S, respectively, of the dB scale. These relations are given by

$$
R_p = -10 \log_{10} \frac{1}{1+\epsilon^2} \implies \epsilon = \sqrt{10^{R_p/10} - 1}
\tag{8.3}
$$

and

$$
A_s = -10 \log_{10} \frac{1}{A^2} \implies A = 10^{A_s/20}
\tag{8.4}
$$

The ripples, δ_1 and δ_2, of the absolute scale are related to ϵ and A by

$$
\frac{1-\delta_1}{1+\delta_1} = \sqrt{\frac{1}{1+\epsilon^2}} \implies \epsilon = \frac{2\sqrt{\delta_1}}{1-\delta_1}
$$

and

$$
\frac{\delta_2}{1+\delta_1} = \frac{1}{A} \implies A = \frac{1+\delta_1}{\delta_2}
$$

Analog filter specifications (8.1), which are given in terms of the magnitude-squared response, contain no phase information. Now to evaluate the s-domain system function $H_a(s)$, consider

$$H_a(j\Omega) = H_a(s)|_{s=j\Omega}$$

Then we have

$$|H_a(j\Omega)|^2 = H_a(j\Omega)H_a^*(j\Omega) = H_a(j\Omega)H_a(-j\Omega) = H_a(s)H_a(-s)|_{s=j\Omega}$$

or

$$H_a(s)H_a(-s) = |H_a(j\Omega)|^2\Big|_{\Omega=s/j} \qquad (8.5)$$

Therefore the poles and zeros of the magnitude-squared function are distributed in a *mirror-image symmetry* with respect to the $j\Omega$ axis. Also for real filters, poles and zeros occur in complex conjugate pairs (or mirror-image symmetry with respect to the real axis). A typical pole-zero pattern of $H_a(s)H_a(-s)$ is shown in Figure 8.2. From this pattern we can construct $H_a(s)$, which is the system function of our analog filter. We want $H_a(s)$ to represent a *causal* and *stable* filter. Then all poles of $H_a(s)$ must lie within the left half-plane. Thus we assign all left-half poles of $H_a(s)H_a(-s)$ to $H_a(s)$. However, zeros of $H_a(s)$ can lie anywhere in the s-plane. Therefore they are not uniquely determined unless they all are on the $j\Omega$ axis. We will choose the zeros of $H_a(s)H_a(-s)$ lying left to or on the $j\Omega$ axis as the zeros of $Ha(s)$. The resulting filter is then called a *minimum-phase* filter.

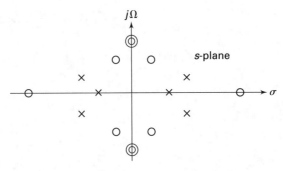

FIGURE 8.2   *Typical pole-zero pattern of $H_a(s)H_a(-s)$*

CHARACTERISTICS OF PROTOTYPE ANALOG FILTERS

IIR filter design techniques rely on existing analog filters to obtain digital filters. We designate these analog filters as *prototype* filters. Three prototypes are widely used in practice. In this section we briefly summarize the characteristics of the lowpass versions of these prototypes: Butterworth lowpass, Chebyshev lowpass (Type I and II), and Elliptic lowpass. Although we will use MATLAB functions to design these filters, it is necessary to learn the characteristics of these filters so that we can use proper parameters in MATLAB functions to obtain correct results.

BUTTER-
WORTH
LOWPASS
FILTERS

This filter is characterized by the property that its magnitude response is flat in both passband and stopband. The magnitude-squared response of an Nth-order lowpass filter is given by

$$|H_a(j\Omega)|^2 = \frac{1}{1 + \left(\dfrac{\Omega}{\Omega_c}\right)^{2N}} \tag{8.6}$$

where N is the order of the filter and Ω_c is the cutoff frequency in rad/sec. The plot of the magnitude-squared response is shown below.

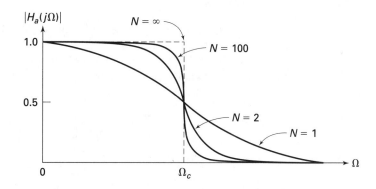

From this plot we can observe the following properties:

- at $\Omega = 0$, $|H_a(j0)|^2 = 1$ for all N.
- at $\Omega = \Omega_c$, $|H_a(j\Omega_c)|^2 = \frac{1}{2}$ for all N, which implies a 3 dB attenuation at Ω_c.
- $|H_a(j\Omega)|^2$ is a monotonically decreasing function of Ω.
- $|H_a(j\Omega)|^2$ approaches an ideal lowpass filter as $N \to \infty$.
- $|H_a(j\Omega)|^2$ is *maximally flat* at $\Omega = 0$ since derivatives of all orders exist and are equal to zero.

To determine the system function $H_a(s)$, we put (8.6) in the form of (8.5) to obtain

$$H_a(s)H_a(-s) = |H_a(j\Omega)|^2\Big|_{\Omega=s/j} = \cfrac{1}{1 + \left(\cfrac{s}{j\Omega_c}\right)^{2N}} = \frac{(j\Omega)^{2N}}{s^{2N} + (j\Omega_c)^{2N}}$$

(8.7)

The roots of the denominator polynomial (or poles of $H_a(s)H_a(-s)$) from (8.7) are given by

$$p_k = (-1)^{\frac{1}{2N}}(j\Omega) = \Omega_c e^{j\frac{\pi}{2N}(2k+N+1)}, \quad k = 0, 1, \ldots, 2N-1 \qquad \textbf{(8.8)}$$

An interpretation of (8.8) is that

- there are $2N$ poles of $H_a(s)H_a(-s)$, which are equally distributed on a circle of radius Ω_c with angular spacing of π/N radians,
- for N odd the poles are given by $p_k = \Omega_c e^{jk\pi/N}$, $k = 0, 1, \ldots, 2N-1$,
- for N even the poles are given by $p_k = \Omega_c e^{j\left(\frac{\pi}{2N} + \frac{k\pi}{N}\right)}$, $k = 0, 1, \ldots, 2N-1$,
- the poles are symmetrically located with respect to the $j\Omega$ axis, and
- a pole never falls on the imaginary axis, and falls on the real axis only if N is odd.

As an example, poles of third- and fourth-order Butterworth filters are shown in Figure 8.3.

A stable and causal filter $H_a(s)$ can now be specified by selecting poles in the left half-plane, and $H_a(s)$ can be written in the form

$$H_a(s) = \frac{\Omega_c^N}{\displaystyle\prod_{\text{LHP poles}} (s - p_k)}$$

(8.9)

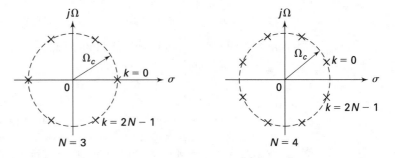

FIGURE 8.3 *Pole plots for Butterworth filters*

□   **EXAMPLE 8.1**   Given that $|H_a(j\Omega)|^2 = \dfrac{1}{1 + 64\Omega^6}$, determine the analog filter system function $H_a(s)$.

Solution   From the given magnitude-squared response,

$$|H_a(j\Omega)|^2 = \frac{1}{1 + 64\Omega^6} = \frac{1}{1 + \left(\dfrac{\Omega}{0.5}\right)^{2(3)}}$$

Comparing this with expression (8.6), we obtain $N = 3$ and $\Omega_c = 0.5$. The poles of $H_a(s)H_a(-s)$ are as shown in Figure 8.4.

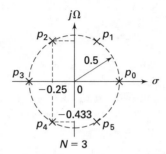

FIGURE 8.4   *Pole plot for Example 8.1*

Hence

$$H_a(j\Omega) = \frac{\Omega_c^3}{(s - s_2)(s - s_3)(s - s_4)}$$

$$= \frac{1/8}{(s + 0.25 - j0.433)(s + 0.5)(s + 0.25 + j0.433)}$$

$$= \frac{0.125}{(s + 0.5)(s^2 + 0.5s + 0.25)}$$ □

MATLAB IMPLEMEN-TATION   MATLAB provides a function called [z,p,k]=buttap(N) to design a *normalized* (i.e., $\Omega_c = 1$) Butterworth analog prototype filter of order N, which returns zeros in z array, poles in p array, and the gain value k. However, we need an unnormalized Butterworth filter with arbitrary Ω_c. From Example 8.1 we observe that there are no zeros and that the poles of the unnormalized filter are on a circle with radius Ω_c instead of on a unit circle. This means that we have to scale the array p of the normalized filter by Ω_c and the gain k by Ω_c^N. In the following function, called U_buttap(N,Omegac), we design the unnormalized Butterworth analog prototype filter.

```
function [b,a] = u_buttap(N,Omegac);
% Unnormalized Butterworth Analog Lowpass Filter Prototype
% ------------------------------------------------------------
% [b,a] = u_buttap(N,Omegac);
%      b = numerator polynomial coefficients of Ha(s)
%      a = denominator polynomial coefficients of Ha(s)
%      N = Order of the Butterworth Filter
% Omegac = Cutoff frequency in radians/sec
%
[z,p,k] = buttap(N);
      p = p*Omegac;
      k = k*Omegac^N;
      B = real(poly(z));
      b0 = k;
      b = k*B;
      a = real(poly(p));
```

The above function provides a direct form (or numerator-denominator) structure. Often we also need a cascade form structure. In Chapter 6 we have already studied how to convert a direct form into a cascade form. The following `sdir2cas` function describes the procedure that is suitable for analog filters.

```
function [C,B,A] = sdir2cas(b,a);
% DIRECT-form to CASCADE-form conversion in s-plane
% ---------------------------------------------------
% [C,B,A] = sdir2cas(b,a)
%  C = gain coefficient
%  B = K by 3 matrix of real coefficients containing bk's
%  A = K by 3 matrix of real coefficients containing ak's
%  b = numerator polynomial coefficients of DIRECT form
%  a = denominator polynomial coefficients of DIRECT form
%
Na = length(a)-1; Nb = length(b)-1;

% compute gain coefficient C
b0 = b(1); b = b/b0;
a0 = a(1); a = a/a0;
 C = b0/a0;
%
% Denominator second-order sections:
p= cplxpair(roots(a)); K = floor(Na/2);
if K*2 == Na      % Computation when Na is even
   A = zeros(K,3);
   for n=1:2:Na
       Arow = p(n:1:n+1,:);
       Arow = poly(Arow);
       A(fix((n+1)/2),:) = real(Arow);
   end
```

```
elseif Na == 1    % Computation when Na = 1
        A = [0 real(poly(p))];

else              % Computation when Na is odd and > 1
    A = zeros(K+1,3);
    for n=1:2:2*K
        Arow = p(n:1:n+1,:);
        Arow = poly(Arow);
        A(fix((n+1)/2),:) = real(Arow);
        end
        A(K+1,:) = [0 real(poly(p(Na)))];
end

% Numerator second-order sections:
z = cplxpair(roots(b)); K = floor(Nb/2);
if Nb == 0           % Computation when Nb = 0
    B = [0 0 poly(z)];

elseif K*2 == Nb     % Computation when Nb is even
    B = zeros(K,3);
    for n=1:2:Nb
        Brow = z(n:1:n+1,:);
        Brow = poly(Brow);
        B(fix((n+1)/2),:) = real(Brow);
    end

elseif Nb == 1       % Computation when Nb = 1
        B = [0 real(poly(z))];

else                 % Computation when Nb is odd and > 1
    B = zeros(K+1,3);
    for n=1:2:2*K
        Brow = z(n:1:n+1,:);
        Brow = poly(Brow);
        B(fix((n+1)/2),:) = real(Brow);
    end
    B(K+1,:) = [0 real(poly(z(Nb)))];
end
```

□   **EXAMPLE 8.2**   Design a third-order Butterworth analog prototype filter with $\Omega_c = 0.5$ given in Example 8.1.

Solution   MATLAB Script _____
```
>> N = 3; OmegaC = 0.5;
>> [b,a] = u_buttap(N,OmegaC);
>> [C,B,A] = sdir2cas(b,a)
C = 0.1250
B = 0    0    1
```

```
A = 1.0000      0.5000      0.2500
        0       1.0000      0.5000
```

The cascade form coefficients agree with those in Example 8.1. □

DESIGN
EQUATIONS

The analog lowpass filter is specified by the parameters Ω_p, R_p, Ω_s, and A_s. Therefore the essence of the design in the case of Butterworth filter is to obtain the order N and the cutoff frequency Ω_c, given these specifications. We want

- at $\Omega = \Omega_p$, $-10 \log_{10} |H_a(j\Omega)|^2 = R_p$ or

$$-10 \log_{10} \left(\frac{1}{1 + \left(\frac{\Omega_p}{\Omega_c} \right)^{2N}} \right) = R_p$$

and

- at $\Omega = \Omega_s$, $-10 \log_{10} |H_a(j\Omega)|^2 = A_s$ or

$$-10 \log_{10} \left(\frac{1}{1 + \left(\frac{\Omega_s}{\Omega_c} \right)^{2N}} \right) = A_s$$

Solving these two equations for N and Ω_c, we have

$$N = \frac{\log_{10} \left[\left(10^{R_p/10} - 1 \right) / \left(10^{A_s/10} - 1 \right) \right]}{2 \log_{10} (\Omega_p / \Omega_s)}$$

In general, the above N will not be an integer. Since we want N to be an integer, we must choose

$$N = \left\lceil \frac{\log_{10} \left[\left(10^{R_p/10} - 1 \right) / \left(10^{A_s/10} - 1 \right) \right]}{2 \log_{10} (\Omega_p / \Omega_s)} \right\rceil \qquad \textbf{(8.10)}$$

where the operation $\lceil x \rceil$ means "choose the smallest integer larger than x"—for example, $\lceil 4.5 \rceil = 5$. Since the actual N chosen is larger than required, specifications can be either met or exceeded either at Ω_p or at Ω_s. To satisfy the specifications exactly at Ω_p,

$$\Omega_c = \frac{\Omega_p}{\sqrt[2N]{\left(10^{R_p/10} - 1 \right)}} \qquad \textbf{(8.11)}$$

or, to satisfy the specifications exactly at Ω_s,

$$\Omega_c = \frac{\Omega_s}{2N\sqrt{\left(10^{A_s/10} - 1\right)}} \tag{8.12}$$

☐ **EXAMPLE 8.3** Design a lowpass Butterworth filter to satisfy

$$\text{Passband cutoff: } \Omega_p = 0.2\pi \ ; \quad \text{Passband ripple: } R_p = 7\text{dB}$$

$$\text{Stopband cutoff: } \Omega_s = 0.3\pi \ ; \quad \text{Stopband ripple: } A_s = 16\text{dB}$$

Solution From (8.10)

$$N = \left\lceil \frac{\log_{10}\left[\left(10^{0.7} - 1\right) / \left(10^{1.6} - 1\right)\right]}{2\log_{10}\left(0.2\pi/0.3\pi\right)} \right\rceil = \lceil 2.79 \rceil = 3$$

To satisfy the specifications exactly at Ω_p, from (8.11) we obtain

$$\Omega_c = \frac{0.2\pi}{\sqrt[6]{\left(10^{0.7} - 1\right)}} = 0.4985$$

To satisfy specifications exactly at Ω_s, from (8.12) we obtain

$$\Omega_c = \frac{0.3\pi}{\sqrt[6]{\left(10^{1.6} - 1\right)}} = 0.5122$$

Now we can choose any Ω_c between the above two numbers. Let us choose $\Omega_c = 0.5$. We have to design a Butterworth filter with $N = 3$ and $\Omega_c = 0.5$, which we did in Example 8.1. Hence

$$H_a(j\Omega) = \frac{0.125}{(s + 0.5)\left(s^2 + 0.5s + 0.25\right)} \qquad ☐$$

MATLAB IMPLEMEN-TATION The above design procedure can be implemented in MATLAB as a simple function. Using the U_buttap function, we provide the afd_butt function to design an analog Butterworth lowpass filter, given its specifications. This function uses (8.11).

```
function [b,a] = afd_butt(Wp,Ws,Rp,As);
% Analog Lowpass Filter Design: Butterworth
% ------------------------------------------
% [b,a] = afd_butt(Wp,Ws,Rp,As);
%  b = Numerator coefficients of Ha(s)
%  a = Denominator coefficients of Ha(s)
% Wp = Passband edge frequency in rad/sec; Wp > 0
% Ws = Stopband edge frequency in rad/sec; Ws > Wp > 0
% Rp = Passband ripple in +dB; (Rp > 0)
```

```
% As = Stopband attenuation in +dB; (As > 0)
%
if Wp <= 0
        error('Passband edge must be larger than 0')
end
if Ws <= Wp
        error('Stopband edge must be larger than Passband edge')
end
if (Rp <= 0) | (As < 0)
        error('PB ripple and/or SB attenuation ust be larger than 0')
end

N = ceil((log10((10^(Rp/10)-1)/(10^(As/10)-1)))/(2*log10(Wp/Ws)));
fprintf('\n*** Butterworth Filter Order = %2.0f \n',N)
OmegaC = Wp/((10^(Rp/10)-1)^(1/(2*N)));
[b,a]=u_buttap(N,OmegaC);
```

To display the frequency-domain plots of analog filters, we provide a function called **freqs_m**, which is a modified version of a function **freqs** provided by MATLAB. This function computes the magnitude response in absolute as well as in relative dB scale and the phase response. This function is similar to the **freqz_m** function discussed earlier. One main difference between them is that in the **freqs_m** function the responses are computed up to a maximum frequency $\Omega_{\max}$.

```
function [db,mag,pha,w] = freqs_m(b,a,wmax);
% Computation of s-domain frequency response: Modified version
% -------------------------------------------------------------
% [db,mag,pha,w] = freqs_m(b,a,wmax);
%   db = Relative magnitude in db over [0 to wmax]
%  mag = Absolute magnitude over [0 to wmax]
%  pha = Phase response in radians over [0 to wmax]
%    w = array of 500 frequency samples between [0 to wmax]
%    b = Numerator polynomial coefficents of Ha(s)
%    a = Denominator polynomial coefficents of Ha(s)
% wmax = Maximum frequency in rad/sec over which response is desired
%
w = [0:1:500]*wmax/500;
H = freqs(b,a,w);
mag = abs(H);
db = 20*log10((mag+eps)/max(mag));
pha = angle(H);
```

The impulse response $h_a(t)$ of the analog filter is computed using MATLAB's **impulse** function.

☐   **EXAMPLE 8.4**   Design the analog Butterworth lowpass filter specified in Example 8.3 using MATLAB.

MATLAB Script

```
>> Wp = 0.2*pi; Ws = 0.3*pi; Rp = 7; As = 16;
>> Ripple = 10 ^ (-Rp/20); Attn = 10 ^ (-As/20);
>> % Analog filter design:
>> [b,a] = afd_butt(Wp,Ws,Rp,As);
*** Butterworth Filter Order =  3
>> % Calculation of second-order sections:
>> [C,B,A] = sdir2cas(b,a)
C = 0.1238
B = 0      0     1
A = 1.0000    0.4985    0.2485
         0    1.0000    0.4985
>> % Calculation of Frequency Response:
>> [db,mag,pha,w] = freqs_m(b,a,0.5*pi);
>> % Calculation of Impulse response:
>> [ha,x,t] = impulse(b,a);
```

The system function is given by

$$H_a\left(s\right) = \frac{0.1238}{\left(s^2 + 0.4985s + 0.2485\right)\left(s + 0.4985\right)}$$

This $H_a\left(s\right)$ is slightly different from the one in Example 8.3 because in that example we used $\Omega_c = 0.5$, while in the afd_butt function Ω_c is chosen to satisfy the specifications at Ω_p. The filter plots are shown in Figure 8.5. □

CHEBYSHEV
LOWPASS
FILTERS

There are two types of Chebyshev filters. The Chebyshev-I filters have *equiripple response in the passband*, while the Chebyshev-II filters have *equiripple response in the stopband*. Butterworth filters have monotonic response in both bands. Recall our discussions regarding equiripple FIR filters. We noted that by choosing a filter that has an equiripple rather than a monotonic behavior, we can obtain a lower-order filter. Therefore Chebyshev filters provide lower order than Butterworth filters for the same specifications.

The magnitude-squared response of a Chebyshev-I filter is

$$|H_a(j\Omega)|^2 = \frac{1}{1 + \epsilon^2 T_N^2\left(\dfrac{\Omega}{\Omega_c}\right)} \tag{8.13}$$

where N is the order of the filter, ϵ is the passband ripple factor, which is related to R_p, and $T_N\left(x\right)$ is the Nth-order Chebyshev polynomial given by

$$T_N(x) = \begin{cases} \cos\left(N\cos^{-1}(x)\right), & 0 \le x \le 1 \\ \cosh\left(\cosh^{-1}(x)\right), & 1 < x < \infty \end{cases} \quad \text{where } x = \frac{\Omega}{\Omega_c}$$

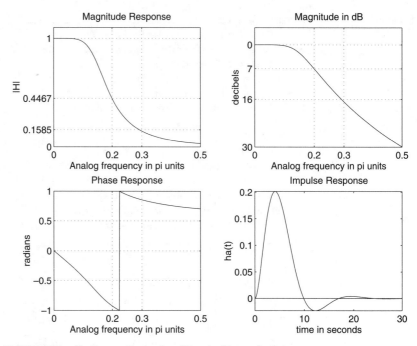

FIGURE 8.5 *Butterworth analog filter in Example 8.4*

The equiripple response of the Chebyshev filters is due to this polynomial $T_N(x)$. Its key properties are (a) for $0 < x < 1$, $T_N(x)$ oscillates between -1 and 1, and (b) for $1 < x < \infty$, $T_N(x)$ increases monotonically to ∞.

There are two possible shapes of $|H_a(j\Omega)|^2$, one for N odd and one for N even as shown below. Note that $x = \Omega/\Omega_c$ is the normalized frequency.

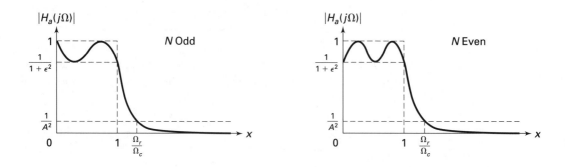

From the above two response plots we observe the following properties:

- At $x = 0$ (or $\Omega = 0$); $|H_a(j0)|^2 = 1$ for N odd.

$$|H_a(j0)|^2 = \frac{1}{1 + \epsilon^2} \quad \text{for } N \text{ even.}$$

- At $x = 1$ (or $\Omega = \Omega_c$); $|H_a(j1)|^2 = \dfrac{1}{1 + \epsilon^2}$ for all N.

- For $0 \le x \le 1$ (or $0 \le \Omega \le \Omega_c$), $|H_a(jx)|^2$ oscillates between 1 and $\dfrac{1}{1 + \epsilon^2}$.

- For $x > 1$ (or $\Omega > \Omega_c$), $|H_a(jx)|^2$ decreases monotonically to 0.

- At $x = \Omega_r$, $|H_a(jx)|^2 = \dfrac{1}{A^2}$.

To determine a causal and stable $H_a(s)$, we must find the poles of $H_a(s)H_a(-s)$ and select the left half-plane poles for $H_a(s)$. The poles of $H_a(s)H_a(-s)$ are obtained by finding the roots of

$$1 + \epsilon^2 T_N^2\left(\frac{s}{j\Omega_c}\right)$$

The solution of this equation is tedious if not difficult to obtain. It can be shown that if $p_k = \sigma_k + j\Omega_k$, $k = 0, \ldots, N-1$ are the (left half-plane) roots of the above polynomial, then

$$
\sigma_k = (a\Omega_c)\cos\left[\frac{\pi}{2} + \frac{(2k+1)\pi}{2N}\right]
$$
$$
\Omega_k = (b\Omega_c)\sin\left[\frac{\pi}{2} + \frac{(2k+1)\pi}{2N}\right]
$$
$$k = 0, \ldots, N-1 \qquad \textbf{(8.14)}$$

where

$$a = \frac{1}{2}\left(\sqrt[N]{\alpha} - \sqrt[N]{1/\alpha}\right), \quad b = \frac{1}{2}\left(\sqrt[N]{\alpha} + \sqrt[N]{1/\alpha}\right), \quad \text{and} \quad \alpha = \frac{1}{\epsilon} + \sqrt{1 + \frac{1}{\epsilon^2}}$$

$$\textbf{(8.15)}$$

These roots fall on an ellipse with major axis $b\Omega_c$ and minor axis $a\Omega_c$. Now the system function is given by

$$H_a(s) = \frac{K}{\prod\limits_k (s - p_k)} \qquad \textbf{(8.16)}$$

where K is a normalizing factor chosen to make

$$H_a(j0) = \begin{cases} 1, & N \text{ odd} \\ \dfrac{1}{\sqrt{1+\epsilon^2}}, & N \text{ even} \end{cases} \qquad \text{(8.17)}$$

MATLAB
IMPLEMEN-
TATION

MATLAB provides a function called `[z,p,k]=cheb1ap(N,Rp)` to design a *normalized* Chebyshev-I analog prototype filter of order `N` and pass-band ripple `Rp` and that returns zeros in `z` array, poles in `p` array, and the gain value `k`. We need an unnormalized Chebyshev-I filter with arbitrary Ω_c. This is achieved by scaling the array `p` of the normalized filter by Ω_c. Similar to the Butterworth prototype, this filter has no zeros. The new gain `k` is determined using (8.17), which is achieved by scaling the old `k` by the ratio of the unnormalized to the normalized denominator polynomials evaluated at $s = 0$. In the following function, called `U_chb1ap(N,Rp,Omegac)`, we design an unnormalized Chebyshev-I analog prototype filter that returns $H_a(s)$ in the direct form.

```
function [b,a] = u_chb1ap(N,Rp,Omegac);
% Unnormalized Chebyshev-1 Analog Lowpass Filter Prototype
% ----------------------------------------------------------
% [b,a] = u_chb1ap(N,Rp,Omegac);
%      b = numerator polynomial coefficients
%      a = denominator polynomial coefficients
%      N = Order of the Elliptic Filter
%     Rp = Passband Ripple in dB; Rp > 0
% Omegac = Cutoff frequency in radians/sec
%
[z,p,k] = cheb1ap(N,Rp);
      a = real(poly(p));
    aNn = a(N+1);
      p = p*Omegac;
      a = real(poly(p));
    aNu = a(N+1);
      k = k*aNu/aNn;
     b0 = k;
      B = real(poly(z));
      b = k*B;
```

DESIGN
EQUATIONS

Given Ω_p, Ω_s, R_p, and A_S, three parameters are required to determine a Chebyshev-I filter: ϵ, Ω_c, and N. From equations (8.3) and (8.4) we obtain

$$\epsilon = \sqrt{10^{0.1R_p} - 1} \qquad \text{and} \qquad A = 10^{A_s/20}$$

From the properties discussed above we have

$$\Omega_c = \Omega_p \qquad \text{and} \qquad \Omega_r = \frac{\Omega_s}{\Omega_p} \tag{8.18}$$

The order N is given by

$$g = \sqrt{(A^2 - 1)/\epsilon^2} \tag{8.19}$$

$$N = \left\lceil \frac{\log_{10}\left[g + \sqrt{g^2 - 1}\right]}{\log_{10}\left[\Omega_r + \sqrt{\Omega_r^2 - 1}\right]} \right\rceil \tag{8.20}$$

Now using (8.15), (8.14), and (8.16), we can determine $H_a(s)$.

☐ **EXAMPLE 8.5** Design a lowpass Chebyshev-I filter to satisfy

$$\text{Passband cutoff: } \Omega_p = 0.2\pi \; ; \quad \text{Passband ripple: } R_p = 1\text{dB}$$
$$\text{Stopband cutoff: } \Omega_s = 0.3\pi \; ; \quad \text{Stopband ripple: } A_s = 16\text{dB}$$

Solution First compute the necessary parameters.

$$\epsilon = \sqrt{10^{0.1(1)} - 1} = 0.5088 \qquad A = 10^{16/20} = 6.3096$$

$$\Omega_c = \Omega_p = 0.2\pi \qquad \Omega_r = \frac{0.3\pi}{0.2\pi} = 1.5$$

$$g = \sqrt{(A^2 - 1)/\epsilon^2} = 12.2429 \qquad N = 4$$

Now we can determine $H_a(s)$.

$$\alpha = \frac{1}{\epsilon} + \sqrt{1 + \frac{1}{\epsilon^2}} = 4.1702$$

$$a = 0.5\left(\sqrt[N]{\alpha} - \sqrt[N]{1/\alpha}\right) = 0.3646$$

$$b = 0.5\left(\sqrt[N]{\alpha} + \sqrt[N]{1/\alpha}\right) = 1.0644$$

There are four poles for $H_a(s)$:

$$p_{0,3} = (a\Omega_c)\cos\left[\frac{\pi}{2} + \frac{\pi}{8}\right] \pm (b\Omega_c)\sin\left[\frac{\pi}{2} + \frac{\pi}{8}\right] = -0.0877 \pm j0.6179$$

$$p_{1,2} = (a\Omega_c)\cos\left[\frac{\pi}{2} + \frac{3\pi}{8}\right] \pm (b\Omega_c)\sin\left[\frac{\pi}{2} + \frac{3\pi}{8}\right] = -0.2117 \pm j0.2559$$

Hence

$$H_a(s) = \frac{K}{\displaystyle\prod_{k=0}^{3}(s - p_k)} = \frac{\overbrace{0.89125 \times .1103 \times .3895}^{0.03829}}{(s^2 + 0.1754s + 0.3895)(s^2 + 0.4234s + 0.1103)}$$

Note that the numerator is such that

$$H_a(j0) = \frac{1}{\sqrt{1+\epsilon^2}} = 0.89125 \qquad\qquad \square$$

MATLAB
IMPLEMEN-
TATION
Using the U_chb1ap function, we provide a function called afd_chb1 to design an analog Chebyshev-II lowpass filter, given its specifications. This is shown below and uses the procedure described in Example 8.5.

```
function [b,a] = afd_chb1(Wp,Ws,Rp,As);
% Analog Lowpass Filter Design: Chebyshev-1
% ----------------------------------------
% [b,a] = afd_chb1(Wp,Ws,Rp,As);
%  b = Numerator coefficients of Ha(s)
%  a = Denominator coefficients of Ha(s)
% Wp = Passband edge frequency in rad/sec; Wp > 0
% Ws = Stopband edge frequency in rad/sec; Ws > Wp > 0
% Rp = Passband ripple in +dB; (Rp > 0)
% As = Stopband attenuation in +dB; (As > 0)
%
if Wp <= 0
        error('Passband edge must be larger than 0')
end
if Ws <= Wp
        error('Stopband edge must be larger than Passband edge')
end
if (Rp <= 0) | (As < 0)
        error('PB ripple and/or SB attenuation ust be larger than 0')
end

ep = sqrt(10^(Rp/10)-1);
A = 10^(As/20);
OmegaC = Wp;
OmegaR = Ws/Wp;
g = sqrt(A*A-1)/ep;
N = ceil(log10(g+sqrt(g*g-1))/log10(OmegaR+sqrt(OmegaR*OmegaR-1)));
fprintf('\n*** Chebyshev-1 Filter Order = %2.0f \n',N)
[b,a]=u_chb1ap(N,Rp,OmegaC);
```

$\square$ **EXAMPLE 8.6**   Design the analog Chebyshev-I lowpass filter given in Example 8.5 using MATLAB.

Solution

MATLAB Script _____
```
>> Wp = 0.2*pi; Ws = 0.3*pi; Rp = 1; As = 16;
>> Ripple = 10 ^ (-Rp/20); Attn = 10 ^ (-As/20);
>> % Analog filter design:
>> [b,a] = afd_chb1(Wp,Ws,Rp,As);
*** Chebyshev-1 Filter Order =  4
```

```
>> % Calculation of second-order sections:
>> [C,B,A] = sdir2cas(b,a)
C = 0.0383
B = 0      0      1
A = 1.0000   0.4233   0.1103
    1.0000   0.1753   0.3895
>> % Calculation of Frequency Response:
>> [db,mag,pha,w] = freqs_m(b,a,0.5*pi);
>> % Calculation of Impulse response:
>> [ha,x,t] = impulse(b,a);
```

The specifications are satisfied by a 4th-order Chebyshev-I filter whose system function is

$$H_a\left(s\right) = \frac{0.0383}{\left(s^2 + 4233s + 0.1103\right)\left(s^2 + 0.1753s + 0.3895\right)}$$

The filter plots are shown in Figure 8.6. □

A Chebyshev-II filter is related to the Chebyshev-I filter through a simple transformation. It has a monotone passband and an equiripple stopband, which implies that this filter has both poles and zeros in the s-plane. Therefore the group delay characteristics are better (and the phase response more linear) in the passband than the Chebyshev-I prototype. If

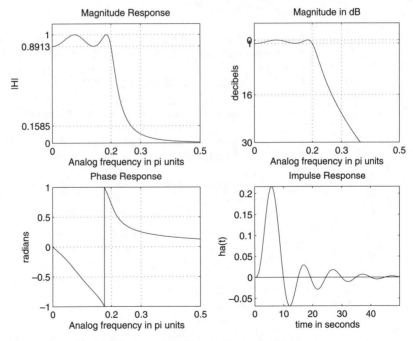

FIGURE 8.6 *Chebyshev-I analog filter in Example 8.6*

we replace the term $\epsilon^2 T_N^2(\Omega/\Omega_c)$ in (8.13) by its reciprocal and also the argument $x = \Omega/\Omega_c$ by its reciprocal, we obtain the magnitude-squared response of Chebyshev-II as

$$|H_a(j\Omega)|^2 = \frac{1}{1 + [\epsilon^2 T_N^2(\Omega_c/\Omega)]^{-1}} \qquad (8.21)$$

One approach to designing a Chebyshev-II filter is to design the corresponding Chebyshev-I first and then apply the above transformations. We will not discuss the details of this filter but will use a function from MATLAB to design a Chebyshev-II filter.

MATLAB IMPLEMENTATION

MATLAB provides a function called `[z,p,k]=cheb2ap(N,As)` to design a *normalized* Chebyshev-II analog prototype filter of order `N` and passband ripple `As` and that returns zeros in `z` array, poles in `p` array, and the gain value `k`. We need an unnormalized Chebyshev-I filter with arbitrary Ω_c. This is achieved by scaling the array `p` of the normalized filter by Ω_c. Since this filter has zeros, we also have to scale the array `z` by Ω_c. The new gain `k` is determined using (8.17), which is achieved by scaling the old `k` by the ratio of the unnormalized to the normalized rational functions evaluated at $s = 0$. In the following function, called `U_chb2ap(N,As,Omegac)`, we design an unnormalized Chebyshev-II analog prototype filter that returns $H_a(s)$ in the direct form.

```
function [b,a] = u_chb2ap(N,As,Omegac);
% Unnormalized Chebyshev-2 Analog Lowpass Filter Prototype
% ----------------------------------------------------------
% [b,a] = u_chb2ap(N,As,Omegac);
%      b = numerator polynomial coefficients
%      a = denominator polynomial coefficients
%      N = Order of the Elliptic Filter
%     As = Stopband Ripple in dB; As > 0
% Omegac = Cutoff frequency in radians/sec
%
[z,p,k] = cheb2ap(N,As);
      a = real(poly(p));
    aNn = a(N+1);
      p = p*Omegac;
      a = real(poly(p));
    aNu = a(N+1);
      b = real(poly(z));
      M = length(b);
    bNn = b(M);
      z = z*Omegac;
      b = real(poly(z));
    bNu = b(M);
```

Chapter 8 ■ IIR FILTER DESIGN

```
      k = k*(aNu*bNn)/(aNn*bNu);
          b0 = k;
           b = k*b;
```

The design equations for the Chebyshev-II prototype are similar to those of the Chebyshev-I except that $\Omega_c = \Omega_s$ since the ripples are in the stopband. Therefore we can develop a MATLAB function similar to the afd_chb1 function for the Chebyshev-II prototype.

```
function [b,a] = afd_chb2(Wp,Ws,Rp,As);
% Analog Lowpass Filter Design: Chebyshev-2
% -----------------------------------------
% [b,a] = afd_chb2(Wp,Ws,Rp,As);
%  b = Numerator coefficients of Ha(s)
%  a = Denominator coefficients of Ha(s)
% Wp = Passband edge frequency in rad/sec; Wp > 0
% Ws = Stopband edge frequency in rad/sec; Ws > Wp > 0
% Rp = Passband ripple in +dB; (Rp > 0)
% As = Stopband attenuation in +dB; (As > 0)
%
if Wp <= 0
        error('Passband edge must be larger than 0')
end
if Ws <= Wp
        error('Stopband edge must be larger than Passband edge')
end
if (Rp <= 0) | (As < 0)
        error('PB ripple and/or SB attenuation ust be larger than 0')
end

ep = sqrt(10^(Rp/10)-1);
A = 10^(As/20);
OmegaC = Wp;
OmegaR = Ws/Wp;
g = sqrt(A*A-1)/ep;
N = ceil(log10(g+sqrt(g*g-1))/log10(OmegaR+sqrt(OmegaR*OmegaR-1)));
fprintf('\n*** Chebyshev-2 Filter Order = %2.0f \n',N)
[b,a]=u_chb2ap(N,As,Ws);
```

□   **EXAMPLE 8.7**   Design a Chebyshev-II analog lowpass filter to satisfy the specifications given in Example 8.5:

$$\text{Passband cutoff: } \Omega_p = 0.2\pi \; ; \quad \text{Passband ripple: } R_p = 1\text{dB}$$

$$\text{Stopband cutoff: } \Omega_s = 0.3\pi \; ; \quad \text{Stopband ripple: } A_s = 16\text{dB}$$

Solution   **MATLAB Script** _____
```
>> Wp = 0.2*pi; Ws = 0.3*pi; Rp = 1; As = 16;
>> Ripple = 10 ^ (-Rp/20); Attn = 10 ^ (-As/20);
```

```
>> % Analog filter design:
>> [b,a] = afd_chb2(Wp,Ws,Rp,As);
*** Chebyshev-2 Filter Order =  4
>> % Calculation of second-order sections:
>> [C,B,A] = sdir2cas(b,a)
C = 0.1585
B = 1.0000          0     6.0654
    1.0000          0     1.0407
A = 1.0000     1.9521     1.4747
    1.0000     0.3719     0.6784
>> % Calculation of Frequency Response:
>> [db,mag,pha,w] = freqs_m(b,a,0.5*pi);
>> % Calculation of Impulse response:
>> [ha,x,t] = impulse(b,a);
```

The specifications are satisfied by a 4th-order Chebyshev-II filter whose system function is

$$H_a(s) = \frac{0.1585 \left(s^2 + 6.0654\right) \left(s^2 + 1.0407\right)}{\left(s^2 + 1.9521s + 1.4747\right) \left(s^2 + 0.3719s + 0.6784\right)}$$

The filter plots are shown in Figure 8.7. ☐

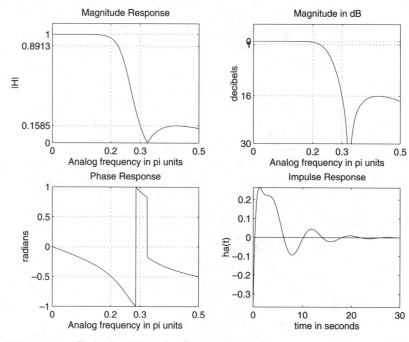

FIGURE 8.7 *Chebyshev-II analog filter in Example 8.7*

Chapter 8 ■ IIR FILTER DESIGN

These filters exhibit equiripple behavior in the passband as well as in the stopband. They are similar in magnitude response characteristics to the FIR equiripple filters. Therefore elliptic filters are optimum filters in that they achieve the minimum order N for the given specifications (or alternately, achieve the sharpest transition band for the given order N). These filters, for obvious reasons, are very difficult to analyze and, therefore, to design. It is not possible to design them using simple tools, and often programs or tables are needed to design them.

The magnitude-squared response of elliptic filters is given by

$$|H_a(j\Omega)|^2 = \frac{1}{1 + \epsilon^2 U_N^2\left(\dfrac{\Omega}{\Omega_c}\right)} \tag{8.22}$$

where N is the order, ϵ is the passband ripple (which is related to R_p), and $U_N(\cdot)$ is the Nth order Jacobian elliptic function. The analysis of this function, even on a superficial level, is beyond the scope of this book. Note the similarity between the above response (8.22) and that of the Chebyshev filters given by (8.13). Typical responses for odd and even N are shown below.

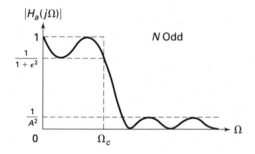
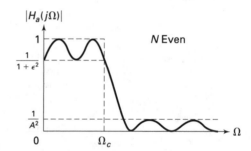

Even though the analysis of (8.22) is difficult, the order calculation formula is very compact and is available in many textbooks [16, 19, 20]. It is given by

$$N = \frac{K(k)K\left(\sqrt{1 - k_1^2}\right)}{K\left(k_1\right)K\left(\sqrt{1 - k^2}\right)} \tag{8.23}$$

where

$$k = \frac{\Omega_p}{\Omega_s}, \quad k_1 = \frac{\epsilon}{\sqrt{A^2 - 1}}$$

and

$$K(x) = \int_0^{\pi/2} \frac{d\theta}{\sqrt{1 - x^2 \sin^2\theta}}$$

is the complete elliptic integral of the first kind. MATLAB provides the function `ellipke` to numerically compute the above integral, which we will use to compute N and to design elliptic filters.

MATLAB
IMPLEMEN-
TATION

MATLAB provides a function called `[z,p,k]=ellipap(N,Rp,As)` to design a *normalized* elliptic analog prototype filter of order N, passband ripple Rp, and stopband attenuation As, and that returns zeros in z array, poles in p array, and the gain value k. We need an unnormalized elliptic filter with arbitrary Ω_c. This is achieved by scaling the arrays p and z of the normalized filter by Ω_c and the gain k by the ratio of the unnormalized to the normalized rational functions evaluated at $s = 0$. In the following function, called `U_elipap(N,Rp,As,Omegac)`, we design an unnormalized elliptic analog prototype filter that returns $H_a(s)$ in the direct form.

```
function [b,a] = u_elipap(N,Rp,As,Omegac);
% Unnormalized Elliptic Analog Lowpass Filter Prototype
% ----------------------------------------------------------
% [b,a] = u_elipap(N,Rp,As,Omegac);
%       b = numerator polynomial coefficients
%       a = denominator polynomial coefficients
%       N = Order of the Elliptic Filter
%      Rp = Passband Ripple in dB; Rp > 0
%      As = Stopband Attenuation in dB; As > 0
% Omegac = Cutoff frequency in radians/sec
%
[z,p,k] = ellipap(N,Rp,As);
      a = real(poly(p));
    aNn = a(N+1);
      p = p*Omegac;
      a = real(poly(p));
    aNu = a(N+1);
      b = real(poly(z));
      M = length(b);
    bNn = b(M);
      z = z*Omegac;
      b = real(poly(z));
    bNu = b(M);
      k = k*(aNu*bNn)/(aNn*bNu);
     b0 = k;
      b = k*b;
```

Using the U_elipap function, we provide a function called afd_elip to design an analog elliptic lowpass filter, given its specifications. This is shown below and uses the filter order computation formula given in (8.23).

```
function [b,a] = afd_elip(Wp,Ws,Rp,As);
% Analog Lowpass Filter Design: Elliptic
% --------------------------------------
% [b,a] = afd_elip(Wp,Ws,Rp,As);
%  b = Numerator coefficients of Ha(s)
%  a = Denominator coefficients of Ha(s)
% Wp = Passband edge frequency in rad/sec; Wp > 0
% Ws = Stopband edge frequency in rad/sec; Ws > Wp > 0
% Rp = Passband ripple in +dB; (Rp > 0)
% As = Stopband attenuation in +dB; (As > 0)
%
if Wp <= 0
        error('Passband edge must be larger than 0')
end
if Ws <= Wp
        error('Stopband edge must be larger than Passband edge')
end
if (Rp <= 0) | (As < 0)
        error('PB ripple and/or SB attenuation ust be larger than 0')
end

ep = sqrt(10^(Rp/10)-1);
A = 10^(As/20);
OmegaC = Wp;
k = Wp/Ws;
k1 = ep/sqrt(A*A-1);
capk = ellipke([k.^2 1-k.^2]); % Version 4.0 code
capk1 = ellipke([(k1 .^2) 1-(k1 .^2)]); % Version 4.0 code
N = ceil(capk(1)*capk1(2)/(capk(2)*capk1(1)));
fprintf('\n*** Elliptic Filter Order = %2.0f \n',N)
[b,a]=u_elipap(N,Rp,As,OmegaC);
```

□   **EXAMPLE 8.8**   Design an analog elliptic lowpass filter to satisfy the following specifications of Example 8.5:

$$\Omega_p = 0.2\pi, \quad R_p = 1\,\text{dB}$$
$$\Omega_s = 0.3\pi, \quad A_s = 16\,\text{db}$$

Solution   **MATLAB Script** _____

```
>> Wp = 0.2*pi; Ws = 0.3*pi; Rp = 1; As = 16;
>> Ripple = 10 ^ (-Rp/20); Attn = 10 ^ (-As/20);
>> % Analog filter design:
>> [b,a] = afd_elip(Wp,Ws,Rp,As);
```

```
*** Elliptic Filter Order =  3
>> % Calculation of second-order sections:
>> [C,B,A] = sdir2cas(b,a)
C = 0.2740
B = 1.0000        0      0.6641
A = 1.0000    0.1696    0.4102
          0    1.0000    0.4435
>> % Calculation of Frequency Response:
>> [db,mag,pha,w] = freqs_m(b,a,0.5*pi);
>> % Calculation of Impulse response:
>> [ha,x,t] = impulse(b,a);
```

The specifications are satisfied by a 3rd-order elliptic filter whose system function is

$$H_a\left(s\right) = \frac{0.274\left(s^2 + 0.6641\right)}{\left(s^2 + 0.1696s + 0.4102\right)\left(s + 0.4435\right)}$$

The filter plots are shown in Figure 8.8. □

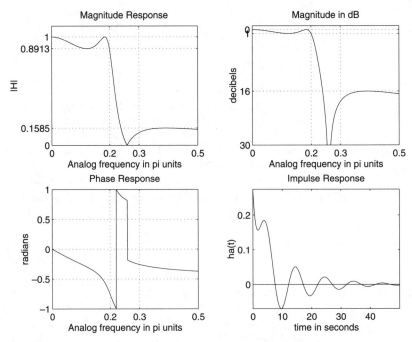

FIGURE 8.8 *Elliptic analog lowpass filter in Example 8.8*

Elliptic filters provide optimal performance in the magnitude-squared response but have highly nonlinear phase response in the passband (which is undesirable in many applications). Even though we decided not to worry about phase response in our designs, phase is still an important issue in the overall system. At the other end of the performance scale are the Butterworth filters, which have maximally flat magnitude response and require a higher-order N (more poles) to achieve the same stopband specification. However, they exhibit a fairly linear phase response in their passband. The Chebyshev filters have phase characteristics that lie somewhere in between. Therefore in practical applications we do consider Butterworth as well as Chebyshev filters, in addition to elliptic filters. The choice depends on both the filter order (which influences processing speed and implementation complexity) and the phase characteristics (which control the distortion).

ANALOG-TO-DIGITAL FILTER TRANSFORMATIONS

After discussing different approaches to the design of analog filters, we are now ready to transform them into digital filters. These transformations are complex-valued mappings that are extensively studied in the literature. These transformations are derived by preserving different aspects of analog and digital filters. If we want to preserve the shape of the impulse response from analog to digital filter, then we obtain a technique called *impulse invariance* transformation. If we want to convert a differential equation representation into a corresponding difference equation representation, then we obtain a *finite difference approximation* technique. Numerous other techniques are also possible. One technique, called *step invariance*, preserves the shape of the step response; this is explored in Problem 9. The most popular technique used in practice is called a *Bilinear* transformation, which preserves the system function representation from analog to digital domain. In this section we will study in detail impulse invariance and bilinear transformations, both of which can be easily implemented in MATLAB.

IMPULSE
INVARIANCE
TRANSFOR-
MATION

In this design method we want the digital filter impulse response to look "similar" to that of a frequency-selective analog filter. Hence we sample $h_a(t)$ at some sampling interval T to obtain $h(n)$; that is,

$$h(n) = h_a(nT)$$

The parameter T is chosen so that the shape of $h_a(t)$ is "captured" by the samples. Since this is a sampling operation, the analog and digital frequencies are related by

$$\omega = \Omega T \text{ or } e^{j\omega} = e^{j\Omega T}$$

Since $z = e^{j\omega}$ on the unit circle and $s = j\Omega$ on the imaginary axis, we have the following transformation from the s-plane to the z-plane:

$$z = e^{sT} \tag{8.24}$$

The system functions $H(z)$ and $H_a(s)$ are related through the frequency-domain aliasing formula (3.27):

$$H(z) = \frac{1}{T} \sum_{k=-\infty}^{\infty} H_a\left(s - j\frac{2\pi}{T}k\right)$$

The complex plane transformation under the mapping (8.24) is shown in Figure 8.9, from which we have the following observations:

1. Using $\sigma = \text{Re}(s)$, we note that

$$\sigma < 0 \quad \text{maps into } |z| < 1 \text{ (inside of the UC)}$$

$$\sigma = 0 \quad \text{maps onto } |z| = 1 \text{ (on the UC)}$$

$$\sigma > 0 \quad \text{maps into } |z| > 1 \text{ (outside of the UC)}$$

2. All semi-infinite strips (shown above) of width $2\pi/T$ map into $|z| < 1$. Thus this mapping is not unique but a *many-to-one* mapping.

3. Since the entire left half of the s-plane maps into the unit circle, a causal and stable analog filter maps into a causal and stable digital filter.

4. If $H_a(j\Omega) = H_a(j\omega/T) = 0$ for $|\Omega| \geq \pi/T$, then

$$H(e^{j\omega}) = \frac{1}{T} H_a(j\omega/T), \quad |\omega| \leq \pi$$

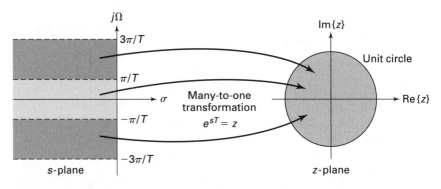

FIGURE 8.9 *Complex-plane mapping in impulse invariance transformation*

and there will be no aliasing. However, no analog filter of finite order can be exactly band-limited. Therefore some aliasing error will occur in this design procedure, and hence the sampling interval T plays a minor role in this design method.

DESIGN
PROCEDURE

Given the digital lowpass filter specifications ω_p, ω_s, R_p, and A_s, we want to determine $H(z)$ by first designing an equivalent analog filter and then mapping it into the desired digital filter. The steps required for this procedure are

1. Choose T and determine the analog frequencies

$$\Omega_p = \frac{\omega_p}{T_p} \quad \text{and} \quad \Omega_s = \frac{\omega_s}{T}$$

2. Design an analog filter $H_a(s)$ using the specifications Ω_p, Ω_s, R_p, and A_s. This can be done using any one of the three (Butterworth, Chebyshev, or elliptic) prototypes of the previous section.

3. Using partial fraction expansion, expand $H_a(s)$ into

$$H_a(s) = \sum_{k=1}^{N} \frac{R_k}{s - p_k}$$

4. Now transform analog poles $\{p_k\}$ into digital poles $\{e^{p_k T}\}$ to obtain the digital filter:

$$H(z) = \sum_{k=1}^{N} \frac{R_k}{1 - e^{p_k T} z^{-1}} \tag{8.25}$$

☐ **EXAMPLE 8.9** Transform

$$H_a(s) = \frac{s+1}{s^2 + 5s + 6}$$

into a digital filter $H(z)$ using the impulse invariance technique in which $T = 0.1$.

Solution

We first expand $H_a(s)$ using partial fraction expansion:

$$H_a(s) = \frac{s+1}{s^2 + 5s + 6} = \frac{2}{s+3} - \frac{1}{s+2}$$

The poles are at $p_1 = -3$ and $p_2 = -2$. Then from (8.25) and using $T = 0.1$, we obtain

$$H(z) = \frac{2}{1 - e^{-3T} z^{-1}} - \frac{1}{1 - e^{-2T} z^{-1}} = \frac{1 - 0.8966 z^{-1}}{1 - 1.5595 z^{-1} + 0.6065 z^{-2}}$$

It is easy to develop a MATLAB function to implement the impulse invariance mapping. Given a rational function description of $H_a(s)$, we can use the residue function to obtain its pole-zero description. Then each analog pole is mapped into a digital pole using (8.24). Finally, the residuez function can be used to convert $H(z)$ into rational function form. This procedure is given in the function imp_invr.

```
function [b,a] = imp_invr(c,d,T)
% Impulse Invariance Transformation from Analog to Digital Filter
% -----------------------------------------------------------------
% [b,a] = imp_invr(c,d,T)
%  b = Numerator polynomial in z^(-1) of the digital filter
%  a = Denominator polynomial in z^(-1) of the digital filter
%  c = Numerator polynomial in s of the analog filter
%  d = Denominator polynomial in s of the analog filter
%  T = Sampling (transformation) parameter
%
[R,p,k] = residue(c,d);
p = exp(p*T);
[b,a] = residuez(R,p,k);
b = real(b'); a = real(a');
```

A similar function called impinvar is available in the new Student Edition of MATLAB. □

☐   **EXAMPLE 8.10**   We demonstrate the use of the imp_invr function on the system function from Example 8.9.

Solution   **MATLAB Script** _____
```
>> c = [1,1]; d = [1,5,6]; T = 0.1;
>> [b,a] = imp_invr(c,d,T)
b =  1.0000   -0.8966
a =  1.0000   -1.5595    0.6065
```
_____

The digital filter is

$$H(z) = \frac{1 - 0.8966z^{-1}}{1 - 1.5595z^{-1} + 0.6065z^{-2}}$$

as expected. In Figure 8.10 we show the impulse responses and the magnitude responses (plotted up to the sampling frequency $1/T$) of the analog and the resulting digital filter. Clearly, the aliasing in the frequency domain is evident. □

In the next several examples we illustrate the impulse invariance design procedure on all three prototypes.

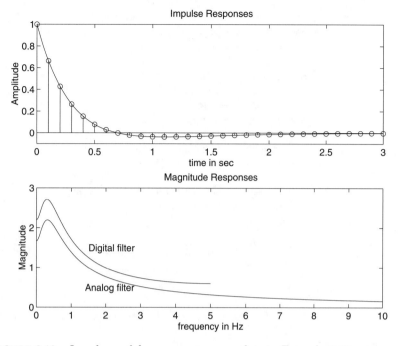

FIGURE 8.10 *Impulse and frequency response plots in Example 8.10*

☐ **EXAMPLE 8.11** Design a lowpass digital filter using a Butterworth prototype to satisfy

$$\omega_p = 0.2\pi, \quad R_p = 1 \text{ dB}$$

$$\omega_s = 0.3\pi, \quad A_s = 15 \text{ dB}$$

Solution The design procedure is described in the following MATLAB script:

```
>> % Digital Filter Specifications:
>> wp = 0.2*pi;                    % digital Passband freq in rad
>> ws = 0.3*pi;                    % digital Stopband freq in rad
>> Rp = 1;                         % Passband ripple in dB
>> As = 15;                        % Stopband attenuation in dB

>> % Analog Prototype Specifications: Inverse mapping for frequencies
>> T = 1;                          % Set T=1
>> OmegaP = wp / T;                % Prototype Passband freq
>> OmegaS = ws / T;                % Prototype Stopband freq

>> % Analog Butterworth Prototype Filter Calculation:
>> [cs,ds] = afd_butt(OmegaP,OmegaS,Rp,As);
*** Butterworth Filter Order =   6
```

Analog-to-Digital Filter Transformations   331

```
>> % Impulse Invariance transformation:
>> [b,a] = imp_invr(cs,ds,T);
>> [C,B,A] = dir2par(b,a)
C = []
B = 1.8557    -0.6304
   -2.1428     1.1454
    0.2871    -0.4466
A = 1.0000    -0.9973     0.2570
    1.0000    -1.0691     0.3699
    1.0000    -1.2972     0.6949
```

The desired filter is a 6th-order Butterworth filter whose system function $H(z)$ is given in the parallel form

$$H(z) = \frac{1.8587 - 0.6304z^{-1}}{1 - 0.9973z^{-1} + 0.257z^{-2}} + \frac{-2.1428 + 1.1454z^{-1}}{1 - 1.0691z^{-1} + 0.3699z^{-2}}$$
$$+ \frac{0.2871 - 0.4463z^{-1}}{1 - 1.2972z^{-1} + 0.6449z^{-2}}$$

The frequency response plots are given in Figure 8.11. □

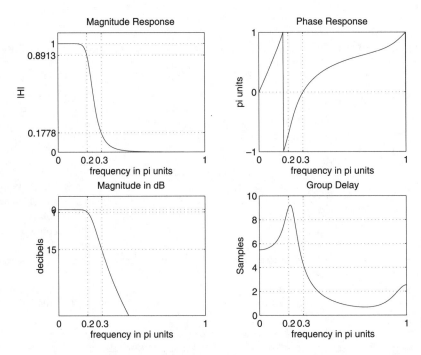

FIGURE 8.11 *Digital Butterworth lowpass filter using impulse invariance design*

Design a lowpass digital filter using a Chebyshev-I prototype to satisfy

$$\omega_p = 0.2\pi, \quad R_p = 1 \text{ dB}$$

$$\omega_s = 0.3\pi, \quad A_s = 15 \text{ dB}$$

Solution   The design procedure is described in the following MATLAB script:

```
>> % Digital Filter Specifications:
>> wp = 0.2*pi;                    % digital Passband freq in rad
>> ws = 0.3*pi;                    % digital Stopband freq in rad
>> Rp = 1;                         % Passband ripple in dB
>> As = 15;                        % Stopband attenuation in dB

>> % Analog Prototype Specifications: Inverse mapping for frequencies
>> T = 1;                          % Set T=1
>> OmegaP = wp / T;                % Prototype Passband freq
>> OmegaS = ws / T;                % Prototype Stopband freq

>> % Analog Chebyshev-1 Prototype Filter Calculation:
>> [cs,ds] = afd_chb1(OmegaP,OmegaS,Rp,As);
*** Chebyshev-1 Filter Order =   4

>> % Impulse Invariance transformation:
>> [b,a] = imp_invr(cs,ds,T);
>> [C,B,A] = dir2par(b,a)
C =   []
B =-0.0833    -0.0246
    0.0833     0.0239
A = 1.0000    -1.4934     0.8392
    1.0000    -1.5658     0.6549
```

The desired filter is a 4th-order Chebyshev-I filter whose system function $H(z)$ is

$$H(z) = \frac{-0.0833 - 0.0246z^{-1}}{1 - 1.4934z^{-1} + 0.8392z^{-2}} + \frac{-0.0833 + 0.0239z^{-1}}{1 - 1.5658z^{-1} + 0.6549z^{-2}}$$

The frequency response plots are given in Figure 8.12.   □

□ **EXAMPLE 8.13**   Design a lowpass digital filter using a Chebyshev-II prototype to satisfy

$$\omega_p = 0.2\pi, \quad R_p = 1 \text{ dB}$$

$$\omega_s = 0.3\pi, \quad A_s = 15 \text{ dB}$$

Solution   Recall that the Chebyshev-II filter is equiripple in the stopband. It means that this analog filter has a response that does not go to zero at high frequencies in the stopband. Therefore after impulse invariance transformation, the aliasing effect will be significant; this can degrade the passband response. The MATLAB script is shown:

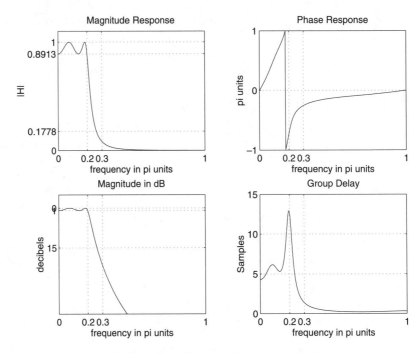

FIGURE 8.12 *Digital Chebyshev-I lowpass filter using impulse invariance design*

```
>> % Digital Filter Specifications:
>> wp = 0.2*pi;                        % digital Passband freq in rad
>> ws = 0.3*pi;                        % digital Stopband freq in rad
>> Rp = 1;                             % Passband ripple in dB
>> As = 15;                            % Stopband attenuation in dB

>> % Analog Prototype Specifications: Inverse mapping for frequencies
>> T = 1;                              % Set T=1
>> OmegaP = wp / T;                    % Prototype Passband freq
>> OmegaS = ws / T;                    % Prototype Stopband freq

>> % Analog Chebyshev-1 Prototype Filter Calculation:
>> [cs,ds] = afd_chb2(OmegaP,OmegaS,Rp,As);
*** Chebyshev-2 Filter Order =  4

>> % Impulse Invariance transformation:
>> [b,a] = imp_invr(cs,ds,T);
>> [C,B,A] = dir2par(b,a);
```

From the frequency response plots in Figure 8.13 we clearly observe the passband as well as stopband degradation. Hence the impulse invariance design technique has failed to produce a desired digital filter. □

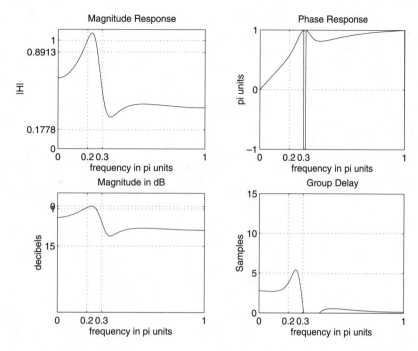

FIGURE 8.13 *Digital Chebyshev-II lowpass filter using impulse invariance design*

☐ **EXAMPLE 8.14** Design a lowpass digital filter using an elliptic prototype to satisfy

$$\omega_p = 0.2\pi, \quad R_p = 1 \text{ dB}$$

$$\omega_s = 0.3\pi, \quad A_s = 15 \text{ dB}$$

Solution The elliptic filter is equiripple in both bands. Hence this situation is similar to that of the Chebyshev-II filter, and we should not expect a good digital filter. The MATLAB script is shown:

```
>> % Digital Filter Specifications:
>> wp = 0.2*pi;                    % digital Passband freq in rad
>> ws = 0.3*pi;                    % digital Stopband freq in rad
>> Rp = 1;                         % Passband ripple in dB
>> As = 15;                        % Stopband attenuation in dB

>> % Analog Prototype Specifications: Inverse mapping for frequencies
>> T = 1;                          % Set T=1
>> OmegaP = wp / T;                % Prototype Passband freq
>> OmegaS = ws / T;                % Prototype Stopband freq

>> % Analog Elliptic Prototype Filter Calculation:
>> [cs,ds] = afd_elip(OmegaP,OmegaS,Rp,As);
*** Elliptic Filter Order =  3
```

Analog-to-Digital Filter Transformations

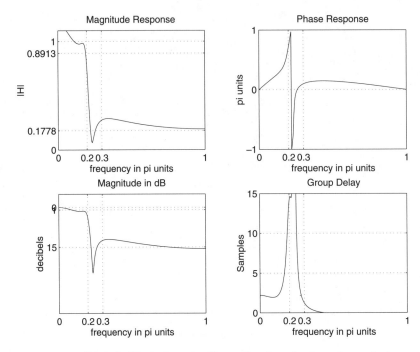

FIGURE 8.14 *Digital elliptic lowpass filter using impulse invariance design*

```
>> % Impulse Invariance transformation:
>> [b,a] = imp_invr(cs,ds,T);
>> [C,B,A] = dir2par(b,a);
```

From the frequency response plots in Figure 8.14 we clearly observe that once again the impulse invariance design technique has failed. □

The advantages of the impulse invariance mapping are that it is a stable design and that the frequencies Ω and ω are linearly related. But the disadvantage is that we should expect some aliasing of the analog frequency response, and in some cases this aliasing is intolerable. Consequently, this design method is useful only when the analog filter is essentially band-limited to a lowpass or bandpass filter in which there are no oscillations in the stopband.

BILINEAR TRANSFOR-MATION

This mapping is the best transformation method; it involves a well-known function given by

$$s = \frac{2}{T} \frac{1 - z^{-1}}{1 + z^{-1}} \Longrightarrow z = \frac{1 + sT/2}{1 - sT/2} \tag{8.26}$$

where T is a parameter. Another name for this transformation is the *linear fractional* transformation because when cleared of fractions, we obtain

$$\frac{T}{2}sz + \frac{T}{2}s - z + 1 = 0$$

which is linear in each variable if the other is fixed, or *bilinear* in s and z. The complex plane mapping under (8.26) is shown in Figure 8.15, from which we have the following observations:

1. Using $s = \sigma + j\Omega$ in (8.26), we obtain

$$z = \left(1 + \frac{\sigma T}{2} + j\frac{\Omega T}{2}\right) \Big/ \left(1 - \frac{\sigma T}{2} - j\frac{\Omega T}{2}\right) \qquad \textbf{(8.27)}$$

Hence

$$\sigma < 0 \implies |z| = \left|\frac{1 + \frac{\sigma T}{2} + j\frac{\Omega T}{2}}{1 - \frac{\sigma T}{2} - j\frac{\Omega T}{2}}\right| < 1$$

$$\sigma = 0 \implies |z| = \left|\frac{1 + j\frac{\Omega T}{2}}{1 - j\frac{\Omega T}{2}}\right| = 1$$

$$\sigma > 0 \implies |z| = \left|\frac{1 + \frac{\sigma T}{2} + j\frac{\Omega T}{2}}{1 - \frac{\sigma T}{2} - j\frac{\Omega T}{2}}\right| > 1$$

2. The entire left half-plane maps into the inside of the unit circle. Hence this is a stable transformation.

3. The imaginary axis maps onto the unit circle in a one-to-one fashion. Hence there is no aliasing in the frequency domain.

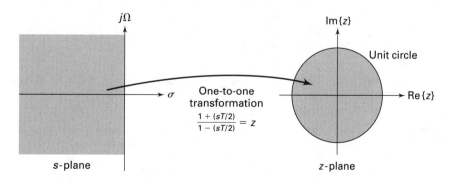

FIGURE 8.15   *Complex-plane mapping in bilinear transformation*

Substituting $\sigma = 0$ in (8.27), we obtain

$$z = \frac{1 + j\frac{\Omega T}{2}}{1 - j\frac{\Omega T}{2}} = e^{j\omega}$$

since the magnitude is 1. Solving for ω as a function of Ω, we obtain

$$\omega = 2\tan^{-1}\left(\frac{\Omega T}{2}\right) \qquad \text{or} \qquad \Omega = \frac{2}{T}\tan\left(\frac{\omega}{2}\right) \qquad (8.28)$$

This shows that Ω is nonlinearly related to (or warped into) ω but that there is no aliasing. Hence in (8.28) we will say that ω is prewarped into Ω.

□ **EXAMPLE 8.15** Transform $H_a(s) = \dfrac{s+1}{s^2 + 5s + 6}$ into a digital filter using the bilinear transformation. Choose $T = 1$.

Solution Using (8.26), we obtain

$$H(z) = H_a\left(\frac{2}{T}\frac{1 - z^{-1}}{1 + z^{-1}}\bigg|_{T=1}\right) = H_a\left(2\frac{1 - z^{-1}}{1 + z^{-1}}\right)$$

$$= \frac{2\dfrac{1 - z^{-1}}{1 + z^{-1}} + 1}{\left(2\dfrac{1 - z^{-1}}{1 + z^{-1}}\right)^2 + 5\left(2\dfrac{1 - z^{-1}}{1 + z^{-1}}\right) + 6}$$

Simplifying,

$$H(z) = \frac{3 + 2z^{-1} - z^{-2}}{20 + 4z^{-1}} = \frac{0.15 + 0.1z^{-1} - 0.05z^{-2}}{1 + 0.2z^{-1}} \qquad \square$$

MATLAB provides a function called `bilinear` to implement this mapping. Its invocation is similar to the `imp_invr` function, but it also takes several forms for different input-output quantities. The Student Edition manual should be consulted for more details. Its use is shown in the following example.

□ **EXAMPLE 8.16** Transform the system function $H_a(s)$ in Example 8.15 using the `bilinear` function.

Solution **MATLAB Script** _____

```
>> c = [1,1]; d = [1,5,6]; T = 1; Fs = 1/T;
>> [b,a] = bilinear(c,d,Fs)
b = 0.1500    0.1000    -0.0500
a = 1.0000    0.2000     0.0000
```

_____

The filter is

$$H(z) = \frac{0.15 + 0.1z^{-1} - 0.05z^{-2}}{1 + 0.2z^{-1}}$$

as before.   □

DESIGN
PROCEDURE

Given digital filter specifications ω_p, ω_s, R_p, and A_s, we want to determine $H(z)$. The design steps in this procedure are the following: ⸙

    1. Choose a value for T. This is arbitrary, and we may set $T = 1$.

    2. Prewarp the cutoff frequencies ω_p and ω_s; that is, calculate Ω_p and Ω_s using (8.28):

$$\Omega_p = \frac{2}{T}\tan\left(\frac{\omega_p}{2}\right), \quad \Omega_s = \frac{2}{T}\tan\left(\frac{\omega_s}{2}\right) \tag{8.29}$$

    3. Design an analog filter $H_a(s)$ to meet the specifications Ω_p, Ω_s, R_p, and A_s. We have already described how to do this in the previous section.

    4. Finally, set

$$H(z) = H_a\left(\frac{2}{T}\frac{1 - z^{-1}}{1 + z^{-1}}\right)$$

and simplify to obtain $H(z)$ as a rational function in z^{-1}.

    In the next several examples we demonstrate this design procedure on our analog prototype filters.

□   **EXAMPLE 8.17**   Design the digital Butterworth filter of Example 8.11. The specifications are

$$\omega_p = 0.2\pi, \quad R_p = 1 \text{ dB}$$
$$\omega_s = 0.3\pi, \quad A_s = 15 \text{ dB}$$

Solution

MATLAB Script _____

```
>> % Digital Filter Specifications:
>> wp = 0.2*pi;                    % digital Passband freq in rad
>> ws = 0.3*pi;                    % digital Stopband freq in rad
>> Rp = 1;                         % Passband ripple in dB
>> As = 15;                        % Stopband attenuation in dB
>> % Analog Prototype Specifications: Inverse mapping for frequencies
>> T = 1; Fs = 1/T;               % Set T=1
>> OmegaP = (2/T)*tan(wp/2);      % Prewarp Prototype Passband freq
>> OmegaS = (2/T)*tan(ws/2);      % Prewarp Prototype Stopband freq
>> % Analog Butterworth Prototype Filter Calculation:
>> [cs,ds] = afd_butt(OmegaP,OmegaS,Rp,As);
*** Butterworth Filter Order =  6
```

```
>> % Bilinear transformation:
>> [b,a] = bilinear(cs,ds,Fs);
>> [C,B,A] = dir2cas(b,a)
C = 5.7969e-004
B = 1.0000    2.0183    1.0186
    1.0000    1.9814    0.9817
    1.0000    2.0004    1.0000
A = 1.0000   -0.9459    0.2342
    1.0000   -1.0541    0.3753
    1.0000   -1.3143    0.7149
```

The desired filter is once again a 6th-order filter and has 6 zeros. Since the 6th-order zero of $H_a(s)$ at $s = -\infty$ is mapped to $z = -1$, these zeros should be at $z = -1$. Due to the finite precision of MATLAB these zeros are not exactly at $z = -1$. Hence the system function should be

$$H(z) = \frac{0.00057969\left(1 + z^{-1}\right)^6}{\left(1 - 0.9459z^{-1} + 0.2342z^{-2}\right)\left(1 - 1.0541z^{-1} + 0.3753z^{-2}\right)\left(1 - 1.3143z^{-1} + 0.7149z^{-2}\right)}$$

The frequency response plots are given in Figure 8.16. Comparing these plots with those in Figure 8.11, we observe that these two designs are very similar.

□

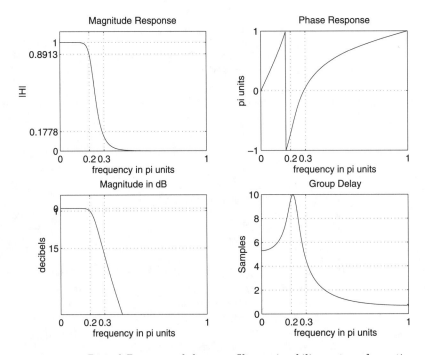

FIGURE 8.16 *Digital Butterworth lowpass filter using bilinear transformation*

Design the digital Chebyshev-I filter of Example 8.12. The specifications are

$$\omega_p = 0.2\pi, \quad R_p = 1 \text{ dB}$$
$$\omega_s = 0.3\pi, \quad A_s = 15 \text{ dB}$$

Solution　　　　**MATLAB Script** _____

```
>> % Digital Filter Specifications:
>> wp = 0.2*pi;                    % digital Passband freq in rad
>> ws = 0.3*pi;                    % digital Stopband freq in rad
>> Rp = 1;                         % Passband ripple in dB
>> As = 15;                        % Stopband attenuation in dB
>> % Analog Prototype Specifications: Inverse mapping for frequencies
>> T = 1; Fs = 1/T;                % Set T=1
>> OmegaP = (2/T)*tan(wp/2);       % Prewarp Prototype Passband freq
>> OmegaS = (2/T)*tan(ws/2);       % Prewarp Prototype Stopband freq
>> % Analog Chebyshev-1 Prototype Filter Calculation:
>> [cs,ds] = afd_chb1(OmegaP,OmegaS,Rp,As);
*** Chebyshev-1 Filter Order =  4
>> % Bilinear transformation:
>> [b,a] = bilinear(cs,ds,Fs);
>> [C,B,A] = dir2cas(b,a)
C = 0.0018
B = 1.0000    2.0000    1.0000
    1.0000    2.0000    1.0000
A = 1.0000   -1.4996    0.8482
    1.0000   -1.5548    0.6493
```

The desired filter is a 4th-order filter and has 4 zeros at $z = -1$. The system function is

$$H(z) = \frac{0.0018 \left(1 + z^{-1}\right)^4}{\left(1 - 1.4996z^{-1} + 0.8482z^{-2}\right)\left(1 - 1.5548z^{-1} + 0.6493z^{-2}\right)}$$

The frequency response plots are given in Figure 8.17 which are similar to those in Figure 8.12.　　　　　　　　　　　　　　　　　　　　　　　　　□

□ **EXAMPLE 8.19**　Design the digital Chebyshev-I filter of Example 8.13. The specifications are

$$\omega_p = 0.2\pi, \quad R_p = 1 \text{ dB}$$
$$\omega_s = 0.3\pi, \quad A_s = 15 \text{ dB}$$

Solution　　　　**MATLAB Script** _____

```
>> % Digital Filter Specifications:
>> wp = 0.2*pi;                    % digital Passband freq in rad
>> ws = 0.3*pi;                    % digital Stopband freq in rad
>> Rp = 1;                         % Passband ripple in dB
>> As = 15;                        % Stopband attenuation in dB
```

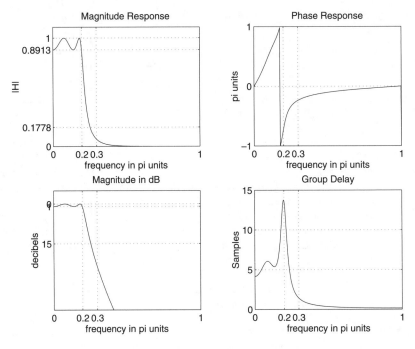

FIGURE 8.17 *Digital Chebyshev-I lowpass filter using bilinear transformation*

```
>> % Analog Prototype Specifications: Inverse mapping for frequencies
>> T = 1; Fs = 1/T;                 % Set T=1
>> OmegaP = (2/T)*tan(wp/2);        % Prewarp Prototype Passband freq
>> OmegaS = (2/T)*tan(ws/2);        % Prewarp Prototype Stopband freq
>> % Analog Chebyshev-2 Prototype Filter Calculation:
>> [cs,ds] = afd_chb2(OmegaP,OmegaS,Rp,As);
*** Chebyshev-2 Filter Order =  4
>> % Bilinear transformation:
>> [b,a] = bilinear(cs,ds,Fs);
>> [C,B,A] = dir2cas(b,a)
C = 0.1797
B = 1.0000    0.5574    1.0000
    1.0000   -1.0671    1.0000
A = 1.0000   -0.4183    0.1503
    1.0000   -1.1325    0.7183
```

The desired filter is again a 4th-order filter with system function

$$H\left(z\right) = \frac{0.1797\left(1 + 0.5574z^{-1} + z^{-2}\right)\left(1 - 1.0671z^{-1} + z^{-2}\right)}{\left(1 - 0.4183z^{-1} + 0.1503z^{-2}\right)\left(1 - 1.1325z^{-1} + 0.7183z^{-2}\right)}$$

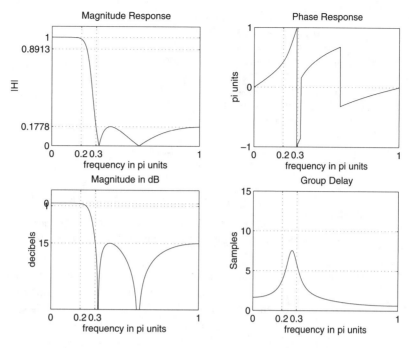

FIGURE 8.18 *Digital Chebyshev-II lowpass filter using bilinear transformation*

The frequency response plots are given in Figure 8.18. Note that the bilinear transformation has properly designed the Chebyshev-II digital filter. □

□ **EXAMPLE 8.20** Design the digital elliptic filter of Example 8.14. The specifications are

$$\omega_p = 0.2\pi, \quad R_p = 1 \text{ dB}$$

$$\omega_s = 0.3\pi, \quad A_s = 15 \text{ dB}$$

Solution

MATLAB Script_____

```
>> % Digital Filter Specifications:
>> wp = 0.2*pi;                      % digital Passband freq in rad
>> ws = 0.3*pi;                      % digital Stopband freq in rad
>> Rp = 1;                           % Passband ripple in dB
>> As = 15;                          % Stopband attenuation in dB
>> % Analog Prototype Specifications: Inverse mapping for frequencies
>> T = 1; Fs = 1/T;                  % Set T=1
>> OmegaP = (2/T)*tan(wp/2);         % Prewarp Prototype Passband freq
>> OmegaS = (2/T)*tan(ws/2);         % Prewarp Prototype Stopband freq
>> % Analog Elliptic Prototype Filter Calculation:
>> [cs,ds] = afd_elip(OmegaP,OmegaS,Rp,As);
*** Elliptic Filter Order =  3
>> % Bilinear transformation:
```

```
>> [b,a] = bilinear(cs,ds,Fs);
>> [C,B,A] = dir2cas(b,a)
C = 0.1214
B = 1.0000    -1.4211    1.0000
    1.0000     1.0000         0
A = 1.0000    -1.4928    0.8612
    1.0000    -0.6183         0
```

The desired filter is a 3rd-order filter with system function

$$H(z) = \frac{0.1214\left(1 - 1.4211z^{-1} + z^{-2}\right)\left(1 + z^{-1}\right)}{\left(1 - 1.4928z^{-1} + 0.8612z^{-2}\right)\left(1 - 0.6183z^{-1}\right)}$$

The frequency response plots are given in Figure 8.19. Note that the bilinear transformation has again properly designed the elliptic digital filter. □

The advantages of this mapping are that (a) it is a stable design, (b) there is no aliasing, and (c) there is no restriction on the type of filter that can be transformed. Therefore this method is used exclusively in computer programs including MATLAB as we shall see next.

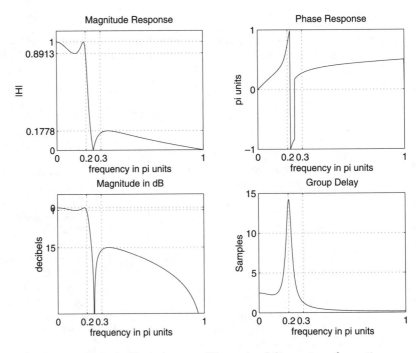

FIGURE 8.19 *Digital elliptic lowpass filter using bilinear transformation*

LOWPASS FILTER DESIGN USING MATLAB

In this section we will demonstrate the use of MATLAB's filter design routines to design digital lowpass filters. These functions use the bilinear transformation because of its desirable advantages as discussed in the previous section. These functions are as follows:

1. [b,a]=butter(N,wn)

This function designs an Nth-order lowpass digital Butterworth filter and returns the filter coefficients in length $N+1$ vectors b and a. The filter order is given by (8.10), and the cutoff frequency wn is determined by the prewarping formula (8.29). However, in MATLAB all digital frequencies are given in *units of* π. Hence wn is computed by using the following relation:

$$\omega_n = \frac{2}{\pi} \tan^{-1} \left(\frac{\Omega_c T}{2} \right)$$

The use of this function is given in Example 8.21.

2. [b,a]=cheby1(N,Rp,wn)

This function designs an Nth-order lowpass digital Chebyshev-I filter with Rp decibels of ripple in the passband. It returns the filter coefficients in length $N + 1$ vectors b and a. The filter order is given by (8.20), and the cutoff frequency wn is the digital passband frequency in units of π; that is,

$$\omega_n = \omega_p/\pi$$

The use of this function is given in Example 8.22.

3. [b,a]=cheby2(N,As,wn)

This function designs an Nth-order lowpass digital Chebyshev-II filter with the stopband attenuation As decibels. It returns the filter coefficients in length $N + 1$ vectors b and a. The filter order is given by (8.20), and the cutoff frequency wn is the digital stopband frequency in units of π; that is,

$$\omega_n = \omega_s/\pi$$

The use of this function is given in Example 8.23.

4. [b,a]=ellip(N,Rp,As,wn)

This function designs an Nth-order lowpass digital elliptic filter with the passband ripple of Rp decibels and a stopband attenuation of As decibels. It returns the filter coefficients in length $N + 1$ vectors b and a. The filter order is given by (8.23), and the cutoff frequency wn is the digital passband frequency in units of π; that is,

$$\omega_n = \omega_p/\pi$$

The use of this function is given in Example 8.24.

All these above functions can also be used to design other frequency-selective filters, such as highpass and bandpass. We will discuss their additional capabilities in Section 8.5.

There is also another set of filter functions, namely the `buttord`, `cheb1ord`, `cheb2ord`, and `ellipord` functions, which can provide filter order N and filter cutoff frequency ω_n, given the specifications. These functions are available in the Signal Processing toolbox but not in the Student Edition, and hence in the examples to follow we will determine these parameters using the formulas given earlier. We will discuss the filter-order functions in the next section.

In the following examples we will redesign the same lowpass filters of previous examples and compare their results. The specifications of the lowpass digital filter are

$$\omega_p = 0.2\pi, \quad R_p = 1 \text{ dB}$$

$$\omega_s = 0.3\pi, \quad A_s = 15 \text{ dB}$$

□   **EXAMPLE 8.21**   Digital Butterworth lowpass filter design:

```
>> % Digital Filter Specifications:
>> wp = 0.2*pi;                    %digital Passband freq in rad
>> ws = 0.3*pi;                    %digital Stopband freq in rad
>> Rp = 1;                         %Passband ripple in dB
>> As = 15;                        %Stopband attenuation in dB

>> % Analog Prototype Specifications:
>> T = 1;                          %Set T=1
>> OmegaP = (2/T)*tan(wp/2);       %Prewarp Prototype Passband freq
>> OmegaS = (2/T)*tan(ws/2);       %Prewarp Prototype Stopband freq

>> % Analog Prototype Order Calculation:
>> N =ceil((log10((10^(Rp/10)-1)/(10^(As/10)-1)))/(2*log10(OmegaP/OmegaS)));
>> fprintf('\n*** Butterworth Filter Order = %2.0f \n',N)
** Butterworth Filter Order =   6
>> OmegaC = OmegaP/((10^(Rp/10)-1)^(1/(2*N))); %Analog BW prototype cutoff
>> wn = 2*atan((OmegaC*T)/2);      %Digital BW cutoff freq

>> % Digital Butterworth Filter Design:
>> wn = wn/pi;                     %Digital Butter cutoff in pi units
>> [b,a]=butter(N,wn);
>> [b0,B,A] = dir2cas(b,a)
```

```
C = 5.7969e-004
B = 1.0000    2.0297    1.0300
    1.0000    1.9997    1.0000
    1.0000    1.9706    0.9709
A = 1.0000   -0.9459    0.2342
    1.0000   -1.0541    0.3753
    1.0000   -1.3143    0.7149
```

The system function is

$$H(z) = \frac{0.00057969 \left(1 + z^{-1}\right)^6}{\left(1 - 0.9459z^{-1} + 0.2342z^{-2}\right)\left(1 - 1.0541z^{-1} + 0.3753z^{-2}\right)\left(1 - 1.3143z^{-1} + 0.7149z^{-2}\right)}$$

which is the same as in Example 8.17. The frequency-domain plots were shown in Figure 8.16. □

□ **EXAMPLE 8.22**   Digital Chebyshev-I lowpass filter design:

```
>> % Digital Filter Specifications:
>> wp = 0.2*pi;                    %digital Passband freq in rad
>> ws = 0.3*pi;                    %digital Stopband freq in rad
>> Rp = 1;                         %Passband ripple in dB
>> As = 15;                        %Stopband attenuation in dB

>> % Analog Prototype Specifications:
>> T = 1;                          %Set T=1
>> OmegaP = (2/T)*tan(wp/2);       %Prewarp Prototype Passband freq
>> OmegaS = (2/T)*tan(ws/2);       %Prewarp Prototype Stopband freq

>> % Analog Prototype Order Calculation:
>> ep = sqrt(10^(Rp/10)-1);        %Passband Ripple Factor
>> A = 10^(As/20);                 %Stopband Attenuation Factor
>> OmegaC = OmegaP;                %Analog Prototype Cutoff freq
>> OmegaR = OmegaS/OmegaP;         %Analog Prototype Transition Ratio
>> g = sqrt(A*A-1)/ep;             %Analog Prototype Intermediate cal.
>> N = ceil(log10(g+sqrt(g*g-1))/log10(OmegaR+sqrt(OmegaR*OmegaR-1)));
>> fprintf('\n*** Chebyshev-1 Filter Order = %2.0f \n',N)
*** Chebyshev-1 Filter Order =  4

>> % Digital Chebyshev-I Filter Design:
>> wn = wp/pi;                     %Digital Passband freq in pi units
>> [b,a]=cheby1(N,Rp,wn);
>> [b0,B,A] = dir2cas(b,a)
b0 = 0.0018
B = 1.0000    2.0000    1.0000
    1.0000    2.0000    1.0000
A = 1.0000   -1.4996    0.8482
    1.0000   -1.5548    0.6493
```

The system function is

$$H(z) = \frac{0.0018\left(1 + z^{-1}\right)^4}{\left(1 - 1.4996z^{-1} + 0.8482z^{-2}\right)\left(1 - 1.5548z^{-1} + 0.6493z^{-2}\right)}$$

which is the same as in Example 8.18. The frequency-domain plots were shown in Figure 8.17. $\square$

□   **EXAMPLE 8.23**   Digital Chebyshev-II lowpass filter design:

```
>> % Digital Filter Specifications:
>> wp = 0.2*pi;                    %digital Passband freq in rad
>> ws = 0.3*pi;                    %digital Stopband freq in rad
>> Rp = 1;                         %Passband ripple in dB
>> As = 15;                        %Stopband attenuation in dB

>> % Analog Prototype Specifications:
>> T = 1;                          %Set T=1
>> OmegaP = (2/T)*tan(wp/2);       %Prewarp Prototype Passband freq
>> OmegaS = (2/T)*tan(ws/2);       %Prewarp Prototype Stopband freq

>> % Analog Prototype Order Calculation:
>> ep = sqrt(10^(Rp/10)-1);        %Passband Ripple Factor
>> A = 10^(As/20);                 %Stopband Attenuation Factor
>> OmegaC = OmegaP;                %Analog Prototype Cutoff freq
>> OmegaR = OmegaS/OmegaP;         %Analog Prototype Transition Ratio
>> g = sqrt(A*A-1)/ep;             %Analog Prototype Intermediate cal.
>> N = ceil(log10(g+sqrt(g*g-1))/log10(OmegaR+sqrt(OmegaR*OmegaR-1)));
>> fprintf('\n*** Chebyshev-2 Filter Order = %2.0f \n',N)
*** Chebyshev-2 Filter Order =  4

>> % Digital Chebyshev-II Filter Design:
>> wn = ws/pi;                           %Digital Stopband freq in pi units
>> [b,a]=cheby2(N,As,wn);
>> [b0,B,A] = dir2cas(b,a)
b0 = 0.1797
B = 1.0000     0.5574     1.0000
    1.0000    -1.0671     1.0000
A = 1.0000    -0.4183     0.1503
    1.0000    -1.1325     0.7183
```

The system function is

$$H(z) = \frac{0.1797\left(1 + 0.5574z^{-1} + z^{-2}\right)\left(1 - 1.0671z^{-1} + z^{-2}\right)}{\left(1 - 0.4183z^{-1} + 0.1503z^{-2}\right)\left(1 - 1.1325z^{-1} + 0.7183z^{-2}\right)}$$

which is the same as in Example 8.19. The frequency-domain plots were shown in Figure 8.18. $\square$

Chapter 8 ■ IIR FILTER DESIGN

□   **EXAMPLE 8.24**   Digital elliptic lowpass filter design:

```
>> % Digital Filter Specifications:
>> wp = 0.2*pi;                    %digital Passband freq in rad
>> ws = 0.3*pi;                    %digital Stopband freq in rad
>> Rp = 1;                         %Passband ripple in dB
>> As = 15;                        %Stopband attenuation in dB

>> % Analog Prototype Specifications:
>> T = 1;                          %Set T=1
>> OmegaP = (2/T)*tan(wp/2);       %Prewarp Prototype Passband freq
>> OmegaS = (2/T)*tan(ws/2);       %Prewarp Prototype Stopband freq

>> % Analog Elliptic Filter order calculations:
>> ep = sqrt(10^(Rp/10)-1);        %Passband Ripple Factor
>> A = 10^(As/20);                 %Stopband Attenuation Factor
>> OmegaC = OmegaP;                %Analog Prototype Cutoff freq
>> k = OmegaP/OmegaS;              %Analog Prototype Transition Ratio;
>> k1 = ep/sqrt(A*A-1);            %Analog Prototype Intermediate cal.
>> capk = ellipke([k.^2 1-k.^2]);
>> capk1 = ellipke([(k1 .^2) 1-(k1 .^2)]);
>> N = ceil(capk(1)*capk1(2)/(capk(2)*capk1(1)));
>> fprintf('\n*** Elliptic Filter Order = %2.0f \n',N)
*** Elliptic Filter Order =  3

>> % Digital Elliptic Filter Design:
>> wn = wp/pi;                     %Digital Passband freq in pi units
>> [b,a]=ellip(N,Rp,As,wn);
>> [b0,B,A] = dir2cas(b,a)
b0 = 0.1214
B = 1.0000   -1.4211    1.0000
    1.0000    1.0000         0
A = 1.0000   -1.4928    0.8612
    1.0000   -0.6183         0
```

The system function is

$$H\left(z\right) = \frac{0.1214\left(1 - 1.4211z^{-1} + z^{-2}\right)\left(1 + z^{-1}\right)}{\left(1 - 1.4928z^{-1} + 0.8612z^{-2}\right)\left(1 - 0.6183z^{-1}\right)}$$

which is the same as in Example 8.20. The frequency-domain plots were shown in Figure 8.19.   □

COMPARISON OF THREE FILTERS

In our examples we designed the same digital filter using four different prototype analog filters. Let us compare their performance. The specifications were $\omega_p = 0.2\pi$, $R_p = 1$ dB, $\omega_s = 0.3\pi$, and $A_s = 15$ dB. This comparison in terms of order N and the minimum stopband attenuations is shown in Table 8.1.

TABLE 8.1 *Comparison of three filters*

Prototype	Order N	Stopband Att.
Butterworth	6	15
Chebyshev-I	4	25
Elliptic	3	27

Clearly, the elliptic prototype gives the best design. However, if we compare their phase responses, then the elliptic design has the most non-linear phase response in the passband.

FREQUENCY-BAND TRANSFORMATIONS

In the preceding two sections we designed digital lowpass filters from their corresponding analog filters. Certainly, we would like to design other types of frequency-selective filters, such as highpass, bandpass, and bandstop. This is accomplished by transforming the frequency axis (or band) of a lowpass filter so that it behaves as another frequency-selective filter. These transformations on the complex variable z are very similar to bilinear transformations, and the design equations are algebraic. The procedure to design a general frequency-selective filter is to first design a *digital prototype* (of fixed bandwidth, say unit bandwidth) lowpass filter and then to apply these algebraic transformations. In this section we will describe the basic philosophy behind these mappings and illustrate their mechanism through examples. MATLAB provides functions that incorporate frequency-band transformation in the s-plane. We will first demonstrate the use of the z-plane mapping and then illustrate the use of MATLAB functions. Typical specifications for most commonly used types of frequency-selective digital filters are shown in Figure 8.20.

Let $H_{LP}(Z)$ be the given prototype lowpass digital filter, and let $H(z)$ be the desired frequency-selective digital filter. Note that we are using two different frequency variables, Z and z, with H_{LP} and H, respectively. Define a mapping of the form

$$Z^{-1} = G(z^{-1})$$

such that

$$H(z) = H_{LP}(Z)|_{Z^{-1}=G(z^{-1})}$$

To do this, we simply replace Z^{-1} everywhere in H_{LP} by the function $G(z^{-1})$. Given that $H_{LP}(Z)$ is a stable and causal filter, we also want $H(z)$ to be stable and causal. This imposes the following requirements:

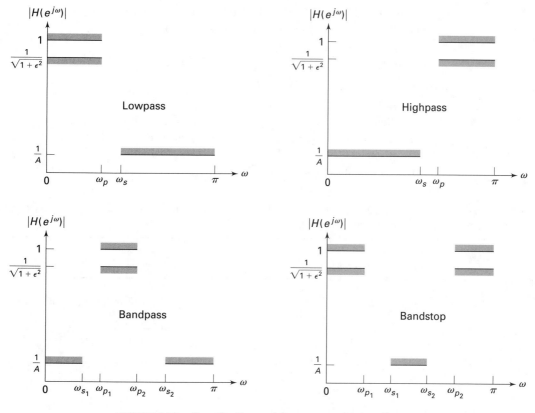

FIGURE 8.20 *Specifications of frequency-selective filters*

1. $G(\cdot)$ must be a rational function in z^{-1} so that $H(z)$ is implementable.

2. The unit circle of the Z-plane must map onto the unit circle of the z-plane.

3. For stable filters, the inside of the unit circle of the Z-plane must also map onto the inside of the unit circle of the z-plane.

Let ω' and ω be the frequency variables of Z and z, respectively—that is, $Z = e^{j\omega'}$ and $z = e^{j\omega}$ on their respective unit circles. Then requirement 2 above implies that

$$\left|Z^{-1}\right| = \left|G(z^{-1})\right| = \left|G(e^{-j\omega})\right| = 1$$

and

$$e^{-j\omega'} = \left|G(e^{-j\omega})\right| e^{j\angle G(e^{-j\omega})}$$

or

$$-\omega' = \angle G(e^{-j\omega})$$

The general form of the function $G(\cdot)$ that satisfies the above requirements is a rational function of the *all-pass* type given by

$$Z^{-1} = G\left(z^{-1}\right) = \pm \prod_{k=1}^{n} \frac{z^{-1} - \alpha_k}{1 - \alpha_k z^{-1}}$$

where $|\alpha_k| < 1$ for stability and to satisfy requirement 3.

Now by choosing an appropriate order n and the coefficients $\{\alpha_k\}$, we can obtain a variety of mappings. The most widely used transformations are given in Table 8.2. We will now illustrate the use of this table for designing a highpass digital filter.

TABLE 8.2 *Frequency transformation for digital filters (prototype lowpass filter has cutoff frequency ω_c')*

Type of Transformation	Transformation	Parameters
Lowpass	$z^{-1} \longrightarrow \dfrac{z^{-1} - \alpha}{1 - \alpha z^{-1}}$	ω_c = cutoff frequency of new filter $\alpha = \dfrac{\sin\left[\left(\omega_c' - \omega_c\right)/2\right]}{\sin\left[\left(\omega_c' + \omega_c\right)/2\right]}$
Highpass	$z^{-1} \longrightarrow -\dfrac{z^{-1} + \alpha}{1 + \alpha z^{-1}}$	ω_c = cutoff frequency of new filter $\alpha = -\dfrac{\cos\left[\left(\omega_c' + \omega_c\right)/2\right]}{\cos\left[\left(\omega_c' - \omega_c\right)/2\right]}$
Bandpass	$z^{-1} \longrightarrow -\dfrac{z^{-2} - \alpha_1 z^{-1} + \alpha_2}{\alpha_2 z^{-2} - \alpha_1 z^{-1} + 1}$	ω_ℓ = lower cutoff frequency ω_u = upper cutoff frequency $\alpha_1 = -2\beta K/(K+1)$ $\alpha_2 = (K-1)/(K+1)$ $\beta = \dfrac{\cos\left[\left(\omega_u + \omega_\ell\right)/2\right]}{\cos\left[\left(\omega_u - \omega_\ell\right)/2\right]}$ $K = \cot\dfrac{\omega_u - \omega_\ell}{2} \tan\dfrac{\omega_c'}{2}$
Bandstop	$z^{-1} \longrightarrow \dfrac{z^{-2} - \alpha_1 z^{-1} + \alpha_2}{\alpha_2 z^{-2} - \alpha_1 z^{-1} + 1}$	ω_ℓ = lower cutoff frequency ω_u = upper cutoff frequency $\alpha_1 = -2\beta/(K+1)$ $\alpha_2 = (K-1)/(K+1)$ $\beta = \dfrac{\cos\left[\left(\omega_u + \omega_\ell\right)/2\right]}{\cos\left[\left(\omega_u - \omega_\ell\right)/2\right]}$ $K = \tan\dfrac{\omega_u - \omega_\ell}{2} \tan\dfrac{\omega_c'}{2}$

☐ **EXAMPLE 8.25** In Example 8.22 we designed a Chebyshev-I lowpass filter with specifications

$$\omega_p' = 0.2\pi, \quad R_p = 1 \text{ dB}$$

$$\omega_s' = 0.3\pi, \quad A_s = 15 \text{ dB}$$

and determined its system function

$$H_{LP}(Z) = \frac{0.001836(1 + Z^{-1})^4}{(1 - 1.4996Z^{-1} + 0.8482Z^{-2})(1 - 1.5548Z^{-1} + 0.6493Z^{-2})}$$

Design a highpass filter with the above tolerances but with passband beginning at $\omega_p = 0.6\pi$.

Solution We want to transform the given lowpass filter into a highpass filter such that the cutoff frequency $\omega_p' = 0.2\pi$ is mapped onto the cutoff frequency $\omega_p = 0.6\pi$. From Table 8.2

$$\alpha = -\frac{\cos[(0.2\pi + 0.6\pi)/2]}{\cos[(0.2\pi - 0.6\pi)/2]} = -0.38197 \tag{8.30}$$

Hence

$$H_{LP}(z) = H(Z)\Big|_{Z = -\frac{z^{-1} - 0.38197}{1 - 0.38197z^{-1}}}$$

$$= \frac{0.02426(1 - z^{-1})^4}{(1 + 0.5661z^{-1} + 0.7657z^{-2})(1 + 1.0416z^{-1} + 0.4019z^{-2})}$$

which is the desired filter. The frequency response plots of the lowpass filter and the new highpass filter are shown in Figure 8.21. ☐

From the above example it is obvious that to obtain the rational function of a new digital filter from the prototype lowpass digital filter, we should be able to implement rational function substitutions from Table 8.2. This appears to be a difficult task, but since these are algebraic functions, we can use the `conv` function repetitively for this purpose. The following **zmapping** function illustrates this approach.

```
function [bz,az] = zmapping(bZ,aZ,Nz,Dz)
% Frequency band Transformation from Z-domain to z-domain
% -----------------------------------------------------------
% [bz,az] = zmapping(bZ,aZ,Nz,Dz)
% performs:
%           b(z)    b(Z)|
%           ---- = ----|        N(z)
%           a(z)    a(Z)|@Z = ----
%                             D(z)
%
bzord = (length(bZ)-1)*(length(Nz)-1);
azord = (length(aZ)-1)*(length(Dz)-1);

bz = zeros(1,bzord+1);
for k = 0:bzord
```

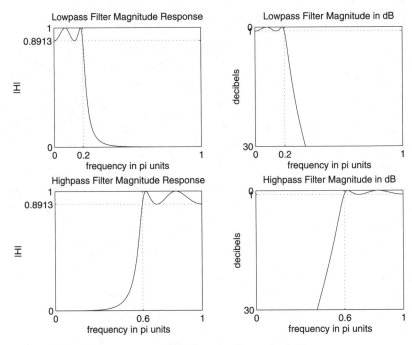

FIGURE 8.21 *Magnitude response plots for Example 8.25*

```
pln = [1];
    for l = 0:k-1
        pln = conv(pln,Nz);
    end
    pld = [1];
    for l = 0:bzord-k-1
        pld = conv(pld,Dz);
    end
    bz = bz+bZ(k+1)*conv(pln,pld);
end

az = zeros(1,azord+1);
for k = 0:azord
    pln = [1];
    for l = 0:k-1
        pln = conv(pln,Nz);
    end
    pld = [1];
    for l = 0:azord-k-1
        pld = conv(pld,Dz);
    end
```

```
      az = az+aZ(k+1)*conv(pln,pld);
      end

      az1 = az(1); az = az/az1; bz = bz/az1;
```

□ **EXAMPLE 8.26**   Use the `zmapping` function to perform the lowpass-to-highpass transformation in Example 8.25.

Solution   First we will design the lowpass digital filter in MATLAB using the bilinear transformation procedure and then use the `zmapping` function.

```
>> % Digital Lowpass Filter Specifications:
>> wplp = 0.2*pi;                        % digital Passband freq in rad
>> wslp = 0.3*pi;                        % digital Stopband freq in rad
>>   Rp = 1;                             % Passband ripple in dB
>>   As = 15;                            % Stopband attenuation in dB

>> % Analog Prototype Specifications: Inverse mapping for frequencies
>> T = 1; Fs = 1/T;                      % Set T=1
>> OmegaP = (2/T)*tan(wplp/2);           % Prewarp Prototype Passband freq
>> OmegaS = (2/T)*tan(wslp/2);           % Prewarp Prototype Stopband freq

>> % Analog Chebyshev Prototype Filter Calculation:
>> [cs,ds] = afd_chb1(OmegaP,OmegaS,Rp,As);
** Chebyshev-1 Filter Order =  4

>> % Bilinear transformation:
>> [blp,alp] = bilinear(cs,ds,Fs);

>> % Digital Highpass Filter Cutoff frequency:
>> wphp = 0.6*pi;                        % Passband edge frequency

>> % LP-to-HP frequency-band transformation:
>> alpha = -(cos((wplp+wphp)/2))/(cos((wplp-wphp)/2))
alpha = -0.3820

>> Nz = -[alpha,1]; Dz = [1,alpha];
>> [bhp,ahp] = zmapping(blp,alp,Nz,Dz);
>> [C,B,A] = dir2cas(bhp,ahp)
C = 0.0243
B = 1.0000   -2.0000    1.0000
    1.0000   -2.0000    1.0000
A = 1.0000    1.0416    0.4019
    1.0000    0.5561    0.7647
```

The system function of the highpass filter is

$$H\left(z\right) = \frac{0.0243(1 - z^{-1})^4}{(1 + 0.5661z^{-1} + 0.7647z^{-2})(1 + 1.0416z^{-1} + 0.4019z^{-2})}$$

which is essentially identical to that in Example 8.25. □

DESIGN
PROCEDURE

In Example 8.26 a lowpass prototype digital filter was available to transform into a highpass filter so that a particular band-edge frequency was properly mapped. In practice we have to first design a prototype lowpass digital filter whose specifications should be obtained from specifications of other frequency-selective filters as given in Figure 8.20. We will now show that the lowpass prototype filter specifications can be obtained from the transformation formulas given in Table 8.2.

Let us use the highpass filter of Example 8.25 as an example. The passband-edge frequencies were transformed using the parameter $\alpha = -0.38197$ in (8.30). What is the stopband-edge frequency of the highpass filter, say ω_s, corresponding to the stopband edge $\omega_s' = 0.3\pi$ of the prototype lowpass filter? This can be answered by (8.30). Since α is fixed for the transformation, we set the equation

$$\alpha = -\frac{\cos[(0.3\pi + \omega_s)/2]}{\cos[(0.3\pi - \omega_s)/2]} = -0.38197$$

This is a transcendental equation whose solution can be obtained iteratively from an initial guess. It can be done using MATLAB, and the solution is

$$\omega_s = 0.4586\pi$$

Now in practice we will know the desired highpass frequencies ω_s and ω_p, and we are required to find the prototype lowpass cutoff frequencies ω_s' and ω_p'. We can choose the passband frequency ω_p' with a reasonable value, say $\omega_p' = 0.2\pi$, and determine α from ω_p using the formula from Table 8.2. Now ω_s' can be determined (for our highpass filter example) from α and

$$Z = -\frac{z^{-1} + \alpha}{1 + \alpha z^{-1}}$$

where $Z = e^{j\omega_s'}$ and $z = e^{j\omega_s}$, or

$$\omega_s' = \angle \left(-\frac{e^{-j\omega_s} + \alpha}{1 + \alpha e^{-j\omega_s}} \right) \qquad \textbf{(8.31)}$$

Continuing our highpass filter example, let $\omega_p = 0.6\pi$ and $\omega_s = 0.4586\pi$ be the band-edge frequencies. Let us choose $\omega_p' = 0.2\pi$. Then $\alpha = -0.38197$

from (8.30), and from (8.31)

$$\omega'_s = \angle \left(-\frac{e^{-j0.4586\pi} - 0.38197}{1 - 0.38197e^{-j-0.38197}} \right) = 0.3\pi$$

as expected. Now we can design a digital lowpass filter and transform it into a highpass filter using the zmapping function to complete our design procedure. For designing a highpass Chebyshev-I digital filter, the above procedure can be incorporated into a MATLAB function called the cheb1hpf function shown below.

```
function [b,a] = cheb1hpf(wp,ws,Rp,As)
% IIR Highpass filter design using Chebyshev-1 prototype
% function [b,a] = cheb1hpf(wp,ws,Rp,As)
%    b = Numerator polynomial of the highpass filter
%    a = Denominator polynomial of the highpass filter
%    wp = Passband frequency in radians
%    ws = Stopband frequency in radians
%    Rp = Passband ripple in dB
%    As = Stopband attenuation in dB
%
% Determine the digital lowpass cutoff frequecies:
wplp = 0.2*pi;
alpha = -(cos((wplp+wp)/2))/(cos((wplp-wp)/2));
wslp = angle(-(exp(-j*ws)+alpha)/(1+alpha*exp(-j*ws)));
%
% Compute Analog lowpass Prototype Specifications:
T = 1; Fs = 1/T;
OmegaP = (2/T)*tan(wplp/2);
OmegaS = (2/T)*tan(wslp/2);

% Design Analog Chebyshev Prototype Lowpass Filter:
[cs,ds] = afd_chb1(OmegaP,OmegaS,Rp,As);

% Perform Bilinear transformation to obtain digital lowpass
[blp,alp] = bilinear(cs,ds,Fs);

% Transform digital lowpass into highpass filter
Nz = -[alpha,1]; Dz = [1,alpha];
[b,a] = zmapping(blp,alp,Nz,Dz);
```

We will demonstrate this procedure in the following example.

☐   **EXAMPLE 8.27**   Design a highpass digital filter to satisfy

$$\omega_p = 0.6\pi, \qquad R_p = 1 \text{ dB}$$

$$\omega_s = 0.4586\pi, \quad A_s = 15 \text{ dB}$$

Use the Chebyshev-I prototype.

```
>> % Digital Highpass Filter Specifications:
>> wp = 0.6*pi;                    % digital Passband freq in rad
>> ws = 0.4586*pi;                 % digital Stopband freq in rad
>> Rp = 1;                         % Passband ripple in dB
>> As = 15;                        % Stopband attenuation in dB

>> [b,a] = cheb1hpf(wp,ws,Rp,As);
>> [C,B,A] = dir2cas(b,a)
C = 0.0243
B = 1.0000   -2.0000    1.0000
    1.0000   -2.0000    1.0000
A = 1.0000    1.0416    0.4019
    1.0000    0.5561    0.7647
```

The system function is

$$H(z) = \frac{0.0243(1 - z^{-1})^4}{(1 + 0.5661z^{-1} + 0.7647z^{-2})(1 + 1.0416z^{-1} + 0.4019z^{-2})}$$

which is identical to that in Example 8.26.   □

The above highpass filter design procedure can be easily extended to other frequency-selective filters using the transformation functions in Table 8.2. These design procedures are explored in Problems 8.17 through 8.21. We now describe MATLAB's filter design functions for designing arbitrary frequency-selective filters.

MATLAB
IMPLEMEN-
TATION

In the preceding section we discussed four MATLAB functions to design digital lowpass filters. These same functions can also be used to design highpass, bandpass, and bandstop filters. The frequency-band transformations in these functions are done in the s-plane, that is, they use Approach-1 discussed on page 301. For the purpose of illustration we will use the function butter. It can be used with the following variations in its input arguments.

• [b,a] = BUTTER(N,wn,'high') designs an Nth-order *highpass* filter with digital 3-dB cutoff frequency wn in units of π.

• [b,a] = BUTTER(N,wn,)designs an order 2N *bandpass* filter if wn is a two-element vector, wn=[w1 w2], with 3-dB passband w1 < w < w2 in units of π.

• [b,a] = BUTTER(N,wn,'stop') is an order 2N *bandstop* filter if wn=[w1 w2] with 3-dB stopband w1 < w < w2 in units of π.

To design any frequency-selective Butterworth filter, we need to know the order N and the 3-dB cutoff frequency vector wn. In this chapter we described how to determine these parameters for lowpass filters. However, these calculations are more complicated for bandpass and bandstop filters. In their Signal Processing toolbox, MATLAB provides a function called `buttord` to compute these parameters. Given the specifications, ω_p, ω_s, R_p, and A_s, this function determines the necessary parameters. Its syntax is

```
[N,wn] = buttord(wp,ws,Rp,As)
```

The parameters `wp` and `ws` have some restrictions, depending on the type of filter:

- for lowpass filters `wp` < `ws`,
- for highpass filters `wp` > `ws`,
- for bandpass filters `wp` and `ws` are two-element vectors, `wp=[wp1, wp2]` and `ws=[ws1,ws2]`, such that `ws1` < `wp1` < `wp2` < `ws2`, and
- for bandstop filters `wp1` < `ws1` < `ws2` < `wp2`.

Now using the `buttord` function in conjunction with the `butter` function, we can design any Butterworth IIR filter. Similar discussions apply for `cheby1`, `cheby2`, and `ellip` functions with appropriate modifications. We illustrate the use of these functions through the following examples.

☐ **EXAMPLE 8.28** In this example we will design a Chebyshev-I highpass filter whose specifications were given in Example 8.27.

Solution **MATLAB Script** _____
```
>> % Digital Filter Specifications:      % Type: Chebyshev-I highpass
>> ws = 0.4586*pi;                       % Dig. stopband edge frequency
>> wp = 0.6*pi;                          % Dig. passband edge frequency
>> Rp = 1;                               % Passband ripple in dB
>> As = 15;                              % Stopband attenuation in dB

>> % Calculations of Chebyshev-I Filter Parameters:
>> [N,wn] = cheb1ord(wp/pi,ws/pi,Rp,As);

>> % Digital Chebyshev-I Highpass Filter Design:
>> [b,a] = cheby1(N,Rp,wn,'high');

>> % Cascade Form Realization:
>> [b0,B,A] = dir2cas(b,a)
b0 = 0.0243
B = 1.0000   -1.9991    0.9991
    1.0000   -2.0009    1.0009
A = 1.0000    1.0416    0.4019
    1.0000    0.5561    0.7647
```
_____

The cascade form system function

$$H(z) = \frac{0.0243(1 - z^{-1})^4}{(1 + 0.5661z^{-1} + 0.7647z^{-2})(1 + 1.0416z^{-1} + 0.4019z^{-2})}$$

is identical to the filter designed in Example 8.27, which demonstrates that the two approaches described on page 301 are identical. The frequency-domain plots are shown in Figure 8.22. □

□ **EXAMPLE 8.29** In this example we will design an elliptic bandpass filter whose specifications are given in the following MATLAB script:

```
>> % Digital Filter Specifications:    % Type: Elliptic Bandpass
>> ws = [0.3*pi 0.75*pi];              % Dig. stopband edge frequency
>> wp = [0.4*pi 0.6*pi];               % Dig. passband edge frequency
>> Rp = 1;                             % Passband ripple in dB
>> As = 40;                            % Stopband attenuation in dB

>> % Calculations of Elliptic Filter Parameters:
>> [N,wn] = ellipord(wp/pi,ws/pi,Rp,As);

>> % Digital Elliptic Bandpass Filter Design:
>> [b,a] = ellip(N,Rp,As,wn);
```

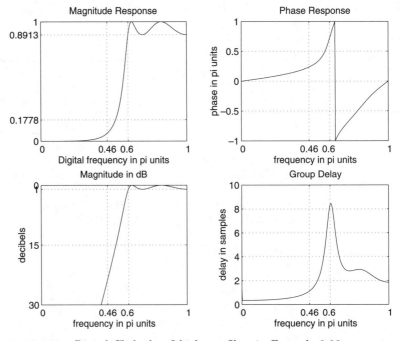

FIGURE 8.22 *Digital Chebyshev-I highpass filter in Example 8.28*

```
>> % Cascade Form Realization:
>> [b0,B,A] = dir2cas(b,a)
b0 = 0.0197
B = 1.0000    1.5066    1.0000
    1.0000    0.9268    1.0000
    1.0000   -0.9268    1.0000
    1.0000   -1.5066    1.0000
A = 1.0000    0.5963    0.9399
    1.0000    0.2774    0.7929
    1.0000   -0.2774    0.7929
    1.0000   -0.5963    0.9399
```

Note that the designed filter is an 8th-order filter. The frequency-domain plots are shown in Figure 8.23. □

□ **EXAMPLE 8.30**   Finally, we will design a Chebyshev-II bandstop filter whose specifications are given in the following MATLAB script.

```
>> % Digital Filter Specifications:     % Type: Chebyshev-II Bandstop
>> ws = [0.4*pi 0.7*pi];                % Dig. stopband edge frequency
>> wp = [0.25*pi 0.8*pi];               % Dig. passband edge frequency
>> Rp = 1;                              % Passband ripple in dB
>> As = 40;                             % Stopband attenuation in dB
```

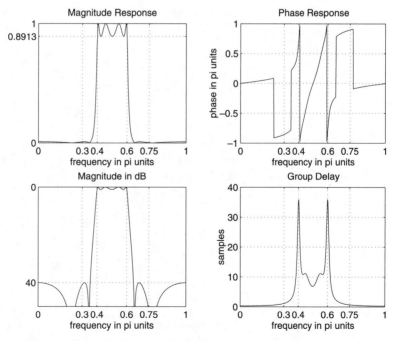

FIGURE 8.23   *Digital elliptic bandpass filter in Example 8.29*

```
>> % Calculations of Chebyshev-II Filter Parameters:
>> [N,wn] = cheb2ord(wp/pi,ws/pi,Rp,As);

>> % Digital Chebyshev-II Bandstop Filter Design:
>> [b,a] = cheby2(N,As,wn,'stop');

>> % Cascade Form Realization:
>> [b0,B,A] = dir2cas(b,a)
b0 = 0.1558
B = 1.0000    1.1456    1.0000
    1.0000    0.8879    1.0000
    1.0000    0.3511    1.0000
    1.0000   -0.2434    1.0000
    1.0000   -0.5768    1.0000
A = 1.0000    1.3041    0.8031
    1.0000    0.8901    0.4614
    1.0000    0.2132    0.2145
    1.0000   -0.4713    0.3916
    1.0000   -0.8936    0.7602
```

This is also a 10th-order filter. The frequency domain plots are shown in Figure 8.24. □

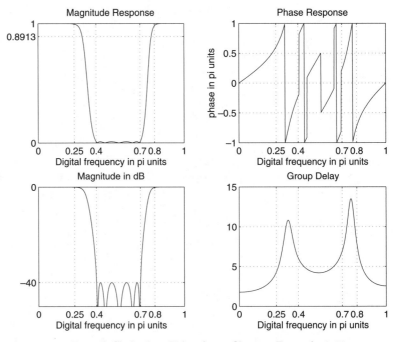

FIGURE 8.24 *Digital Chebyshev-II bandstop filter in Example 8.30*

COMPARISON OF FIR VS. IIR FILTERS

So far we have seen many techniques for designing both FIR and IIR filters. In practice one wonders about which filter (FIR or IIR) should be chosen for a given application and which method should be used to design it. Because these design techniques involve different methodologies, it is difficult to compare them. However, some meaningful comparisons can be attempted if we focus on the minimax optimal (or equiripple) filters. In the case of FIR filters these optimal filters are the equiripple filters designed via the Parks-McClellan algorithm (or Remez Exchange Algorithm), while in the case of IIR filters these are the elliptic filters.

One basis of comparison is the number of multiplications required to compute one output sample in the standard realization of these filters. For FIR filters the standard realization is the linear-phase direct form, while for elliptic filters cascade forms are widely used. Let M be the *length* of a linear phase FIR filter (assume M odd). Then we need

$$\frac{M+1}{2} \simeq \frac{M}{2} \quad \text{for large } M$$

multiplications per output sample. Let N (assume N even) be the *order* of an elliptic filter with the cascade form realization. Then there are $N/2$ second-order sections, each requiring 3 multiplications (in the most efficient implementation). There are an additional three multiplications in the overall structure for a total of

$$3\frac{N}{2} + 3 \simeq \frac{3N}{2} \quad \text{(for large } N)$$

multiplications per output sample.

Now if we assume that each filter meets exactly the same specifications: (e.g., ω_p, ω_s, δ_1 (or passband ripple R_p), and δ_2 (or stopband attenuation A_s) for a lowpass filter), then these two filters are equivalent if

$$\frac{M+1}{2} = \frac{3N+3}{2} \implies \frac{M}{N} = 3 + \frac{1}{N} \simeq 3 \quad \text{for large } N$$

This means that if the ratio $M/N = 3$, then two filters are roughly efficient. However, an equiripple FIR filter is more efficient if $M/N < 3$, or an elliptic IIR filter is more efficient if $M/N > 3$.

It has been shown experimentally that

- for $\omega_p \geq 0.3$, $M/N \geq 3$ for all δ_1, δ_2, N
- for $N \geq 10$, $M/N \geq 3$ for all δ_1, δ_2, N
- for large N, $M/N \approx$ in 100's.

This shows that for most applications IIR elliptic filters are desirable from the computational point of view. The most favorable conditions for FIR filters are

- large values of δ_1,
- small values of δ_2, and
- large transition width.

Furthermore, if we take into account the phase equalizers (which are all-pass filters) connected in cascade with elliptic filters that are needed for linear-phase characteristics, then FIR equiripple filter designs look good because of their exact linear-phase characteristics.

PROBLEMS

P8.1 Design an analog Butterworth lowpass filter that has a 1-dB or better ripple at 30 rad/sec and at least 30 dB of attenuation at 40 rad/sec. Determine the system function in a cascade form. Plot the magnitude response, the log-magnitude response in dB, the phase response, and the impulse response of the filter.

P8.2 Design a lowpass analog elliptic filter with the following characteristics:
- an acceptable passband ripple of 1 dB,
- passband cutoff frequency of 10 rad/sec, and
- stopband attenuation of 40 dB or greater beyond 15 rad/sec.

Determine the system function in a rational function form. Plot the magnitude response, the log-magnitude response in dB, the phase response, and the impulse response of the filter.

P8.3 A signal $x_a(t)$ contains two frequencies, 100 Hz and 130 Hz. We want to suppress the 130-Hz component to 50-dB attenuation while passing the 100-Hz component with less than 2-dB attenuation. Design a minimum-order Chebyshev-I analog filter to perform this filtering operation. Plot the log-magnitude response and verify the design.

P8.4 Design an analog Chebyshev-II lowpass filter that has a 0.5 dB or better ripple at 250 Hz and at least 45 dB of attenuation at 300 Hz. Plot the magnitude response, the log-magnitude response in dB, the phase response, and the impulse response of the filter.

P8.5 Write a MATLAB function to design analog lowpass filters. The format of this function should be

```
function [b,a] =afd(type,Fp,Fs,Rp,As)
%
% function [b,a] =afd(type,Fp,Fs,Rp,As)
%    Designs analog lowpass filters
% type = 'butter' or 'cheby1' or 'cheby2' or 'ellip'
%    Fp = passband cutoff in Hz
%    Fs = stopband cutoff in Hz
%    Rp = passband ripple in dB
%    As = stopband attenuation in dB
```

Use the `afd_butt`, `afd_chb1`, `afd_chb2`, and `afd_elip` functions developed in this chapter. Check your function on specifications given in Problems 8.1 through 8.4.

P8.6 Design a lowpass digital filter to be used in a structure

$$x_a(t) \longrightarrow \boxed{\text{A/D}} \longrightarrow \boxed{H(z)} \longrightarrow \boxed{\text{D/A}} \longrightarrow y_a(t)$$

to satisfy the following requirements:

- sampling rate of 8000 sam/sec,
- passband edge of 1500 Hz with ripple of 3dB,
- stopband edge of 2000 Hz with attenuation of 40 dB,
- equiripple passband but monotone stopband, and
- impulse invariance method.

a. Choose $T = 1$ in the impulse invariance method and determine the system function $H(z)$ in parallel form. Plot the log-magnitude response in dB and the impulse response $h(n)$.

b. Choose $T = 1/8000$ in the impulse invariance method and determine the system function $H(z)$ in parallel form. Plot the log-magnitude response in dB and the impulse response $h(n)$. Compare this design with the above one and comment on the effect of T on the impulse invariance design.

P8.7 Design a Butterworth digital lowpass filter to satisfy these specifications:

$$\text{passband edge:} \quad 0.4\pi, \quad R_p = 0.5 \text{ dB}$$

$$\text{stopband edge:} \quad 0.6\pi, \quad As = 50 \text{ dB}$$

Use the impulse invariance method with $T = 2$. Determine the system function in the rational form and plot the log-magnitude response in dB. Plot the impulse response $h(n)$ and the impulse response $h_a(t)$ of the analog prototype and compare their shapes.

P8.8 Write a MATLAB function to design digital lowpass filters based on the impulse invariance transformation. The format of this function should be

```
function [b,a] =dlpfd_ii(type,wp,ws,Rp,As,T)
%
% function [b,a] =dlpfd_ii(type,wp,ws,Rp,As,T)
%   Designs digital lowpass filters using impulse invariance
% type = 'butter' or 'cheby1'
%   wp = passband cutoff in Hz
%   ws = stopband cutoff in Hz
%   Rp = passband ripple in dB
%   As = stopband attenuation in dB
%    T = sampling interval
```

Use the `afd` function developed in Problem 8.5. Check your function on specifications given in Problems 8.6 and 8.7.

P8.9 In this problem we will develop a technique called the *step invariance* transformation. In this technique the step response of an analog prototype filter is preserved in the resulting digital filter; that is, if $\xi_a(t)$ is the step response of the prototype and if $\xi(n)$ is the step response of the digital filter, then

$$\xi(n) = \xi_a(t = nT), \quad T : \text{sampling interval}$$

Note that the frequency-domain quantities are related by

$$\Xi_a(s) \triangleq \mathcal{L}[\xi_a(t)] = H_a(s)/s$$

and

$$\Xi(z) \triangleq \mathcal{Z}[\xi(n)] = H(z)\frac{1}{1-z^{-1}}$$

Hence the step invariance transformation steps are as follows: Given $H_a(s)$,

- Divide $H_a(s)$ by s to obtain $\Xi_a(s)$.
- Find residues $\{R_k\}$ and poles $\{p_k\}$ of $\Xi_a(s)$.
- Transform analog poles $\{p_k\}$ into digital poles $\{e^{p_k T}\}$, where T is arbitrary.
- Determine $\Xi(z)$ from residues $\{R_k\}$ and poles $\{e^{p_k T}\}$.
- Determine $H(z)$ by multiplying $\Xi(z)$ by $(1-z^{-1})$.

Use the above procedure to develop a MATLAB function to implement the step invariance transformation. The format of this function should be

```
function [b,a] =stp_invr(c,d,T)

% Step Invariance Transformation from Analog to Digital Filter
%  [b,a] =stp_invr(c,d,T)
%  b = Numerator polynomial in z^(-1) of the digital filter
%  a = Denominator polynomial in z^(-1) of the digital filter
%  c = Numerator polynomial in s of the analog filter
%  d = Denominator polynomial in s of the analog filter
%  T = Sampling (transformation) parameter
```

P8.10 Design the lowpass Butterworth digital filter of Problem 8.7 using the step invariance method. Plot the log-magnitude response in dB and compare it with that in Problem 8.7. Plot the step response $\xi(n)$ and the impulse response $\xi_a(t)$ of the analog prototype and compare their shapes.

P8.11 Consider the design of the lowpass Butterworth filter of Problem 8.7.

a. Use the bilinear transformation technique outlined in this chapter and the **bilinear** function. Plot the log-magnitude response in dB. Compare the impulse responses of the analog prototype and the digital filter.

b. Use the **butter** function and compare this design with the above one.

P8.12 Design the analog Chebyshev-I filter of Problem 8.6 using the bilinear transformation method. Compare the two designs.

P8.13 Design a digital lowpass filter using elliptic prototype to satisfy these requirements:

$$\text{passband edge:} \quad 0.4\pi, \quad R_p = 1 \text{ dB}$$

$$\text{stopband edge:} \quad 0.5\pi, \quad As = 60 \text{ dB}$$

Use the **bilinear** as well as the **ellip** function and compare your designs.

P8.14 Design a digital lowpass filter to satisfy these specifications:

$$\text{passband edge:} \quad 0.3\pi, \quad R_p = 0.5 \text{ dB}$$

$$\text{stopband edge:} \quad 0.4\pi, \quad As = 50 \text{ dB}$$

a. Use the `butter` function and determine the order N and the actual minimum stopband attenuation in dB.

b. Use the `cheby1` function and determine the order N and the actual minimum stopband attenuation in dB.

c. Use the `cheby2` function and determine the order N and the actual minimum stopband attenuation in dB.

d. Use the `ellip` function and determine the order N and the actual minimum stopband attenuation in dB.

e. Compare the orders, the actual minimum stopband attenuations, and the group delays in each of the above designs.

P8.15 Write a MATLAB function to determine the lowpass prototype digital filter frequencies from a highpass digital filter's specifications using the procedure outlined in this chapter. The format of this function should be

```
function [wpLP,wsLP,alpha] = hp2lpfre(wphp,wshp)
% Band-edge frequency conversion from highpass to lowpass digital filter
% [wpLP,wsLP,a] = hp2lpfre(wphp,wshp)
%  wpLP = passband edge for the lowpass prototype
%  wsLP = stopband edge for the lowpass prototype
% alpha = lowpass to highpass transformation parameter
%  wphp = passband edge for the highpass
%  wshp = stopband edge for the highpass
```

Using this function, develop a MATLAB function to design a highpass digital filter using the bilinear transformation. The format of this function should be

```
function [b,a] = dhpfd_bl(type,wp,ws,Rp,As)
% IIR Highpass filter design using bilinear transformation
% [b,a] = dhpfd_bl(type,wp,ws,Rp,As)
% type = 'butter' or 'cheby1' or 'chevy2' or 'ellip'
%    b = Numerator polynomial of the highpass filter
%    a = Denominator polynomial of the highpass filter
%   wp = Passband frequency in radians
%   ws = Stopband frequency in radians (wp < ws)

%   Rp = Passband ripple in dB
%   As = Stopband attenuation in dB
```

Verify your function using the specifications in Example 8.27.

P8.16 Design a highpass filter to satisfy these specifications:

$$\text{stopband edge:} \quad 0.4\pi, \quad A_s = 60 \text{ dB}$$

$$\text{passband edge:} \quad 0.6\pi, \quad R_p = 0.5 \text{ dB}$$

a. Use the `dhpfd_bl` function of Problem 8.15 and the elliptic prototype to design this filter. Plot the log-magnitude response in dB of the designed filter.

b. Use the `ellip` function for design and plot the log-magnitude response in dB. Compare these two designs.

P8.17 Write a MATLAB function to determine the lowpass prototype digital filter frequencies from an arbitrary lowpass digital filter's specifications using the functions given in Table 8.2 and the procedure outlined for highpass filters. The format of this function should be

```
function [wpLP,wsLP,alpha] = lp2lpfre(wplp,wslp)
% Band-edge frequency conversion from lowpass to lowpass digital filter
% [wpLP,wsLP,a] = lp2lpfre(wplp,wslp)
%  wpLP = passband edge for the lowpass prototype
%  wsLP = stopband edge for the lowpass prototype
% alpha = lowpass to highpass transformation parameter
%  wplp = passband edge for the lowpass
%  wslp = stopband edge for the lowpass
```

Using this function, develop a MATLAB function to design a bandpass filter from a prototype lowpass digital filter using the bilinear transformation. The format of this function should be

```
function [b,a] = dbpfd_bl(type,wp,ws,Rp,As)
% IIR bandpass filter design using bilinear transformation
% [b,a] = dbpfd_bl(type,wp,ws,Rp,As)
% type = 'butter' or 'cheby1' or 'chevy2' or 'ellip'
%    b = Numerator polynomial of the bandpass filter
%    a = Denominator polynomial of the bandpass filter
%   wp = Passband frequency vector [wp_lower, wp_upper] in radians
%   ws = Stopband frequency vector [wp_lower, wp_upper] in radians
%   Rp = Passband ripple in dB
%   As = Stopband attenuation in dB
```

Verify your function using the designs in Problem 8.14.

P8.18 Design a bandpass digital filter using the `Cheby2` function. The specifications are

$$
\begin{array}{ll}
\text{lower stopband edge:} & 0.3\pi \\
\text{upper stopband edge:} & 0.6\pi
\end{array} \quad A_s = 50 \text{ dB}
$$

$$
\begin{array}{ll}
\text{lower passband edge:} & 0.4\pi \\
\text{upper passband edge:} & 0.5\pi
\end{array} \quad R_p = 0.5 \text{ dB}
$$

Plot the impulse response and the log-magnitude response in dB of the designed filter.

P8.19 Write a MATLAB function to determine the lowpass prototype digital filter frequencies from a bandpass digital filter's specifications using the functions given in Table 8.2 and the procedure outlined for highpass filters. The format of this function should be

```
function [wpLP,wsLP,alpha] = bp2lpfre(wpbp,wsblp)
% Band-edge frequency conversion from bandpass to lowpass digital filter
% [wpLP,wsLP,a] = bp2lpfre(wpbp,wsbp)
```

```
%  wpLP = passband edge for the lowpass prototype
%  wsLP = stopband edge for the lowpass prototype
% alpha = lowpass to highpass transformation parameter
%  wpbp = passband edge frequency vector [wp_lower, wp_upper] for the bandpass
%  wsbp = stopband edge frequency vector [ws_lower, ws_upper] for the bandpass
```

Using this function, develop a MATLAB function to design a bandpass filter from a prototype lowpass digital filter using the bilinear transformation. The format of this function should be

```
function [b,a] = dbpfd_bl(type,wp,ws,Rp,As)
% IIR bandpass filter design using bilinear transformation
% [b,a] = dbpfd_bl(type,wp,ws,Rp,As)
% type = 'butter' or 'cheby1' or 'chevy2' or 'ellip'
%     b = Numerator polynomial of the bandpass filter
%     a = Denominator polynomial of the bandpass filter
%    wp = Passband frequency vector [wp_lower, wp_upper] in radians
%    ws = Stopband frequency vector [wp_lower, wp_upper] in radians
%    Rp = Passband ripple in dB
%    As = Stopband attenuation in dB
```

Verify your function using the design in Problem 8.18.

P8.20 We wish to use the Chebyshev-I prototype to design a bandpass IIR digital filter that meets the following specifications:

$$0.95 \leq \left| H\left(e^{j\omega}\right) \right| \leq 1.05, \qquad 0 \leq |\omega| \leq 0.25\pi$$

$$0 \leq \left| H\left(e^{j\omega}\right) \right| \leq 0.01, \quad 0.35\pi \leq |\omega| \leq 0.65\pi$$

$$0.95 \leq \left| H\left(e^{j\omega}\right) \right| \leq 1.05, \quad 0.75\pi \leq |\omega| \leq \pi$$

Use the **cheby1** function and determine the system function $H(z)$ of such a filter. Provide a plot containing subplots of the log-magnitude response in dB and the impulse response.

P8.21 Write a MATLAB function to determine the lowpass prototype digital filter frequencies from a bandstop digital filter's specifications using the functions given in Table 8.2 and the procedure outlined for highpass filters. The format of this function should be

```
function [wpLP,wsLP,alpha] = bs2lpfre(wpbp,wsblp)
% Band-edge frequency conversion from bandstop to lowpass digital filter
% [wpLP,wsLP,a] = bs2lpfre(wpbp,wsbp)
%  wpLP = passband edge for the lowpass prototype
%  wsLP = stopband edge for the lowpass prototype
% alpha = lowpass to highpass transformation parameter
%  wpbp = passband edge frequency vector [wp_lower, wp_upper] for the bandstop
%  wsbp = stopband edge frequency vector [ws_lower, ws_upper] for the bandstop
```

Using this function, develop a MATLAB function to design a bandstop filter from a prototype lowpass digital filter using the bilinear transformation. The format of this function should be

```
function [b,a] = dbsfd_bl(type,wp,ws,Rp,As)
% IIR bandstop filter design using bilinear transformation
% [b,a] = dbsfd_bl(type,wp,ws,Rp,As)
% type = 'butter' or 'cheby1' or 'chevy2' or 'ellip'
%    b = Numerator polynomial of the bandstop filter
%    a = Denominator polynomial of the bandstop filter
%   wp = Passband frequency vector [wp_lower, wp_upper] in radians
%   ws = Stopband frequency vector [wp_lower, wp_upper] in radians
%   Rp = Passband ripple in dB
%   As = Stopband attenuation in dB
```

Verify your function using the design in Problem 8.20.

P8.22 An analog signal $x_a(t) = 5\sin(200\pi t) + 2\cos(300\pi t)$ is to be processed by a

$$x_a(t) \longrightarrow \boxed{\text{A/D}} \longrightarrow \boxed{H(z)} \longrightarrow \boxed{\text{D/A}} \longrightarrow y_a(t)$$

system in which the sampling frequency is 1000 sam/sec.

a. Design a minimum-order IIR digital filter that will pass the 150-Hz component with attenuation of less than 1 dB and suppress the 100-Hz component to at least 40 dB. The filter should have a monotone passband and an equiripple stopband. Determine the system function in rational function form and plot the log-magnitude response.

b. Generate 300 samples (sampled at 1000 sam/sec) of the above signal $x_a(t)$ and process through the designed filter to obtain the output sequence. Interpolate this sequence (using any one of the interpolating techniques discussed in Chapter 3) to obtain $y_a(t)$. Plot the input and the output signals and comment on your results.

P8.23 Using the bilinear transformation method, design a tenth-order elliptic bandstop filter to remove the digital frequency $\omega = 0.44\pi$ with bandwidth of 0.08π. Choose a reasonable value for the stopband attenuation. Plot the magnitude response. Generate 201 samples of the sequence

$$x(n) = \sin[0.44\pi n], \quad n = 0, \ldots, 200$$

and process thorough the bandstop filter. Comment on your results.

P8.24 Design a digital highpass filter $H(z)$ to be used in a

$$x_a(t) \longrightarrow \boxed{\text{A/D}} \longrightarrow \boxed{H(z)} \longrightarrow \boxed{\text{D/A}} \longrightarrow y_a(t)$$

structure to satisfy the following requirements

- sampling rate of 10 Khz,
- stopband edge of 1.5 Khz with attenuation of 40 dB,
- passband edge of 2 Khz with ripple of 3dB,
- monotone passband and stopband, and
- impulse invariance transformation method.

a. Plot magnitude response of the overall analog filter over the $[0, 5\,\text{Khz}]$ interval.

b. Plot the magnitude response of the digital lowpass prototype.

c. Plot the magnitude response of the analog lowpass prototype.

d. What limitations must be placed on the input signals so that the above structure truly acts like a highpass filter to them?

9

APPLICATIONS IN ADAPTIVE FILTERING

In Chapters 7 and 8 we described methods for designing FIR and IIR digital filters to satisfy some desired specifications. Our goal was to determine the coefficients of the digital filter that met the desired specifications.

In contrast to the filter design techniques considered in those two chapters, there are many digital signal processing applications in which the filter coefficients cannot be specified a priori. For example, let us consider a high-speed modem that is designed to transmit data over telephone channels. Such a modem employs a filter called a channel equalizer to compensate for the channel distortion. The modem must effectively transmit data through communication channels that have different frequency response characteristics and hence result in different distortion effects. The only way in which this is possible is if the channel equalizer has *adjustable coefficients* that can be optimized to minimize some measure of the distortion, on the basis of measurements performed on the characteristics of the channel. Such a filter with adjustable parameters is called an *adaptive filter*, in this case an *adaptive equalizer*.

Numerous applications of adaptive filters have been described in the literature. Some of the more noteworthy applications include (1) adaptive antenna systems, in which adaptive filters are used for beam steering and for providing nulls in the beam pattern to remove undesired interference [23]; (2) digital communication receivers, in which adaptive filters are used to provide equalization of intersymbol interference and for channel identification [18]; (3) adaptive noise canceling techniques, in which an adaptive filter is used to estimate and eliminate a noise component in some desired signal [22, 10, 14]; and (4) system modeling, in which an adaptive filter is used as a model to estimate the characteristics of an unknown system. These are just a few of the best known examples on the use of adaptive filters.

Although both IIR and FIR filters have been considered for adaptive filtering, the FIR filter is by far the most practical and widely used. The reason for this preference is quite simple. The FIR filter has only adjustable zeros, and hence it is free of stability problems associated with adaptive IIR filters that have adjustable poles as well as zeros. We should not conclude, however, that adaptive FIR filters are always stable. On the contrary, the stability of the filter depends critically on the algorithm for adjusting its coefficients.

Of the various FIR filter structures that we may use, the direct form and the lattice form are the ones often used in adaptive filtering applications. The direct form FIR filter structure with adjustable coefficients $h(0), h(1), \ldots, h(N-1)$ is illustrated in Figure 9.1. On the other hand, the adjustable parameters in an FIR lattice structure are the reflection coefficients K_n shown in Figure 6.18.

An important consideration in the use of an adaptive filter is the criterion for optimizing the adjustable filter parameters. The criterion must not only provide a meaningful measure of filter performance, but it must also result in a practically realizable algorithm.

One criterion that provides a good measure of performance in adaptive filtering applications is the least-squares criterion, and its counterpart in a statistical formulation of the problem, namely, the mean-square-error (MSE) criterion. The least squares (and MSE) criterion results in a quadratic performance index as a function of the filter coefficients, and hence it possesses a single minimum. The resulting algorithms for adjusting the coefficients of the filter are relatively easy to implement.

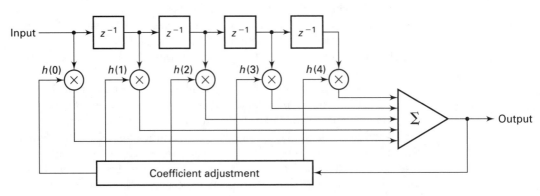

FIGURE 9.1 *Direct form adaptive FIR filter*

In this chapter we describe a basic algorithm, called the *least-mean-square* (LMS) *algorithm*, to adaptively adjust the coefficients of an FIR filter. The adaptive filter structure that will be implemented is the direct form FIR filter structure with adjustable coefficients $h(0), h(1), \ldots,$ $h(N-1)$, as illustrated in Figure 9.1. After we describe the LMS algorithm, we apply it to several practical systems in which adaptive filters are employed.

LMS ALGORITHM FOR COEFFICIENT ADJUSTMENT

Suppose we have an FIR filter with adjustable coefficients $\{h(k), 0 \leq k \leq N-1\}$. Let $\{x(n)\}$ denote the input sequence to the filter, and let the corresponding output be $\{y(n)\}$, where

$$y(n) = \sum_{k=0}^{N-1} h(k) x(n-k), \quad n = 0, \ldots, M \tag{9.1}$$

Suppose that we also have a desired sequence $\{d(n)\}$ with which we can compare the FIR filter output. Then we can form the error sequence $\{e(n)\}$ by taking the difference between $d(n)$ and $y(n)$. That is,

$$e(n) = d(n) - y(n), \quad n = 0, \ldots, M \tag{9.2}$$

The coefficients of the FIR filter will be selected to minimize the sum of squared errors. Thus we have

$$\mathcal{E} = \sum_{n=0}^{M} e^2(n) = \sum_{n=0}^{M} \left[d(n) - \sum_{k=0}^{N-1} h(k) x(n-k) \right] \tag{9.3}$$

$$= \sum_{n=0}^{M} d^2(n) - 2 \sum_{k=0}^{N-1} h(k) r_{dx}(k) + \sum_{k=0}^{N-1} \sum_{\ell=0}^{N-1} h(k) h(\ell) r_{xx}(k-\ell)$$

where, by definition,

$$r_{dx}(k) = \sum_{n=0}^{M} d(n) x(n-k), \quad 0 \leq k \leq N-1 \tag{9.4}$$

$$r_{xx}(k) = \sum_{n=0}^{M} x(n) x(n+k), \quad 0 \leq k \leq N-1 \tag{9.5}$$

We call $\{r_{dx}(k)\}$ the crosscorrelation between the desired output sequence $\{d(n)\}$ and the input sequence $\{x(n)\}$, and $\{r_{xx}(k)\}$ is the autocorrelation sequence of $\{x(n)\}$.

The sum of squared errors $\mathcal{E}$ is a quadratic function of the FIR filter coefficients. Consequently, the minimization of $\mathcal{E}$ with respect to the filter coefficients $\{h(k)\}$ results in a set of linear equations. By differentiating $\mathcal{E}$ with respect to each of the filter coefficients, we obtain

$$\frac{\partial \mathcal{E}}{\partial h(m)} = 0, \quad 0 \leq m \leq N - 1 \tag{9.6}$$

and, hence

$$\sum_{k=0}^{N-1} h(k)\, r_{xx}(k - m) = r_{dx}(m), \quad 0 \leq m \leq N - 1 \tag{9.7}$$

This is the set of linear equations that yield the optimum filter coefficients.

To solve the set of linear equations directly, we must first compute the autocorrelation sequence $\{r_{xx}(k)\}$ of the input signal and the cross-correlation sequence $\{r_{dx}(k)\}$ between the desired sequence $\{d(n)\}$ and the input sequence $\{x(n)\}$.

The LMS algorithm provides an alternative computational method for determining the optimum filter coefficients $\{h(k)\}$ without explicitly computing the correlation sequences $\{r_{xx}(k)\}$ and $\{r_{dx}(k)\}$. The algorithm is basically a recursive gradient (steepest-descent) method that finds the minimum of $\mathcal{E}$ and thus yields the set of optimum filter coefficients.

We begin with any arbitrary choice for the initial values of $\{h(k)\}$, say $\{h_0(k)\}$. For example, we may begin with $h_0(k) = 0, \quad 0 \leq k \leq N-1$. Then after each new input sample $\{x(n)\}$ enters the adaptive FIR filter, we compute the corresponding output, say $\{y(n)\}$, form the error signal $e(n) = d(n) - y(n)$, and update the filter coefficients according to the equation

$$h_n(k) = h_{n-1}(k) + \triangle \cdot e(n) \cdot x(n - k), \quad 0 \leq k \leq N - 1, \quad n = 0, 1, \dots \tag{9.8}$$

where $\triangle$ is called the step size parameter, $x(n-k)$ is the sample of the input signal located at the kth tap of the filter at time n, and $e(n)\, x(n - k)$ is an approximation (estimate) of the negative of the gradient for the kth filter coefficient. This is the LMS recursive algorithm for adjusting the filter coefficients adaptively so as to minimize the sum of squared errors $\mathcal{E}$.

The step size parameter $\triangle$ controls the rate of convergence of the algorithm to the optimum solution. A large value of $\triangle$ leads to large step size adjustments and thus to rapid convergence, while a small value of $\triangle$ results in slower convergence. However, if $\triangle$ is made too large the algorithm becomes unstable. To ensure stability, $\triangle$ must be chosen [18]

Chapter 9 ■ APPLICATIONS IN ADAPTIVE FILTERING

to be in the range

$$0 < \triangle < \frac{1}{10NP_x} \qquad (9.9)$$

where N is the length of the adaptive FIR filter and P_x is the power in the input signal, which can be approximated by

$$P_x \approx \frac{1}{1+M} \sum_{n=0}^{M} x^2(n) = \frac{r_{xx}(0)}{M+1} \qquad (9.10)$$

The mathematical justification of equations (9.9) and (9.10) and the proof that the LMS algorithm leads to the solution for the optimum filter coefficients is given in more advanced treatments of adaptive filters. The interested reader may refer to the books by Haykin [9] and Proakis [18].

MATLAB IMPLEMEN-TATION

The LMS algorithm (9.8) can easily be implemented in MATLAB. Given the input sequence $\{x(n)\}$, the desired sequence $\{d(n)\}$, step size $\triangle$, and the desired length of the adaptive FIR filter N, we can use (9.1), (9.2), and (9.8) to determine the adaptive filter coefficients $\{h(n), 0 \le n \le N-1\}$ recursively. This is shown in the following function called lms.

```
function [h,y] = lms(x,d,delta,N)
% LMS Algorithm for Coefficient Adjustment
% ----------------------------------------
% [h,y] = lms(x,d,delta,N)
%      h = estimated FIR filter
%      y = output array y(n)
%      x = input array x(n)
%      d = desired array d(n), length must be same as x
% delta = step size
%      N = length of the FIR filter
%
M = length(x); y = zeros(1,M);
h = zeros(1,N);
for n = N:M
    x1 = x(n:-1:n-N+1);
     y = h * x1';
     e = d(n) - y;
     h = h + delta*e*x1;
end
```

In addition, the lms function provides the output $\{y(n)\}$ of the adaptive filter.

Below, we apply the LMS algorithm to several practical applications involving adaptive filtering.

SYSTEM IDENTIFICATION OR SYSTEM MODELING

To formulate the problem, let us refer to Figure 9.2. We have an unknown linear system that we wish to identify. The unknown system may be an all-zero (FIR) system or a pole-zero (IIR) system. The unknown system will be approximated (modeled) by an FIR filter of length N. Both the unknown system and the FIR model are connected in parallel and are excited by the same input sequence $\{x(n)\}$. If $\{y(n)\}$ denotes the output of the model and $\{d(n)\}$ denotes the output of the unknown system, the error sequence is $\{e(n) = d(n) - y(n)\}$. If we minimize the sum of squared errors, we obtain the same set of linear equations as in (9.7). Therefore the LMS algorithm given by (9.8) may be used to adapt the coefficients of the FIR model so that its output approximates the output of the unknown system.

PROJECT 9.1:
SYSTEM
IDENTIFI-
CATION

There are three basic modules that are needed to perform this project.

1. A noise signal generator that generates a sequence of random numbers with zero mean value. For example, we may generate a sequence of uniformly distributed random numbers over the interval $[-a, a]$. Such a sequence of uniformly distributed numbers has an average value of zero and a variance of $a^2/3$. This signal sequence, call it $\{x(n)\}$, will be used as the input to the unknown system and the adaptive FIR model. In this case the input signal $\{x(n)\}$ has power $P_x = a^2/3$. In MATLAB this can be implemented using the **rand** function.

2. An unknown system module that may be selected is an IIR filter and implemented by its difference equation. For example, we may select an IIR filter specified by the second-order difference equation

$$d(n) = a_1 d(n-1) + a_2 d(n-2) + x(n) + b_1 x(n-1) + b_2 x(n-2) \quad \textbf{(9.11)}$$

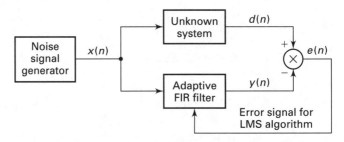

FIGURE 9.2 *Block diagram of system identification or system modeling problem*

where the parameters $\{a_1, a_2\}$ determine the positions of the poles and $\{b_1, b_2\}$ determine the positions of the zeros of the filter. These parameters are input variables to the program. This can be implemented by the `filter` function.

3. An adaptive FIR filter module where the FIR filter has N tap coefficients that are adjusted by means of the LMS algorithm. The length N of the filter is an input variable to the program. This can be implemented using the `lms` function given in the previous section.

The three modules are configured as shown in Figure 9.2. From this project we can determine how closely the impulse response of the FIR model approximates the impulse response of the unknown system after the LMS algorithm has converged.

To monitor the convergence rate of the LMS algorithm, we may compute a short-term average of the squared error $e^2(n)$ and plot it. That is, we may compute

$$\text{ASE}(m) = \frac{1}{K} \sum_{k=n+1}^{n+K} e^2(k) \tag{9.12}$$

where $m = n/K = 1, 2, \ldots$. The averaging interval K may be selected to be (approximately) $K = 10N$. The effect of the choice of the step size parameter $\triangle$ on the convergence rate of the LMS algorithm may be observed by monitoring the $\text{ASE}(m)$.

Besides the main part of the program, you should also include, as an aside, the computation of the impulse response of the unknown system, which can be obtained by exciting the system with a unit sample sequence $\delta(n)$. This actual impulse response can be compared with that of the FIR model after convergence of the LMS algorithm. The two impulse responses can be plotted for the purpose of comparison.

SUPPRESSION OF NARROWBAND INTERFERENCE IN A WIDEBAND SIGNAL

Let us assume that we have a signal sequence $\{x(n)\}$ that consists of a desired wideband signal sequence, say $\{w(n)\}$, corrupted by an additive narrowband interference sequence $\{s(n)\}$. The two sequences are uncorrelated. This problem arises in digital communications and in signal detection, where the desired signal sequence $\{w(n)\}$ is a spread-spectrum signal, while the narrowband interference represents a signal from another user of the frequency band or some intentional interference from a jammer who is trying to disrupt the communication or detection system.

From a filtering point of view, our objective is to design a filter that suppresses the narrowband interference. In effect, such a filter should place a notch in the frequency band occupied by the interference. In practice, however, the frequency band of the interference might be unknown. Moreover, the frequency band of the interference may vary slowly in time.

The narrowband characteristics of the interference allow us to estimate $s(n)$ from past samples of the sequence $x(n) = s(n) + w(n)$ and to subtract the estimate from $x(n)$. Since the bandwidth of $\{s(n)\}$ is narrow compared to the bandwidth of $\{w(n)\}$, the samples of $\{s(n)\}$ are highly correlated. On the other hand, the wideband sequence $\{w(n)\}$ has a relatively narrow correlation.

The general configuration of the interference suppression system is shown in Figure 9.3. The signal $x(n)$ is delayed by D samples, where the delay D is chosen sufficiently large so that the wideband signal components $w(n)$ and $w(n-D)$, which are contained in $x(n)$ and $x(n-D)$, respectively, are uncorrelated. The output of the adaptive FIR filter is the estimate

$$\hat{s}(n) = \sum_{k=0}^{N-1} h(k)x(n-k-D) \tag{9.13}$$

The error signal that is used in optimizing the FIR filter coefficients is $e(n) = x(n) - \hat{s}(n)$. The minimization of the sum of squared errors again leads to a set of linear equations for determining the optimum coefficients. Due to the delay D, the LMS algorithm for adjusting the coefficients recursively becomes

$$h_n(k) = h_{n-1}(k) + \triangle e(n)x(n-k-D)\ , \quad \begin{matrix} k = 0, 1, \ldots, N-1 \\ n = 1, 2, \ldots \end{matrix} \tag{9.14}$$

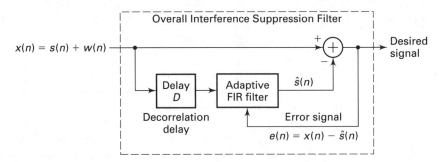

FIGURE 9.3 *Adaptive filter for estimating and suppressing a narrowband interference*

Chapter 9 ▪ APPLICATIONS IN ADAPTIVE FILTERING

PROJECT 9.2: SUPPRESSION OF SINUSOIDAL INTERFERENCE

There are three basic modules required to perform this project.

1. A noise signal generator module that generates a wideband sequence $\{w(n)\}$ of random numbers with zero mean value. In particular, we may generate a sequence of uniformly distributed random numbers using the **rand** function as previously described in the project on system identification. The signal power is denoted as P_w.

2. A sinusoidal signal generator module that generates a sine wave sequence $s(n) = A \sin \omega_0 n$, where $0 < \omega_0 < \pi$ and A is the signal amplitude. The power of the sinusoidal sequence is denoted as P_s.

3. An adaptive FIR filter module using the **lms** function, where the FIR filter has N tap coefficients that are adjusted by the LMS algorithm. The length N of the filter is an input variable to the program.

The three modules are configured as shown in Figure 9.4. In this project the delay $D = 1$ is sufficient, since the sequence $\{w(n)\}$ is a white noise (spectrally flat or uncorrelated) sequence. The objective is to adapt the FIR filter coefficients and then to investigate the characteristics of the adaptive filter.

It is interesting to select the interference signal to be much stronger than the desired signal $w(n)$, for example, $P_s = 10P_w$. Note that the power P_x required in selecting the step size parameter in the LMS algorithm is $P_x = P_s + P_w$. The frequency response characteristic $H(e^{j\omega})$ of the adaptive FIR filter with coefficients $\{h(k)\}$ should exhibit a resonant peak at the frequency of the interference. The frequency response of the interference suppression filter is $H_s(e^{j\omega}) = 1 - H(e^{j\omega})$, which should then exhibit a notch at the frequency of the interference.

It is interesting to plot the sequences $\{w(n)\}$, $\{s(n)\}$, and $\{x(n)\}$. It is also interesting to plot the frequency responses $H(e^{j\omega})$ and $H_s(e^{j\omega})$ after the LMS algorithm has converged. The short-time average squared error $\mathrm{ASE}(m)$, defined by (9.12), may be used to monitor the convergence characteristics of the LMS algorithm. The effect of the length of the adaptive filter on the quality of the estimate should be investigated.

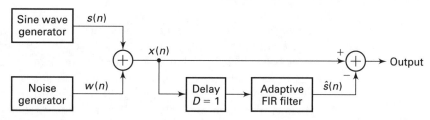

FIGURE 9.4 *Configuration of modules for experiment on interference suppression*

The project may be generalized by adding a second sinusoid of a different frequency. Then $H\left(e^{j\omega}\right)$ should exhibit two resonant peaks, provided the frequencies are sufficiently separated. Investigate the effect of the filter length N on the resolution of two closely spaced sinusoids.

ADAPTIVE LINE ENHANCEMENT

In the preceding section we described a method for suppressing a strong narrowband interference from a wideband signal. An adaptive line enhancer (ALE) has the same configuration as the interference suppression filter in Figure 9.3, except that the objective is different.

In the adaptive line enhancer, $\{s(n)\}$ is the desired signal and $\{w(n)\}$ represents a wideband noise component that masks $\{s(n)\}$. The desired signal $\{s(n)\}$ may be a spectral line (a pure sinusoid) or a relatively narrowband signal. Usually, the power in the wideband signal is greater than that in the narrowband signal—that is, $P_w > P_s$. It is apparent that the ALE is a self-tuning filter that has a peak in its frequency response at the frequency of the input sinusoid or in the frequency band occupied by the narrowband signal. By having a narrow bandwidth FIR filter, the noise outside the frequency band of the signal is suppressed, and thus the spectral line is enhanced in amplitude relative to the noise power in $\{w(n)\}$.

PROJECT 9.3:
ADAPTIVE
LINE
ENHANCEMENT

This project requires the same software modules as those used in the project on interference suppression. Hence the description given in the preceding section applies directly. One change is that in the ALE, the condition is that $P_w > P_s$. Secondly, the output signal from the ALE is $\{s(n)\}$. Repeat the project described in the previous section under these conditions.

ADAPTIVE CHANNEL EQUALIZATION

The speed of data transmission over telephone channels is usually limited by channel distortion that causes intersymbol interference (ISI). At data rates below 2400 bits the ISI is relatively small and is usually not a problem in the operation of a modem. However, at data rates above 2400 bits, an adaptive equalizer is employed in the modem to compensate for the channel distortion and thus to allow for highly reliable high-speed data transmission. In telephone channels, filters are used throughout the system to separate signals in different frequency bands. These filters cause amplitude and phase distortion. The adaptive equalizer is basically an adaptive FIR filter with coefficients that are adjusted by means of the LMS algorithm to correct for the channel distortion.

A block diagram showing the basic elements of a modem transmitting data over a channel is given in Figure 9.5. Initially, the equalizer coefficients are adjusted by transmitting a short training sequence, usually less than one second in duration. After the short training period, the transmitter begins to transmit the data sequence $\{a(n)\}$. To track the possible slow time variations in the channel, the equalizer coefficients must continue to be adjusted in an adaptive manner while receiving data. This is usually accomplished, as illustrated in Figure 9.5, by treating the decisions at the output of the decision device as correct, and using the decisions in place of the reference $\{d(n)\}$ to generate the error signal. This approach works quite well when decision errors occur infrequently, such as less than one error in 100 data symbols. The occasional decision errors cause only a small misadjustment in the equalizer coefficients.

PROJECT 9.4: ADAPTIVE CHANNEL EQUALIZATION

The objective of this project is to investigate the performance of an adaptive equalizer for data transmission over a channel that causes intersymbol interference. The basic configuration of the system to be simulated is shown in Figure 9.6. As we observe, five basic modules are required. Note that we have avoided carrier modulation and demodulation, which is required in a telephone channel modem. This is done to simplify the simulation program. However, all processing involves complex arithmetic operations.

The five modules are as follows:

1. The data generator module is used to generate a sequence of complex-valued information symbols $\{a(n)\}$. In particular, employ four equally probable symbols $s+js$, $s-js$, $-s+js$, and $-s-js$, where s is a scale factor that may be set to $s = 1$, or it can be an input parameter.

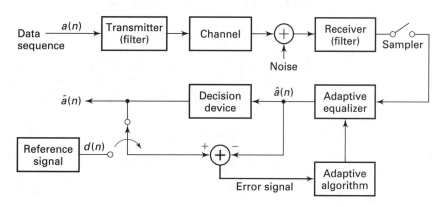

FIGURE 9.5 *Application of adaptive filtering to adaptive channel equalization*

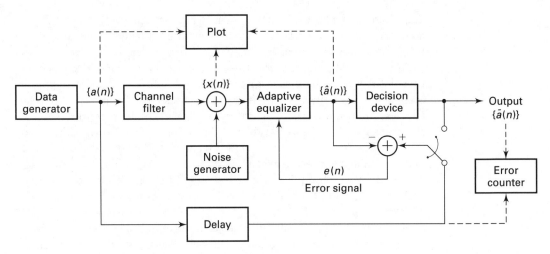

FIGURE 9.6 *Experiment for investigating the performance of an adaptive equal-izer*

2. The channel filter module is an FIR filter with coefficients $\{c(n),\ 0 \le n \le K - 1\}$ that simulates the channel distortion. For distortionless transmission, set $c(0) = 1$ and $c(n) = 0$ for $1 \le n \le K - 1$. The length K of the filter is an input parameter.

3. The noise generator module is used to generate additive noise that is usually present in any digital communication system. If we are modeling noise that is generated by electronic devices, the noise distribution should be Gaussian with zero mean. Use the `randu` function.

4. The adaptive equalizer module is an FIR filter with tap coefficients $\{h(k),\ 0 < k < N - 1\}$, which are adjusted by the LMS algorithm. However, due to the use of complex arithmetic, the recursive equation in the LMS algorithm is slightly modified to

$$h_n(k) = h_{n-1}(k) + \triangle\, e(n) x^*(n - k) \qquad (9.15)$$

where the asterisk denotes the complex conjugate.

5. The decision device module takes the estimate $\hat{a}(n)$ and quantizes it to one of the four possible signal points on the basis of the following decision rule:

$$\mathrm{Re}\,[\hat{a}\,(n)] > 0 \quad \text{and} \quad \mathrm{Im}\,[\hat{a}\,(n)] > 0 \quad \longrightarrow \quad 1 + j$$
$$\mathrm{Re}\,[\hat{a}\,(n)] > 0 \quad \text{and} \quad \mathrm{Im}\,[\hat{a}\,(n)] < 0 \quad \longrightarrow \quad 1 - j$$
$$\mathrm{Re}\,[\hat{a}\,(n)] < 0 \quad \text{and} \quad \mathrm{Im}\,[\hat{a}\,(n)] > 0 \quad \longrightarrow \quad -1 + j$$
$$\mathrm{Re}\,[\hat{a}\,(n)] < 0 \quad \text{and} \quad \mathrm{Im}\,[\hat{a}\,(n)] < 0 \quad \longrightarrow \quad -1 - j$$

The effectiveness of the equalizer in suppressing the ISI introduced by the channel filter may be seen by plotting the following relevant sequences in a two-dimensional (real–imaginary) display. The data generator output $\{a(n)\}$ should consist of four points with values $\pm 1 \pm j$. The effect of channel distortion and additive noise may be viewed by displaying the sequence $\{x(n)\}$ at the input to the equalizer. The effectiveness of the adaptive equalizer may be assessed by plotting its output $\{\hat{a}(n)\}$ after convergence of its coefficients. The short-time average squared error $ASE(n)$ may also be used to monitor the convergence characteristics of the LMS algorithm. Note that a delay must be introduced into the output of the data generator to compensate for the delays that the signal encounters due to the channel filter and the adaptive equalizer. For example, this delay may be set to the largest integer closest to $(N + K)/2$. Finally, an error counter may be used to count the number of symbol errors in the received data sequence, and the ratio for the number of errors to the total number of symbols (error rate) may be displayed. The error rate may be varied by changing the level of the ISI and the level of the additive noise.

It is suggested that simulations be performed for the following three channel conditions:

a. No ISI: $c(0) = 1$, $c(n) = 0$, $1 \leq n \leq K - 1$
b. Mild ISI: $c(0) = 1$, $c(1) = 0.2$, $c(2) = -0.2$, $c(n) = 0$, $3 \leq n \leq K - 1$
c. Strong ISI: $c(0) = 1$, $c(1) = 0.5$, $c(2) = 0.5$, $c(n) = 0$, $3 \leq n \leq K-1$

The measured error rate may be plotted as a function of the signal-to-noise ratio (SNR) at the input to the equalizer, where SNR is defined as P_s/P_n, where P_s is the signal power, given as $P_s = s^2$, and P_n is the noise power of the sequence at the output of the noise generator.

SUMMARY

In this chapter we introduced the reader to the theory and implementation of adaptive FIR filters with applications to system identification, interference suppression, narrowband frequency enhancement, and adaptive equalization. Projects were formulated involving these applications of adaptive filtering; these can be implemented using MATLAB.

APPLICATIONS IN COMMUNICATIONS

Today MATLAB finds widespread use in the simulation of a variety of communication systems. In this chapter we shall focus on several applications dealing with waveform representation and coding, especially speech coding, and with digital communications. In particular, we shall describe several methods for digitizing analog waveforms, with specific application to speech coding and transmission. These methods are pulse-code modulation (PCM), differential PCM and adaptive differential PCM (ADPCM), delta modulation (DM) and adaptive delta modulation (ADM), and linear predictive coding (LPC). A project is formulated involving each of these waveform encoding methods for simulation using MATLAB.

The last three topics treated in this chapter deal with signal-detection applications that are usually encountered in the implementation of a receiver in a digital communication system. For each of these topics we describe a project that involves the implementations via simulation of the detection scheme in MATLAB.

PULSE-CODE MODULATION

Pulse-code modulation is a method for quantizing an analog signal for the purpose of transmitting or storing the signal in digital form. PCM is widely used for speech transmission in telephone communications and for telemetry systems that employ radio transmission. We shall concentrate our attention on the application of PCM to speech signal processing.

Speech signals transmitted over telephone channels are usually limited in bandwidth to the frequency range below 4kHz. Hence the Nyquist rate for sampling such a signal is less than 8kHz. In PCM the analog speech signal is sampled at the nominal rate of 8kHz (samples per second), and each sample is quantized to one of 2^b levels, and represented digitally by a sequence of b bits. Thus the bit rate required to transmit the digitized speech signal is $8000\,b$ bits per second.

The quantization process may be modeled mathematically as

$$\tilde{s}(n) = s(n) + q(n) \tag{10.1}$$

where $\tilde{s}(n)$ represents the quantized value of $s(n)$, and $q(n)$ represents the quantization error, which we treat as an additive noise. Assuming that a uniform quantizer is used and the number of levels is sufficiently large, the quantization noise is well characterized statistically by the uniform probability density function,

$$p(q) = \frac{1}{\Delta}, \quad -\frac{\Delta}{2} \leq q \leq \frac{\Delta}{2} \tag{10.2}$$

where the step size of the quantizer is $\Delta = 2^{-b}$. The mean square value of the quantization error is

$$E\left(q^2\right) = \frac{\Delta^2}{12} = \frac{2^{-2b}}{12} \tag{10.3}$$

Measured in decibels, the mean square value of the noise is

$$10\log\left(\frac{\Delta^2}{12}\right) = 10\log\left(\frac{2^{-2b}}{12}\right) = -6b - 10.8 \text{ dB} \tag{10.4}$$

We observe that the quantization noise decreases by 6 dB/bit used in the quantizer. High-quality speech requires a minimum of 12 bits per sample and hence a bit rate of 96,000 bits per second (bps).

Speech signals have the characteristic that small signal amplitudes occur more frequently than large signal amplitudes. However, a uniform quantizer provides the same spacing between successive levels throughout the entire dynamic range of the signal. A better approach is to use a nonuniform quantizer, which provides more closely spaced levels at the low signal amplitudes and more widely spaced levels at the large signal amplitudes. For a nonuniform quantizer with b bits, the resulting quantization error has a mean square value that is smaller than that given by (10.4). A nonuniform quantizer characteristic is usually obtained by passing the signal through a nonlinear device that compresses the signal amplitude, followed by a uniform quantizer. For example, a logarithmic compressor employed in U.S. and Canadian telecommunications systems, called a μ-law compressor, has an input-output magnitude characteristic of the form

$$y = \frac{\ln\left(1 + \mu\,|s|\right)}{\ln(1 + \mu)}\, \text{sgn}\left(s\right); \quad |s| \leq 1, |y| \leq 1 \tag{10.5}$$

where s is the normalized input, y is the normalized output, sgn $(\cdot)$ is the sign function, and μ is a parameter that is selected to give the desired compression characteristic.

In the encoding of speech waveforms the value of $\mu = 255$ has been adopted as a standard in the U.S. and Canada. This value results in about a 24-dB reduction in the quantization noise power relative to uniform quantization. Consequently, an 8-bit quantizer used in conjunction with a $\mu = 255$ logarithmic compressor produces the same quality speech as a 12-bit uniform quantizer with no compression. Thus the compressed PCM speech signal has a bit rate of 64,000 bps.

The logarithmic compressor standard used in European telecommunication systems is called A-law and is defined as

$$
y = \begin{cases}
\dfrac{1 + \ln\left(A\,|s|\right)}{1 + \ln A}\,\operatorname{sgn}\left(s\right), & \frac{1}{A} \le |s| \le 1 \\[3mm]
\dfrac{A\,|s|}{1 + \ln A}\,\operatorname{sgn}\left(s\right), & 0 \le |s| \le \frac{1}{A}
\end{cases}
\tag{10.6}
$$

where A is chosen as 87.56. Although (10.5) and (10.6) are different nonlinear functions, the two compression characteristics are very similar. Figure 10.1 illustrates these two compression functions. Note their strong similarity.

In the reconstruction of the signal from the quantized values, the decoder employs an inverse logarithmic relation to expand the signal am-

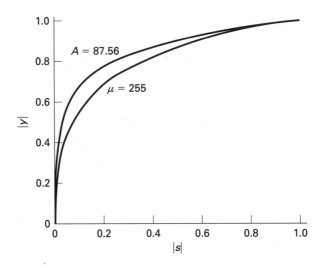

FIGURE 10.1 *Comparison of μ-law and A-law nonlinearities*

plitude. For example, in μ-law the inverse relation is given by

$$|s| = \frac{(1+\mu)^{|y|} - 1}{\mu}; \quad |y| \le 1, |s| \le 1 \tag{10.7}$$

The combined compressor-expander pair is termed a *compander*.

PROJECT 10.1:
PCM

The purpose of this project is to gain an understanding of PCM compression (linear-to-logarithmic) and PCM expansion (logarithmic-to-linear). Write the following three MATLAB functions for this project:

1. a μ-law compressor function to implement (10.5) that accepts a zero-mean normalized ($|s| \le 1$) signal and produces a compressed zero-mean signal with μ as a free parameter that can be specified,
2. a quantizer function that accepts a zero-mean input and produces an integer output after b-bit quantization that can be specified, and
3. a μ-law expander to implement (10.7) that accepts an integer input and produces a zero-mean output for a specified μ parameter.

For simulation purposes generate a large number of samples (10,000 or more) of the following sequences: (a) a sawtooth sequence, (b) an exponential pulse train sequence, (c) a sinusoidal sequence, and (d) a random sequence with small variance. Care must be taken to generate nonperiodic sequences by choosing their normalized frequencies as irrational numbers (i.e., sample values should not repeat). For example, a sinusoidal sequence can be generated using

$$s(n) = 0.5 \sin(n/33), \quad 0 \le n \le 10{,}000$$

From our discussions in Chapter 2 this sequence is nonperiodic, yet it has a periodic envelope. Other sequences can also be generated in a similar fashion. Process these signals through the above μ-law compressor, quantizer, and expander functions as shown in Figure 10.2, and compute the

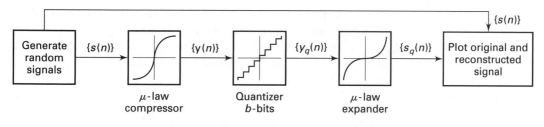

FIGURE 10.2 *PCM project*

signal-to-quantization noise ratio (SQNR) in dB as

$$\text{SQNR} = 10 \log_{10} \left(\frac{\displaystyle\sum_{n=1}^{N} s^2(n)}{\displaystyle\sum_{n=1}^{N} (s(n) - s_q(n))^2} \right).$$

For different b-bit quantizers, systematically determine the value of μ that maximizes the SQNR. Also plot the input and output waveforms and comment on the results.

DIFFERENTIAL PCM (DPCM)
■

In PCM each sample of the waveform is encoded independently of all the other samples. However, most signals, including speech, sampled at the Nyquist rate or faster exhibit significant correlation between successive samples. In other words, the average change in amplitude between successive samples is relatively small. Consequently, an encoding scheme that exploits the redundancy in the samples will result in a lower bit rate for the speech signal.

A relatively simple solution is to encode the differences between successive samples rather than the samples themselves. Since differences between samples are expected to be smaller than the actual sampled amplitudes, fewer bits are required to represent the differences. A refinement of this general approach is to predict the current sample based on the previous p samples. To be specific, let $s(n)$ denote the current sample of speech and let $\hat{s}(n)$ denote the predicted value of $s(n)$, defined as

$$\hat{s}(n) = \sum_{i=1}^{p} a(i) s(n-i) \tag{10.8}$$

Thus $\hat{s}(n)$ is a weighted linear combination of the past p samples, and the $a(i)$ are the predictor (filter) coefficients. The $a(i)$ are selected to minimize some function of the error between $s(n)$ and $\hat{s}(n)$.

A mathematically and practically convenient error function is the sum of squared errors. With this as the performance index for the predictor, we select the $a(i)$ to minimize

$$\mathcal{E}_p \triangleq \sum_{n=1}^{N} e^2(n) = \sum_{n=1}^{N} \left[s(n) - \sum_{i=1}^{p} a(i) s(n-i) \right]^2 \tag{10.9}$$

$$= r_{ss}(0) - 2\sum_{i=1}^{p} a(i) r_{ss}(i) + \sum_{i=1}^{p}\sum_{j=1}^{p} a(i) a(j) r_{ss}(i-j)$$

where $r_{ss}(m)$ is the autocorrelation function of the sampled signal sequence $s(n)$, defined as

$$r_{ss}(m) = \sum_{i=1}^{N} s(i) s(i+m) \qquad (10.10)$$

Minimization of $\mathcal{E}_p$ with respect to the predictor coefficients $\{a_i(n)\}$ results in the set of linear equations, called the normal equations,

$$\sum_{i=1}^{p} a(i) r_{ss}(i-j) = r_{ss}(j), \quad j = 1, 2, \ldots, p \qquad (10.11)$$

or in the matrix form,

$$\mathbf{Ra} = \mathbf{r} \Longrightarrow \mathbf{a} = \mathbf{R}^{-1}\mathbf{r} \qquad (10.12)$$

where $\mathbf{R}$ is the autocorrelation matrix, $\mathbf{a}$ is the coefficient vector, and $\mathbf{r}$ is the autocorrelation vector. Thus the values of the predictor coefficients are established.

  Having described the method for determining the predictor coefficients, let us now consider the block diagram of a practical DPCM system, shown in Figure 10.3. In this configuration the predictor is implemented with the feedback loop around the quantizer. The input to the predictor is denoted as $\tilde{s}(n)$, which represents the signal sample $s(n)$ modified by the quantization process, and the output of the predictor is

$$\widehat{\tilde{s}} = \sum_{i=1}^{p} a(i)\, \tilde{s}(n-i) \qquad (10.13)$$

The difference

$$e(n) = s(n) - \widehat{\tilde{s}}(n) \qquad (10.14)$$

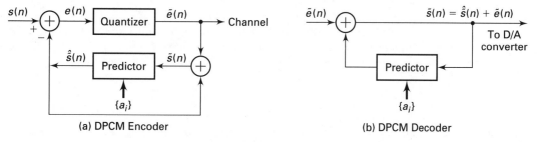

(a) DPCM Encoder   (b) DPCM Decoder

FIGURE 10.3   *Block diagram of a DPCM transcoder: (a) Encoder, (b) Decoder*

is the input to the quantizer, and $\tilde{e}(n)$ denotes the output. Each value of the quantized prediction error $\tilde{e}(n)$ is encoded into a sequence of binary digits and transmitted over the channel to the receiver. The quantized error $\tilde{e}(n)$ is also added to the predicted value $\widehat{\tilde{s}}(n)$ to yield $\tilde{s}(n)$.

At the receiver the same predictor that was used at the transmitting end is synthesized, and its output $\widehat{\tilde{s}}(n)$ is added to $\tilde{e}(n)$ to yield $\tilde{s}(n)$. The signal $\tilde{s}(n)$ is the desired excitation for the predictor and also the desired output sequence from which the reconstructed signal $\tilde{s}(t)$ is obtained by filtering, as shown in Figure 10.3b.

The use of feedback around the quantizer, as described above, ensures that the error in $\tilde{s}(n)$ is simply the quantization error $q(n) = \tilde{e}(n) - e(n)$ and that there is no accumulation of previous quantization errors in the implementation of the decoder. That is,

$$q(n) = \tilde{e}(n) - e(n) = \tilde{e}(n) - s(n) + \widehat{\tilde{s}}(n) = \tilde{s}(n) - s(n) \qquad \textbf{(10.15)}$$

Hence $\tilde{s}(n) = s(n) + q(n)$. This means that the quantized sample $\tilde{s}(n)$ differs from the input $s(n)$ by the quantization error $q(n)$ independent of the predictor used. Therefore the quantization errors do not accumulate.

In the DPCM system illustrated in Figure 10.3, the estimate or predicted value $\tilde{s}(n)$ of the signal sample $s(n)$ is obtained by taking a linear combination of past values $\tilde{s}(n-k)$, $k = 1, 2, \ldots, p$, as indicated by (10.13). An improvement in the quality of the estimate is obtained by including linearly filtered past values of the quantized error. Specifically, the estimate of $s(n)$ may be expressed as

$$\widehat{\tilde{s}}(n) = \sum_{i=1}^{p} a(i)\,\tilde{s}(n-i) + \sum_{i=1}^{m} b(i)\,\tilde{e}(n-i) \qquad \textbf{(10.16)}$$

where $b(i)$ are the coefficients of the filter for the quantized error sequence $\tilde{e}(n)$. The block diagram of the encoder at the transmitter and the decoder at the receiver are shown in Figure 10.4. The two sets of coefficients $a(i)$ and $b(i)$ are selected to minimize some function of the error $e(n) = \tilde{s}(n) - s(n)$, such as the sum of squared errors.

By using a logarithmic compressor and a 4-bit quantizer for the error sequence $e(n)$, DPCM results in high-quality speech at a rate of 32,000 bps, which is a factor of two lower than logarithmic PCM.

PROJECT 10.2: DPCM

The objective of this project is to gain understanding of the DPCM encoding and decoding operations. For simulation purposes, generate correlated

Chapter 10 ■ APPLICATIONS IN COMMUNICATIONS

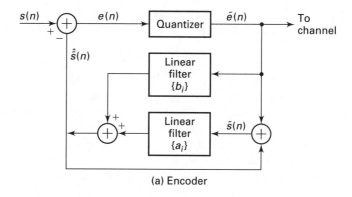

(a) Encoder

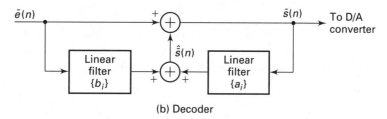

(b) Decoder

FIGURE 10.4 *DPCM modified by the linearly filtered error sequence*

random sequences using a pole-zero signal model of the form

$$s(n) = a(1)s(n-1) + b_0 x(n) + b_1 x(n-1) \tag{10.17}$$

where $x(n)$ is a zero-mean unit variance Gaussian sequence. This can be done using the `filter` function. The sequences developed in Project 10.1 can also be used for simulation. Develop the following three MATLAB modules for this project:

1. a model predictor function to implement (10.12), given the input signal $s(n)$;

2. a DPCM encoder function to implement the block diagram of Figure 10.3a, which accepts a zero-mean input sequence and produces a quantized b-bit integer error sequence, where b is a free parameter; and

3. a DPCM decoder function of Figure 10.3b, which reconstructs the signal from the quantized error sequence.

Experiment with several p-order prediction models for a given signal and determine the optimum order. Compare this DPCM implementation with the PCM system of Project 10.1 and comment on the results. Extend this implementation to include an mth-order moving average filter as indicated in (10.16).

ADAPTIVE PCM (ADPCM) AND DPCM

In general, the power in a speech signal varies slowly with time. PCM and DPCM encoders, however, are designed on the basis that the speech signal power is constant, and hence the quantizer is fixed. The efficiency and performance of these encoders can be improved by having them adapt to the slowly time-variant power level of the speech signal.

In both PCM and DPCM the quantization error $q(n)$ resulting from a uniform quantizer operating on a slowly varying power level input signal will have a time-variant variance (quantization noise power). One improvement that reduces the dynamic range of the quantization noise is the use of an adaptive quantizer.

Adaptive quantizers can be classified as feedforward or feedback. A feedforward adaptive quantizer adjusts its step size for each signal sample, based on a measurement of the input speech signal variance (power). For example, the estimated variance, based as a sliding window estimator, is

$$\hat{\sigma}_{n+1}^2 = \frac{1}{M} \sum_{k=n+1-M}^{n+1} s^2(k) \tag{10.18}$$

Then the step size for the quantizer is

$$\Delta(n+1) = \Delta(n)\,\hat{\sigma}_{n+1} \tag{10.19}$$

In this case it is necessary to transmit $\Delta(n+1)$ to the decoder in order for it to reconstruct the signal.

A feedback adaptive quantizer employs the output of the quantizer in the adjustment of the step size. In particular, we may set the step size as

$$\Delta(n+1) = \alpha(n)\,\Delta(n) \tag{10.20}$$

where the scale factor $\alpha(n)$ depends on the previous quantizer output. For example, if the previous quantizer output is small, we may select $\alpha(n) < 1$ in order to provide for finer quantization. On the other hand, if the quantizer output is large, then the step size should be increased to reduce the possibility of signal clipping. Such an algorithm has been successfully used in the encoding of speech signals. Figure 10.5 illustrates such a (3-bit) quantizer in which the step size is adjusted recursively according to the relation

$$\Delta(n+1) = \Delta(n) \cdot M(n)$$

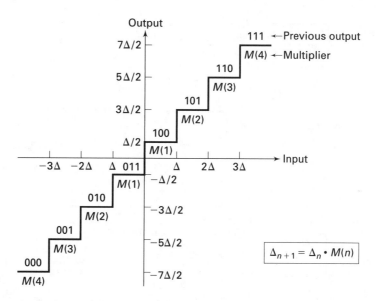

FIGURE 10.5 *Example of a quantizer with an adaptive step size ([10])*

where $M(n)$ is a multiplication factor whose value depends on the quantizer level for the sample $s(n)$, and $\Delta(n)$ is the step size of the quantizer for processing $s(n)$. Values of the multiplication factors optimized for speech encoding have been given by [13]. These values are displayed in Table 10.1 for 2-, 3-, and 4-bit quantization for PCM and DPCM.

In DPCM the predictor can also be made adaptive. Thus in ADPCM the coefficients of the predictor are changed periodically to reflect the changing signal statistics of the speech. The linear equations given by (10.11) still apply, but the short-term autocorrelation function of $s(n)$, $r_{ss}(m)$ changes with time.

TABLE 10.1 *Multiplication factors for adaptive step size adjustment ([10])*

	PCM			DPCM		
	2	*3*	*4*	*2*	*3*	*4*
$M(1)$	0.60	0.85	0.80	0.80	0.90	0.90
$M(2)$	2.20	1.00	0.80	1.60	0.90	0.90
$M(3)$		1.00	0.80		1.25	0.90
$M(4)$		1.50	0.80		1.70	0.90
$M(5)$			0.80			1.20
$M(6)$			0.80			1.60
$M(7)$			0.80			2.00
$M(8)$			0.80			2.40

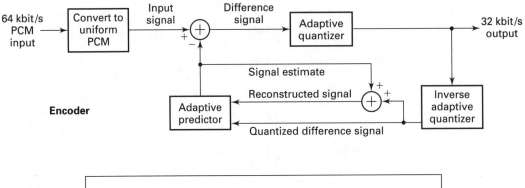

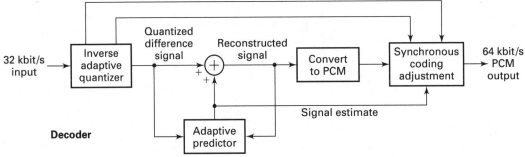

FIGURE 10.6 *ADPCM block diagram*

ADPCM
STANDARD

Figure 10.6 illustrates, in block diagram form, a 32,000 bps ADPCM encoder and decoder that has been adopted as an international (CCITT) standard for speech transmission over telephone channels. The ADPCM encoder is designed to accept 8-bit PCM compressed signal samples at 64,000 bps, and by means of adaptive prediction and adaptive 4-bit quantization to reduce the bit rate over the channel to 32,000 bps. The ADPCM decoder accepts the 32,000 bps data stream and reconstructs the signal in the form of an 8-bit compressed PCM at 64,000 bps. Thus we have a configuration shown in Figure 10.7, where the ADPCM encoder/

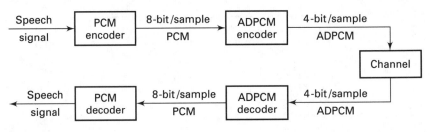

FIGURE 10.7 *ADPCM interface to PCM system*

decoder is embedded into a PCM system. Although the ADPCM encoder/decoder could be used directly on the speech signal, the interface to the PCM system is necessary in practice in order to maintain compatibility with existing PCM systems that are widely used in the telephone network.

The ADPCM encoder accepts the 8-bit PCM compressed signal and expands it to a 14-bit-per-sample linear representation for processing. The predicted value is subtracted from this 14-bit linear value to produce a difference signal sample that is fed to the quantizer. Adaptive quantization is performed on the difference signal to produce a 4-bit output for transmission over the channel.

Both the encoder and decoder update their internal variables, based only on the ADPCM values that are generated. Consequently, an ADPCM decoder including an inverse adaptive quantizer is embedded in the encoder so that all internal variables are updated, based on the same data. This ensures that the encoder and decoder operate in synchronism without the need to transmit any information on the values of internal variables.

The adaptive predictor computes a weighted average of the last six dequantized difference values and the last two predicted values. Hence this predictor is basically a two-pole ($p = 2$) and six-zero ($m = 6$) filter governed by the difference equation given by (10.16). The filter coefficients are updated adaptively for every new input sample.

At the receiving decoder and at the decoder that is embedded in the encoder, the 4-bit transmitted ADPCM value is used to update the inverse adaptive quantizer, whose output is a dequantized version of the difference signal. This dequantized value is added to the value generated by the adaptive predictor to produce the reconstructed speech sample. This signal is the output of the decoder, which is converted to compressed PCM format at the receiver.

PROJECT 10.3: ADPCM

The objective of this project is to gain familiarity with, and understanding of, ADPCM and its interface with a PCM encoder/decoder (transcoder). As described above, the ADPCM transcoder is inserted between the PCM compressor and the PCM expander as shown in Figure 10.7. Use the already developed MATLAB PCM and DPCM modules for this project.

The input to the PCM-ADPCM transcoder system can be supplied from internally generated waveform data files, just as in the case of the PCM project. The output of the transcoder can be plotted. Comparisons should be made between the output signal from the PCM-ADPCM transcoder with the signal from the PCM transcoder (PCM project 10.1), and with the original input signal.

DELTA MODULATION (DM)

Delta modulation may be viewed as a simplified form of DPCM in which a two-level (1-bit) quantizer is used in conjunction with a fixed first-order predictor. The block diagram of a DM encoder-decoder is shown in Figure 10.8. We note that

$$\widehat{\tilde{s}}(n) = \tilde{s}(n-1) = \widehat{\tilde{s}}(n-1) + \tilde{e}(n-1) \tag{10.21}$$

Since

$$q(n) = \tilde{e}(n) - e(n) = \tilde{e}(n) - \left[s(n) - \widehat{\tilde{s}}(n)\right]$$

it follows that

$$\widehat{\tilde{s}}(n) = s(n-1) + q(n-1) \tag{10.22}$$

Thus the estimated (predicted) value of $s(n)$ is really the previous sample $s(n-1)$ modified by the quantization noise $q(n-1)$. We also note that the difference equation in (10.21) represents an integrator with an input $\tilde{e}(n)$. Hence an equivalent realization of the one-step predictor is an accumulator with an input equal to the quantized error signal $\tilde{e}(n)$. In general, the quantized error signal is scaled by some value, say Δ_1, which is called the step size. This equivalent realization is illustrated in Figure 10.9. In effect, the encoder shown in Figure 10.9 approximates a waveform $s(t)$ by a linear staircase function. In order for the approximation to be relatively good, the waveform $s(t)$ must change slowly relative to

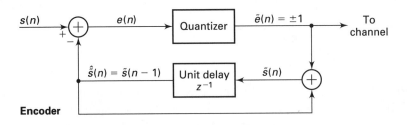

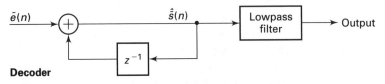

FIGURE 10.8 *Block diagram of a delta modulation system*

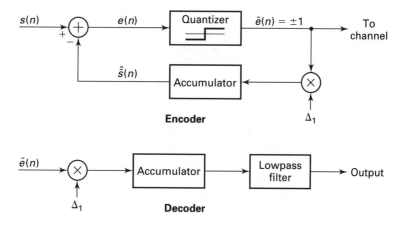

FIGURE 10.9 *An equivalent realization of a delta modulation system*

the sampling rate. This requirement implies that the sampling rate must be several (a factor of at least 5) times the Nyquist rate. A lowpass filter is usually incorporated into the decoder to smooth out discontinuities in the reconstructed signal.

ADAPTIVE DELTA MODULATION (ADM)

At any given sampling rate, the performance of the DM encoder is limited by two types of distortion as shown in Figure 10.10. One is called slope-overload distortion. It is due to the use of a step size Δ_1 that is too small to follow portions of the waveform that have a steep slope. The second type of distortion, called granular noise, results from using a step size that is too large in parts of the waveform having a small slope. The need to minimize both of these two types of distortion results in conflicting requirements in the selection of the step size Δ_1.

An alternative solution is to employ a variable size that adapts itself to the short-term characteristics of the source signal. That is, the step size is increased when the waveform has a steep slope and decreased when the waveform has a relatively small slope.

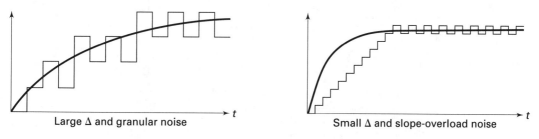

FIGURE 10.10 *Two types of distortion in the DM encoder*

A variety of methods can be used to set adaptively the step size in every iteration. The quantized error sequence $\tilde{e}(n)$ provides a good indication of the slope characteristics of the waveform being encoded. When the quantized error $\tilde{e}(n)$ is changing signs between successive iterations, this is an indication that the slope of the waveform in the locality is relatively small. On the other hand, when the waveform has a steep slope, successive values of the error $\tilde{e}(n)$ are expected to have identical signs. From these observations it is possible to devise algorithms that decrease or increase the step size, depending on successive values of $\tilde{e}(n)$. A relatively simple rule devised by [12] is to vary adaptively the step size according to the relation

$$\Delta(n) = \Delta(n-1)\, K^{\tilde{e}(n)\tilde{e}(n-1)}, \quad n = 1, 2, \ldots \tag{10.23}$$

where $K \geq 1$ is a constant that is selected to minimize the total distortion. A block diagram of a DM encoder-decoder that incorporates this adaptive algorithm is illustrated in Figure 10.11.

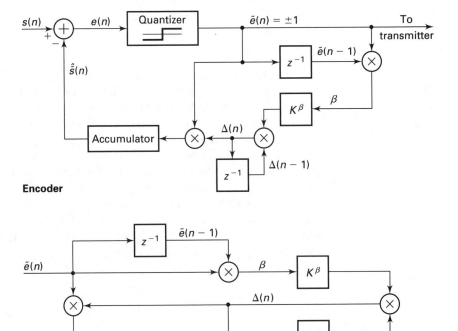

FIGURE 10.11 *An example of a delta modulation system with adaptive step size*

Several other variations of adaptive DM encoding have been investigated and described in the technical literature. A particularly effective and popular technique first proposed by [8] is called *continuously variable slope delta modulation* (CVSD). In CVSD the adaptive step size parameter may be expressed as

$$\Delta(n) = \alpha\Delta(n-1) + k_1 \tag{10.24}$$

if $\tilde{e}(n)$, $\tilde{e}(n-1)$, and $\tilde{e}(n-2)$ have the same sign; otherwise

$$\Delta(n) = \alpha\Delta(n-1) + k_2 \tag{10.25}$$

The parameters α, k_1, and k_2 are selected such that $0 < \alpha < 1$ and $k_1 > k_2 > 0$. For more discussion on this and other variations of adaptive DM, the interested reader is referred to the papers by [13] and [6] and to the extensive references contained in these papers.

PROJECT 10.4:
DM AND ADM

The purpose of this project is to gain an understanding of delta modulation and adaptive delta modulation for coding of waveforms. This project involves writing MATLAB functions for the DM encoder and decoder as shown in Figure 10.9, and for the ADM encoder and decoder shown in Figure 10.11. The lowpass filter at the decoder can be implemented as a linear-phase FIR filter. For example, a Hanning filter that has the impulse response

$$h(n) = \frac{1}{2}\left[1 - \cos\left(\frac{2\pi n}{N-1}\right)\right], \quad 0 \leq n \leq N-1 \tag{10.26}$$

may be used, where the length N may be selected in the range $5 \leq N \leq 15$.

The input to the DM and ADM systems can be supplied from the waveforms generated in Project 10.1 except that the sampling rate should be higher by a factor of 5 to 10. The output of the decoder can be plotted. Comparisons should be made between the output signal from the DM and ADM decoders and the original input signal.

LINEAR PREDICTIVE CODING (LPC) OF SPEECH

The linear predictive coding (LPC) method for speech analysis and synthesis is based on modeling the vocal tract as a linear all-pole (IIR) filter having the system function

$$H(z) = \frac{G}{1 + \displaystyle\sum_{k=1}^{p} a_p(k)z^{-k}} \tag{10.27}$$

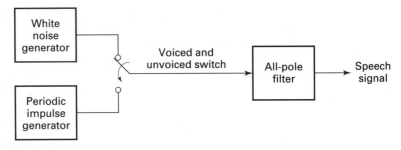

FIGURE 10.12 *Block diagram model for the generation of a speech signal*

where p is the number of poles, G is the filter gain, and $\{a_p(k)\}$ are the parameters that determine the poles. There are two mutually exclusive excitation functions to model voiced and unvoiced speech sounds. On a short-time basis, voiced speech is periodic with a fundamental frequency F_0, or a pitch period $1/F_0$, which depends on the speaker. Thus voiced speech is generated by exciting the all-pole filter model by a periodic impulse train with a period equal to the desired pitch period. Unvoiced speech sounds are generated by exciting the all-pole filter model by the output of a random-noise generator. This model is shown in Figure 10.12.

Given a short-time segment of a speech signal, usually about 20 ms or 160 samples at an 8 kHz sampling rate, the speech encoder at the transmitter must determine the proper excitation function, the pitch period for voiced speech, the gain parameter G, and the coefficients $a_p(k)$. A block diagram that illustrates the speech encoding system is given in Figure 10.13. The parameters of the model are determined adaptively from the data and encoded into a binary sequence and transmitted to the receiver.

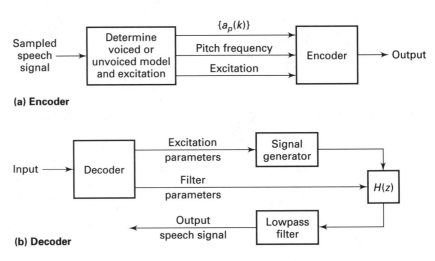

FIGURE 10.13 *Encoder and decoder for LPC*

At the receiver the speech signal is synthesized from the model and the excitation signal.

The parameters of the all-pole filter model are easily determined from the speech samples by means of linear prediction. To be specific, the output of the FIR linear prediction filter is

$$\hat{s}(n) = -\sum_{k=1}^{p} a_p(k) s(n-k) \tag{10.28}$$

and the corresponding error between the observed sample $s(n)$ and the predicted value $\hat{s}(n)$ is

$$e(n) = s(n) + \sum_{k=1}^{p} a_p(k) s(n-k) \tag{10.29}$$

By minimizing the sum of squared errors, that is,

$$\mathcal{E} = \sum_{n=0}^{N} e^2(n) = \sum_{n=0}^{N} \left[s(n) + \sum_{k=1}^{p} a_p(k) s(n-k) \right]^2 \tag{10.30}$$

we can determine the pole parameters $\{a_p(k)\}$ of the model. The result of differentiating $\mathcal{E}$ with respect to each of the parameters and equating the result to zero, is a set of p linear equations

$$\sum_{k=1}^{p} a_p(k) r_{ss}(m-k) = -r_{ss}(m), \quad m = 1, 2, \ldots, p \tag{10.31}$$

where $r_{ss}(m)$ is the autocorrelation of the sequence $s(n)$ defined as

$$r_{ss}(m) = \sum_{n=0}^{N} s(n) s(n+m) \tag{10.32}$$

The linear equations (10.31) can be expressed in matrix form as

$$\mathbf{R}_{ss}\mathbf{a} = -\mathbf{r}_{ss} \tag{10.33}$$

where $\mathbf{R}_{ss}$ is a $p \times p$ autocorrelation matrix, $\mathbf{r}_{ss}$ is a $p \times 1$ autocorrelation vector, and $\mathbf{a}$ is a $p \times 1$ vector of model parameters. Hence

$$\mathbf{a} = -\mathbf{R}_{ss}^{-1}\mathbf{r}_{ss} \tag{10.34}$$

These equations can also be solved recursively and most efficiently, without resorting to matrix inversion, by using the Levinson-Durbin algorithm [19]. However, in MATLAB it is convenient to use the matrix inversion. The all-pole filter parameters $\{a_p(k)\}$ can be converted to the all-pole lattice parameters $\{K_i\}$ (called the reflection coefficients) using the MATLAB function dir2latc developed in Chapter 6.

The gain parameter of the filter can be obtained by noting that its input-output equation is

$$s(n) = -\sum_{k=1}^{p} a_p(k) s(n-k) + Gx(n) \qquad (10.35)$$

where $x(n)$ is the input sequence. Clearly,

$$Gx(n) = s(n) + \sum_{k=1}^{p} a_p(k) s(n-k) = e(n)$$

Then

$$G^2 \sum_{n=0}^{N-1} x^2(n) = \sum_{n=0}^{N-1} e^2(n) \qquad (10.36)$$

If the input excitation is normalized to unit energy by design, then

$$G^2 = \sum_{n=0}^{N-1} e^2(n) = r_{ss}(0) + \sum_{k=1}^{p} a_p(k) r_{ss}(k) \qquad (10.37)$$

Thus G^2 is set equal to the residual energy resulting from the least-squares optimization.

Once the LPC coefficients are computed, we can determine whether the input speech frame is voiced, and if so, what the pitch is. This is accomplished by computing the sequence

$$r_e(n) = \sum_{k=1}^{p} r_a(k) r_{ss}(n-k) \qquad (10.38)$$

where $r_a(k)$ is defined as

$$r_a(k) = \sum_{i=1}^{p} a_p(i) a_p(i+k) \qquad (10.39)$$

which is the autocorrelation sequence of the prediction coefficients. The pitch is detected by finding the peak of the normalized sequence $r_e(n)/r_e(0)$ in the time interval that corresponds to 3 to 15 ms in the 20-ms sampling frame. If the value of this peak is at least 0.25, the frame of speech is considered voiced with a pitch period equal to the value of $n = N_p$, where $r_e(N_p)/r_e(0)$ is a maximum. If the peak value is less than 0.25, the frame of speech is considered unvoiced and the pitch is zero.

The values of the LPC coefficients, the pitch period, and the type of excitation are transmitted to the receiver, where the decoder synthesizes the speech signal by passing the proper excitation through the all-pole filter model of the vocal tract. Typically, the pitch period requires 6 bits, and the gain parameter may be represented by 5 bits after its dynamic range is compressed logarithmically. If the prediction coefficients were to be coded, they would require between 8 to 10 bits per coefficient for accurate representation. The reason for such high accuracy is that relatively small changes in the prediction coefficients result in a large change in the pole positions of the filter model. The accuracy requirements are lessened by transmitting the reflection coefficients $\{K_i\}$, which have a smaller dynamic range—that is, $|K_i| < 1$. These are adequately represented by 6 bits per coefficient. Thus for a 10th-order predictor the total number of bits assigned to the model parameters per frame is 72. If the model parameters are changed every 20 milliseconds, the resulting bit rate is 3,600 bps. Since the reflection coefficients are usually transmitted to the receiver, the synthesis filter at the receiver is implemented as an all-pole lattice filter, described in Chapter 6.

PROJECT 10.5:   The objective of this project is to analyze a speech signal through an
LPC   LPC coder and then to synthesize it through the corresponding PLC decoder. Use several `.wav` sound files (sampled at 8000 sam/sec rate), which are available in MATLAB for this purpose. Divide speech signals into short-time segments (with lengths between 120 and 150 samples) and process each segment to determine the proper excitation function (voiced or unvoiced), the pitch period for voiced speech, the coefficients $\{a_p(k)\}$ $(p \le 10)$, and the gain G. The decoder that performs the synthesis is an all-pole lattice filter whose parameters are the reflection coefficients that can be determined from $\{a_p(k)\}$. The output of this project is a synthetic speech signal that can be compared with the original speech signal. The distortion effects due to LPC analysis/synthesis may be assessed qualitatively.

DUAL-TONE MULTIFREQUENCY (DTMF) SIGNALS

DTMF is the generic name for push-button telephone signaling that is equivalent to the Touch Tone system in use within the Bell System. DTMF also finds widespread use in electronic mail systems and telephone banking systems in which the user can select options from a menu by sending DTMF signals from a telephone.

In a DTMF signaling system a combination of a high-frequency tone and a low-frequency tone represent a specific digit or the characters * and #. The eight frequencies are arranged as shown in Figure 10.14, to

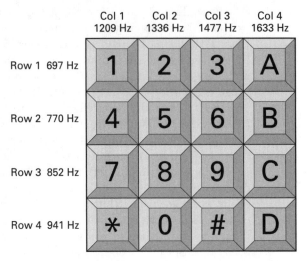

Col 1 1209 Hz | Col 2 1336 Hz | Col 3 1477 Hz | Col 4 1633 Hz

Row 1 697 Hz — 1 2 3 A
Row 2 770 Hz — 4 5 6 B
Row 3 852 Hz — 7 8 9 C
Row 4 941 Hz — * 0 # D

DTMF digit = row tone + column tone

FIGURE 10.14 *DTMF digits*

accommodate a total of 16 characters, 12 of which are assigned as shown, while the other four are reserved for future use.

DTMF signals are easily generated in software and detected by means of digital filters, also implemented in software, that are tuned to the eight frequency tones. Usually, DTMF signals are interfaced to the analog world via a *codec* (coder/decoder) chip or by linear A/D and D/A converters. Codec chips contain all the necessary A/D and D/A, sampling, and filtering circuitry for a bi-directional analog/digital interface.

The DTMF tones may be generated either mathematically or from a look-up table. In a hardware implementation (e.g., in a digital signal processor), digital samples of two sine waves are generated mathematically, scaled, and added together. The sum is logarithmically compressed and sent to the codec for conversion to an analog signal. At an 8 kHz sampling rate the hardware must output a sample every 125 ms. In this case a sine look-up table is not used because the values of the sine wave can be computed quickly without using the large amount of data memory that a table look-up would require. For simulation and investigation purposes the look-up table might be a good approach in MATLAB.

At the receiving end the logarithmically compressed, 8-bit digital data words from the codec are received, logarithmically expanded to their 16-bit linear format, and then the tones are detected to decide on the transmitted digit. The detection algorithm can be a DFT implementation using the FFT algorithm or a filter bank implementation. For the relatively small number of tones to be detected, the filter bank implementation is more efficient. Below, we describe the use of the Goertzel algorithm to implement the eight tuned filters.

Recall from the discussion in Chapter 5 that the DFT of an N-point data sequence $\{x(n)\}$ is

$$X(k) = \sum_{n=0}^{N-1} x(n) W_N^{nk}, \quad k = 0, 1, \ldots, N-1 \qquad \text{(10.40)}$$

If the FFT algorithm is used to perform the computation of the DFT, the number of computations (complex multiplications and additions) is $N \log_2 N$. In this case we obtain all N values of the DFT at once. However, if we desire to compute only M points of the DFT, where $M < \log_2 N$, then a direct computation of the DFT is more efficient. The Goertzel algorithm, which is described below, is basically a linear filtering approach to the computation of the DFT, and provides an alternative to direct computation.

THE GOERTZEL ALGORITHM

The Goertzel algorithm exploits the periodicity of the phase factors $\{W_N^k\}$ and allows us to express the computation of the DFT as a linear filtering operation. Since $W_N^{-kN} = 1$, we can multiply the DFT by this factor. Thus

$$X(k) = W_N^{-kN} X(k) = \sum_{m=0}^{N-1} x(m) W_N^{-k(N-m)} \qquad \text{(10.41)}$$

We note that (10.41) is in the form of a convolution. Indeed, if we define the sequence $y_k(n)$ as

$$y_k(n) = \sum_{m=0}^{N-1} x(m) W_N^{-k(n-m)} \qquad \text{(10.42)}$$

then it is clear that $y_k(n)$ is the convolution of the finite-duration input sequence $x(n)$ of length N with a filter that has an impulse response

$$h_k(n) = W_N^{-kn} u(n) \qquad \text{(10.43)}$$

The output of this filter at $n = N$ yields the value of the DFT at the frequency $\omega_k = 2\pi k/N$. That is,

$$X(k) = y_k(n)|_{n=N} \qquad \text{(10.44)}$$

as can be verified by comparing (10.41) with (10.42).

The filter with impulse response $h_k(n)$ has the system function

$$H_k(z) = \frac{1}{1 - W_N^{-k} z^{-1}} \qquad (10.45)$$

This filter has a pole on the unit circle at the frequency $\omega_k = 2\pi k/N$. Thus the entire DFT can be computed by passing the block of input data into a parallel bank of N single-pole filters (resonators), where each filter has a pole at the corresponding frequency of the DFT.

Instead of performing the computation of the DFT as in (10.42), via convolution, we can use the difference equation corresponding to the filter given by (10.45) to compute $y_k(n)$ recursively. Thus we have

$$y_k(n) = W_N^{-k} y_k(n-1) + x(n), \quad y_k(-1) = 0 \qquad (10.46)$$

The desired output is $X(k) = y_k(N)$. To perform this computation, we can compute once and store the phase factor W_N^{-k}.

The complex multiplications and additions inherent in (10.46) can be avoided by combining the pairs of resonators possessing complex conjugate poles. This leads to two-pole filters with system functions of the form

$$H_k(z) = \frac{1 - W_N^k z^{-1}}{1 - 2\cos(2\pi k/N) z^{-1} + z^{-2}} \qquad (10.47)$$

The realization of the system illustrated in Figure 10.15 is described by the difference equations

$$v_k(n) = 2\cos\frac{2\pi k}{N} v_k(n-1) - v_k(n-2) + x(n) \qquad (10.48)$$

$$y_k(n) = v_k(n) - W_N^k v_k(n-1) \qquad (10.49)$$

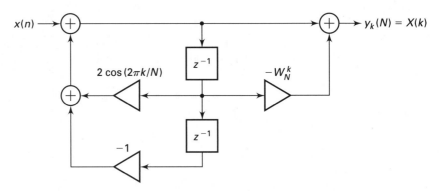

$x(n) \longrightarrow$ ... $y_k(N) = X(k)$

$2\cos(2\pi k/N)$

z^{-1}

$-W_N^k$

z^{-1}

-1

FIGURE 10.15 *Realization of two-pole resonator for computing the DFT*

with initial conditions $v_k(-1) = v_k(-2) = 0$. This is the Goertzel algorithm.

The recursive relation in (10.48) is iterated for $n = 0, 1, \ldots, N$, but the equation in (10.49) is computed only once, at time $n = N$. Each iteration requires one real multiplication and two additions. Consequently, for a real input sequence $x(n)$, this algorithm requires $N + 1$ real multiplications to yield not only $X(k)$ but also, due to symmetry, the value of $X(N - k)$.

We can now implement the DTMF decoder by use of the Goertzel algorithm. Since there are eight possible tones to be detected, we require eight filters of the type given by (10.47), with each filter tuned to one of the eight frequencies. In the DTMF detector, there is no need to compute the complex value $X(k)$; only the magnitude $|X(k)|$ or the magnitude-squared value $|X(k)|^2$ will suffice. Consequently, the final step in the computation of the DFT value involving the numerator term (feedforward part of the filter computation) can be simplified. In particular, we have

$$|X(k)|^2 = |y_k(N)|^2 = \left| v_k(N) - W_N^k v_k(N-1) \right|^2 \tag{10.50}$$

$$= v_k^2(N) + v_k^2(N-1) - \left(2\cos\frac{2\pi k}{N} \right) v_k(N) v_k(N-1)$$

Thus complex-valued arithmetic operations are completely eliminated in the DTMF detector.

PROJECT 10.6: DTMF SIGNALING

The objective of this project is to gain an understanding of the DTMF tone generation software and the DTMF decoding algorithm (the Goertzel algorithm). Design the following MATLAB modules:

1. a tone generation function that accepts an array containing dialing digits and produces a signal containing appropriate tones (from Figure 10.14) of one-half-second duration for each digit at 8 kHz sampling frequency,

2. a dial-tone generator generating samples of $(350 + 440)$ Hz frequency at 8 kHz sampling interval for a specified amount of duration, and

3. a decoding function to implement (10.50) that accepts a DTMF signal and produces an array containing dialing digits.

Generate several dialing list arrays containing a mix of digits and dial tones. Experiment with the tone generation and detection modules and comment on your observations. Use MATLAB's sound generation capabilities to listen to the tones and to observe the frequency components of the generated tones.

BINARY DIGITAL COMMUNICATIONS

Digitized speech signals that have been encoded via PCM, ADPCM, DM, and LPC are usually transmitted to the decoder by means of digital modulation. A binary digital communications system employs two signal waveforms, say $s_1(t) = s(t)$ and $s_2(t) = -s(t)$, to transmit the binary sequence representing the speech signal. The signal waveform $s(t)$, which is nonzero over the interval $0 \leq t \leq T$, is transmitted to the receiver if the data bit is a 1, and the signal waveform $-s(t)$, $0 \leq t \leq T$ is transmitted if the data bit is a 0. The time interval T is called the signal interval, and the bit rate over the channel is $R = 1/T$ bits per second. A typical signal waveform $s(t)$ is a rectangular pulse—that is, $s(t) = A$, $0 \leq t \leq T$—which has energy $A^2 T$.

In practice the signal waveforms transmitted over the channel are corrupted by additive noise and other types of channel distortions that ultimately limit the performance of the communications system. As a measure of performance we normally use the average probability of error, which is often called the bit error rate.

PROJECT 10.7:
BINARY DATA
COMMUNI-
CATIONS
SYSTEM

The purpose of this project is to investigate the performance of a binary data communications system on an additive noise channel by means of simulation. The basic configuration of the system to be simulated is shown in Figure 10.16. Five MATLAB functions are required.

1. A binary data generator module that generates a sequence of independent binary digits with equal probability.

2. A modulator module that maps a binary digit 1 into a sequence of M consecutive +1's, and maps a binary digit 0 into a sequence of M consecutive −1's. Thus the M consecutive +1's represent a sampled version of the rectangular pulse.

3. A noise generator that generates a sequence of uniformly distributed numbers over the interval $(-a, a)$. Each noise sample is added to a corresponding signal sample.

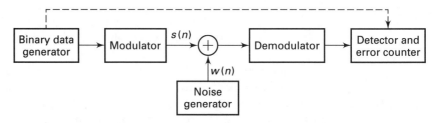

FIGURE 10.16 *Model of binary data communications system*

4. A demodulator module that sums the M successive outputs of the noise corrupted sequence $+1$'s or -1's received from the channel. We assume that the demodulator is time synchronized so that it knows the beginning and end of each waveform.

5. A detector and error-counting module. The detector compares the output of the modulator with zero and decides in favor of 1 if the output is greater than zero and in favor of 0 if the output is less than zero. If the output of the detector does not agree with the transmitted bit from the transmitter, an error is counted by the counter. The error rate depends on the ratio (called signal-to-noise ratio) of the size of M to the additive noise power, which is $P_n = a^2/3$.

The measured error rate can be plotted for different signal-to-noise ratios, either by changing M and keeping P_n fixed or vice versa.

SPREAD-SPECTRUM COMMUNICATIONS

Spread-spectrum signals are often used in the transmission of digital data over communication channels that are corrupted by interference due to intentional jamming or from other users of the channel (e.g., cellular telephones and other wireless applications). In applications other than communications, spread-spectrum signals are used to obtain accurate range (time delay) and range rate (velocity) measurements in radar and navigation. For the sake of brevity we shall limit our discussion to the use of spread spectrum for digital communications. Such signals have the characteristic that their bandwidth is much greater than the information rate in bits per second.

In combatting intentional interference (jamming), it is important to the communicators that the jammer who is trying to disrupt their communication does not have prior knowledge of the signal characteristics. To accomplish this, the transmitter introduces an element of unpredictability or randomness (pseudo-randomness) in each of the possible transmitted signal waveforms, which is known to the intended receiver, but not to the jammer. As a consequence, the jammer must transmit an interfering signal without knowledge of the pseudo-random characteristics of the desired signal.

Interference from other users arises in multiple-access communications systems in which a number of users share a common communications channel. At any given time a subset of these users may transmit information simultaneously over a common channel to corresponding receivers. The transmitted signals in this common channel may be distinguished from one another by superimposing a different pseudo-random pattern, called a *multiple-access code*, in each transmitted signal. Thus a particular

---

receiver can recover the transmitted data intended for it by knowing the pseudo-random pattern, that is, the key used by the corresponding transmitter. This type of communication technique, which allows multiple users to simultaneously use a common channel for data transmission, is called *code division multiple access* (CDMA).

The block diagram shown in Figure 10.17 illustrates the basic elements of a spread-spectrum digital communications system. It differs from a conventional digital communications system by the inclusion of two identical pseudo-random pattern generators, one that interfaces with the modulator at the transmitting end, and the second that interfaces with the demodulator at the receiving end. The generators generate a pseudo-random or *pseudo-noise* (PN) binary-valued sequence (±1's), which is impressed on the transmitted signal at the modulator and removed from the received signal at the demodulator.

Synchronization of the PN sequence generated at the demodulator with the PN sequence contained in the incoming received signal is required in order to demodulate the received signal. Initially, prior to the transmission of data, synchronization is achieved by transmitting a short fixed PN sequence to the receiver for purposes of establishing synchronization. After time synchronization of the PN generators is established, the transmission of data commences.

PROJECT 10.8: BINARY SPREAD-SPECTRUM COMMUNICATIONS

The objective of this project is to demonstrate the effectiveness of a PN spread-spectrum signal in suppressing sinusoidal interference. Let us consider the binary communication system described in Project 10.7, and let us multiply the output of the modulator by a binary (±1) PN sequence. The same binary PN sequence is used to multiply the input to the demodulator and thus to remove the effect of the PN sequence in the desired signal. The channel corrupts the transmitted signal by the addition of a wideband noise sequence $\{w(n)\}$ and a sinusoidal interference sequence of the form $i(n) = A \sin \omega_0 n$, where $0 < \omega_0 < \pi$. We may assume that $A \geq M$, where M is the number of samples per bit from the modulator. The basic binary spread spectrum-system is shown in Figure 10.18. As can

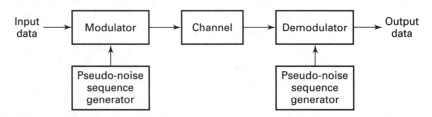

FIGURE 10.17 *Basic spread spectrum digital communications system*

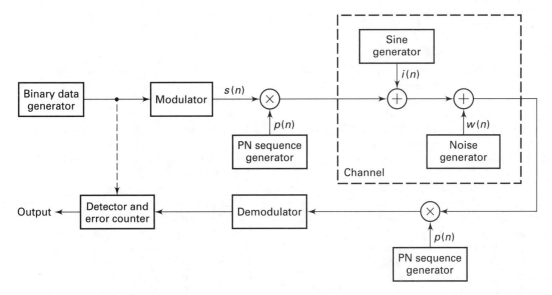

FIGURE 10.18 *Block diagram of binary PN spread-spectrum system for simulation experiment*

be observed, this is just the binary digital communication system shown in Figure 10.16, to which we have added the sinusoidal interference and the PN sequence generators. The PN sequence may be generated by using a random-number generator to generate a sequence of equally probable ±1's.

Execute the simulated system with and without the use of the PN sequence, and measure the error rate under the condition that $A \geq M$ for different values of M, such as $M = 50, 100, 500, 1000$. Explain the effect of the PN sequence on the sinusoidal interference signal. Thus explain why the PN spread-spectrum system outperforms the conventional binary communication system in the presence of the sinusoidal jamming signal.

SUMMARY

In this chapter we focused on applications to waveform representation and coding. In particular, we described several methods for digitizing an analog waveform, including PCM, DPCM, ADPCM, DM, ADM, and LPC. These methods have been widely used for speech coding and transmission. Projects involving these waveform encoding methods were formulated for implementation via simulation in MATLAB.

We also described signal-detection and communication systems where MATLAB may be used to perform the signal processing tasks. Projects were also devised for these applications.

BIBLIOGRAPHY

[1] *MATLAB Reference Guide: High-Performance Numeric computation and Visualization Software.* The MathWorks, Inc., South Natick, MA, 1984–1994.

[2] *MATLAB User's Guide: High Performance Numeric Computation and Visualization Software.* The MathWorks, Inc., South Natick, MA, 1984–1994.

[3] *The MathWorks, Inc.: The Student Edition of MATLAB.* Prentice Hall, Englewood Cliffs, NJ, version 4 edition, 1995.

[4] J. W. Cooley and J. W. Tukey. An algorithm for the machine computation of complex Fourier series. *Mathematical Computations*, 19:297–301, April 1965.

[5] C. de Boor. *A Practical Guide to Splines.* Springer-Verlag, 1978.

[6] J. L. Flanagan et al. Speech coding. *IEEE Transactions on Communications*, COM-27:710–736, April 1979.

[7] D. A. George, R. R. Bowen, and J. R. Storey. An adaptive decision feedback equalizer. *IEEE Transactions on Communications Technology*, pages 281–293, June 1971.

[8] J. A. Greefkes. A digitally companded delta modulation modem for speech transmission. In *Proceedings of IEEE International Conference on Communications*, pages 7.33–7.48, June 1970.

[9] S. Haykin. *Adaptive Filter Theory.* Prentice Hall, Englewood Cliffs, NJ, 1986.

[10] F. M. Hsu and A. A. Giordano. Digital whitening techniques for improving spread spectrum communications performance in the presence of narrowband jamming and interference. *IEEE Transactions on Communications*, COM-26:209–216, February 1978.

[11] V. K. Ingle and J. G. Proakis. *Digital Signal Processing using the ADSP-2101.* Prentice Hall, Englewood Cliffs, NJ, 1991.

[12] N. S. Jayant. Adaptive delta modulation with one-bit memory. *Bell System Technical Journal*, pages 321–342, March 1970.

[13] N. S. Jayant. Digital coding of speech waveforms: Pcm, dpcm and dm quantizers. *Proceedings of the IEEE*, 62:611–632, May 1974.

[14] J. W. Ketchum and J. G. Proakis. Adaptive algorithms for estimation and suppression of narrowband interference in pn spread-spectrum systems. *IEEE Transactions on Communications*, COM-30:913–922, May 1982.

[15] N. Levinson. The wiener rms (root-mean-square) error criterion in filter design and prediction. *Journal of Mathematical Physics*, 25:261–278, 1947.

[16] A. V. Oppenheim and R. W. Schafer. *Discrete-Time Signal Processing.* Prentice Hall, Englewood Cliffs, New Jersey, 1989.

[17] T. W. Parks and J. H. McClellan. A program for the design of linear-phase finite impulse response digital filters. *IEEE Transactions on Audio and Electroacoustics*, AU-20:195–199, August 1972.

[18] J. G. Proakis. *Digital Communications.* McGraw-Hill, New York, NY, third edition, 1995.

[19] J. G. Proakis and D. G. Manolakis. *Digital Signal Processing: Principles, Algorithms and Applications.* Macmillan, New York, NY, third edition, 1996.

[20] L. R. Rabiner and B. Gold. *Theory and Applications in Digital Signal Processing.* Prentice Hall, Englewood Cliffs, NJ, 1975.

[21] L. R. Rabiner, R. W. Schafer, and C. A. McGonegal. An approach to the approximation problem for nonrecursive digital filters. *IEEE Transactions on Audio and Electroacoustics*, AU-18:83–106, June 1970.

[22] B. Widrow et al. Adaptive noise cancelling: Principles and applications. *Proceedings of the IEEE*, 63:1692–1716, December 1975.

[23] B. Widrow, P. Manley, and L. J. Griffiths. Adaptive antenna systems. *Proceedings of the IEEE*, 55:2143–2159, December 1967.

INDEX

A-law, 386
Absolute specifications, 225
Absolutely summable, 22
Accumulated amplitude response, 245
Adaptive channel equalizer, 380
  project in, 381
Adaptive delta modulation (ADM), 397
  project in, 399
Adaptive differential PCM (ADPCM), 392
  project in, 395
  standard, 394
Adaptive FIR filter, direct form, 372
Adaptive line enhancement, 380
  project in, 380
Adder, 183
Advantages of DSP over ASP, 3
afd, 364
afd_butt, 311
afd_chb1, 318
afd_chb2, 321
afd_elip, 325
Aliasing formula, 61
All-pole lattice filter, 212
All-zero lattice filter, 208
Alternation theorem, 284
Ampl_Res, 295
Amplitude response, 231
  accumulated, 245
Analog filter design (AFD), 301
Analog lowpass filter design
  (see Analog to digital filter
  transformations)
Analog prototype filters, 305
  characteristics, 305
Analog signal processing (ASP), 2
Analog signals, 2, 7
  reconstruction, 66
  sampling, 61
Analog to digital conversion (ADC), 3, 60
Analog to digital filter
  transformation, 327
Attenuation parameter,
  stopband, 302
Autocorrelation, 20, 27
  in communications, 374, 389
  in LPC speech analysis,
  synthesis, 401
Autoregressive (AR) filter, 34
Autoregressive moving average
  (ARMA) filter, 35

Band-limited signal, 62
Bartlett (triangular), 248
Basic elements of filter structures, 183
  adder, 183
  delay element (shifter), 183
  multiplier, 183
Bessel function, modified
  zero-order, 252
bilinear, 338
Bilinear transformation, 327, 336
  design procedure, 339
Binary digital communication, 408
  project in, 408
Binary spread spectrum
  communication, 409
  project in, 410
Biquad section, 186, 190
blackman, 253
Blackman window, 250
Block convolutions, 157–158
Bowen, R. R.
  (see George, D. A.)
boxcar, 253
bp2lpfre, 370
bs2lpfre, 370
buttapp, 307
butter, 345, 358
Butterworth filter, 302
  design equations, 310
  analog lowpass, 305
buttord, 346, 359

cas2dir, 188
Cascade form, FIR filter
  structure, 197, 198
Cascade form, IIR filter
  structure, 184, 185
casfiltr, 188
Causal sequence, 22
Causality, 22
  in *z*-domain, 102
ceiling, 169
Characteristics of prototype
  analog filters, 305
cheb1ap, 316
cheb1hpf, 357
cheb1ord, 346
cheb2ap, 320
cheb2ord, 346
cheby1, 345
cheby2, 345

Chebyshev error
  (see Minimax approximation
  error)
Chebyshev filter, 302
  analog lowpass, 313
  design equations, 316
  type-I, 313
  type-II, 313, 319
circevod, 143, 175
circonvf, 177
circonvt, 151
Circulant matrix, 177
Circular-even component, 143
Circular-odd component, 143
Circular conjugate symmetry, 142
Circular convolution, 148
circular shift, 146
circulnt, 177
cirshftf, 176
cirshftt, 146
clock, 168
Column vector, 43
Compandor, 387
Comparison of FIR vs. IIR filters, 363
Complex frequency, 81
Conjugate-antisymmetric, 36, 75
Conjugate-symmetric, 35, 42, 75
Constraints on the number of
  extrema, 282
conv, 25
  in polynomial multiplication, 85
conv_m, 26
  in polynomial multiplication, 85
conv_tp, 38
Convergence (ROC), region of, 81
Convolution, 22
  block, 157
  circular, 148
  fast, 169
  high-speed block, 170
  linear, 21
  linear, properties of, 37
  overlap-add, 160
  overlap-save, 158
  sum, 21
Cooley and Tukey, 160, 415
Correlation, 20, 27
  cross-, 20, 27, 376
  (see also Autocorrelation)
cplxcomp, 192
cplxpair, 188

Cross-correlation, 20, 27, 374
Cubic spline interpolation, 69
Cubic splines, 69
Cutoff frequency, 243
   passband, 302

`dbpfd_bl`, 368
`dbpfd_bl`, 369
DC gain, 57
Decimation, 36
Decimation-in-frequency
   (see Fast Fourier transform)
Decimation-in-time
   (see Fast Fourier transform)
`deconv`, 86, 112
   in polynomial division, 89
`deconv_m`, 112
Deconvolution, 86
Delay element, 183
Delta modulation (DM), 396
   adaptive, 397
   project in, 399
Denominator polynomial, 82
Design
   analog filter (AFD), 301
   analog lowpass filters, 302
   FIR filter, 224
   frequency sampling, 264
   IIR filter, 301
   optimal equiripple, 277
   problem statement, 227
   window technique, 243
`dfs`, 119
`dft`, 131
`dhpfd_bl`, 368
Difference equation, 29
   FIR, 34
   IIR, 34
   solutions of, 105
   system representation from, 96
Differential PCM (DPCM), 388
   project in, 390
Differentiator
   (see Digital differentiator)
Digital differentiator, 39, 274, 291
   ideal, 262
Digital filters, 34
   FIR, 34
   IIR, 34
   structures, 182
Digital frequency, 41
Digital prototype filter, 350
Digital signal processing (DSP), 1
   overview of, 2
Digital signal processor, 3
Digital *sinc* function, 128
Digital to analog converter
   (DAC), 3, 61

practical, 67
Dilation, signal, 36
`dir2cas`, 187
`dir2fs`, 204
   modified, 222
`dir2ladr`, 215
`dir2latc`, 210
`dir2par`, 191
Direct form, FIR filter structures,
   197, 198
Direct form, IIR filter structure,
   184
   form I, 184
   form II, 185
Direct form adaptive FIR filter,
   372
Discrete-time
   Fourier transform (DTFT), 40
   inverse Fourier transform
     (IDTFT), 41
   signals, 7
   systems, 20
Discrete-time Fourier transform
   interpolation formula, 127
Discrete-time Fourier transform
   properties, 47
   conjugation, 48
   convolution, 48
   energy, 49
   folding, 48
   frequency-shifting, 48
   linearity, 48
   multiplication, 49
   periodicity, 41
   symmetry, 42
   time-shifting, 48
Discrete Fourier series (DFS), 116
   definition, 117
   matrix, 119
   relation to the DTFT, 123
   relation to the z-transform, 121
Discrete Fourier transform
   (DFT), 116, 129
   definition, 130
   matrix, 131
Discrete Fourier transform (DFT)
   properties, 139
   circular convolution, 148
   circular folding, 139–140
   circular shift in the
     frequency-domain, 148
   circular shift in the
     time-domain, 145
   conjugation, 142
   frequency leakage, 179, 180
   linearity, 139
   multiplication, 153
   Parseval's relation, 153
   symmetry, 142

Discrete systems, 20
`dlpfd_bl`, 369
`dlpfd_ii`, 365
Dot-product, 163
Down-sampling, 36
`dtft`, 44, 74
Dual tone multi-frequency
   (DTMF), 403
   project in, 407
Durbin, J., 401

Efficient computation, goal of,
   161
`ellip`, 345
`ellipap`, 324
`ellipord`, 346
Elliptic filter, 302
   analog lowpass, 323
   computation of filter order,
     323
Energy density spectrum, 49
Equalizer, adaptive channel, 380
   project in, 381
Equiripple design technique,
   277
   problem statement, 281
Equiripple filters, 278
Error analysis, 155
`etime`, 168
Even and odd synthesis, 17
`evenodd`, 18
Excitation, 20
Exponential sequence
   complex-valued, 9
   real-valued, 9
Extra-ripple filters, 284

Fast convolution, 169
Fast Fourier transform (FFT),
   160
   mixed radix, 165
   radix-2
     decimation-in-frequency
     (DIF), 167
   radix-2 decimation-in-time
     (DIT), 165, 166
   radix-R, 165
`fft`, 167
Filter, 2, 182
   analog, prototype, 305
   approximations, 224
   autoregressive, 34
   Butterworth analog lowpass,
     305
   Chebyshev analog lowpass,
     313
   digital, 34

digital prototype, 350
Elliptic analog lowpass, 323
equiripple, 278
extra-ripple, 284
FIR, 34
ideal bandpass, 258
ideal highpass, 77
ideal lowpass, 76
IIR, 34
implementation, 225
linear phase FIR, 231–234
minimum-phase, 304
moving average, 34
nonrecursive, 34
recursive, 34
specifications, 224
staircase, 290
filter, 30
  with initial conditions, 108
Filter transformations, analog to
    digital, 327
  bilinear transformation, 327
  finite difference approximation,
    327
  impulse invariance, 327
  step invariance, 327, 365
filtic, 109
Finite-duration impulse response
    (FIR) filters, 5, 34
  adaptive, 372
  cascade form, 197, 198
  design, 224
  difference equation, 34
  direct form, 197, 198
  frequency sampling design
    technique, 264
  frequency sampling form, 197,
    202
  linear-phase form, 197, 199
  structures, 197
Finite-duration sequence, 7
Finite difference approximation
    technique, 327
First-order hold (FOH)
    interpolation, 69
formants, 208
Fourier transform
  discrete, 130
  discrete-time, 40
  fast, 160
  inverse discrete-time, 41
freqresp, 77
freqs_m, 312
Frequency
  complex, 81
  cutoff, 243
  digital, 41
  natural, 30
  resolution, 123

response, 54
response, linear-phase, 230
sampling theorem, 125
Frequency-band transformations,
    350
  design procedure for lowpass to
    highpass, 356
Frequency-domain representation
    of LTI systems, 53
Frequency response function from
    difference equations, 57
Frequency sampling design
    technique, 264
  basic idea, 266
  naive design method, 267
  optimum design method, 268
Frequency sampling form, 197,
    202
freqz, 45
freqz_m, 254
Fundamental period, 10

Geometric series, 19
George, D. A., 413
Gibbs phenomenon, 247
Giordano, A. A.
  (see Hsu, F. M.)
Goal of an efficient computation,
    161
Goertzel algorithm, 161, 405
Gold, B.
  (see Rabiner, L. R.)
Greefkes, J. A., 413
Griffiths, L. J.
  (see Widrow, B.)
Group delay, constant, 229

hamming, 253
Hamming window, 249
hanning, 253
Hanning window, 249
Haykin, S., 415
High-speed block convolution,
    170
High-speed convolution
  (see Fast convolution)
Hilbert transformer, 263, 275,
    292
Homogeneous solution, 29, 105,
    107
hp21pfre, 368
Hr_Type1, 234
Hr_Type2, 235
Hr_Type3, 235
Hr_Type4, 235
hsolpsav, 171
Hsu, F. M., 413

Ideal
  bandpass filter, 258
  digital differentiator, 262
  highpass filter, 77
  lowpass filter, 76
ideal_lp, 253
Identification, system, 376
idfs, 120
idft, 131
ifft, 168
imp_invr, 330
impseq, 8
impulse, 312
Impulse invariance
    transformation, 327
  design procedure, 329
Impulse response, 21
  antisymmetric, 200
  symmetric, 199
  time-varying, 21
Infinite-duration impulse
    response (IIR) filters, 5, 34
  cascade form, 184, 185
  design, 301
  difference equation, 34
  direct form, 184
  parallel form, 184, 190
  structures, 183
Initial-condition input, 108
Interpolation
  cubic spline, 69
  first-order hold (FOH), 69
  formula (DTFT), 127
  formula (time-domain), 67
  zero-order hold (ZOH), 68
Intersymbol interference, 381
Inverse
  discrete-time Fourier transform
    (IDTFT), 41
  DFT (IDFT), 130
  FFT (IFFT), 168
  z-transform, 81, 89

Jayant, N. S., 413

kai_bpf, 297
kai_bsf, 297
kai_hpf, 297
kai_lpf, 297
kaiser, 253
Kaiser window, 250
  design equations, 253
Ketchum, J. W., 413

Ladder coefficients, 215
ladr2dir, 216

ladrfilt, 216
latc2dir, 211
latcfilt, 210
Lattice-ladder filter, 214
  structure, 215
Lattice filter structures, 208
  FIR, 208
  IIR, 212, 214
Levinson, N., 401, 413
Levinson-Durbin recursion, 401
Linear-phase FIR filters
  advantages, 227
  frequency response, 230
  properties, 228
  Type-1, 231
  Type-2, 232
  Type-3, 233
  Type-4, 234
  zero constellation, 236
  zero locations, 236
Linear-phase form, 197, 199
Linear convolution, 21
  properties of, 37
  using the DFT, 154
Linear fractional transformation
  (see Bilinear transformation)
Linear predictive coding (LPC) of
  speech, 399
  project in, 403
Linear systems, 20
Linear time-invariant (LTI)
  system, 21
  frequency-domain
    representation, 53
LMS algorithm, 373
Lowpass filter design
  analog prototype (see
    Analog-to-digital filter
    transformations)
  digital, using Matlab, 345
Lowpass filters
  (see Filters)
lp2lpfre, 369

M-fold periodicity, 173
Magnitude-only specifications,
  225
Magnitude (or gain) response,
  54
Manley, P.
  (see Widrow, B.)
Manolakis, D. G.
  (see Proakis, J. G.)
Matlab
  a few words about, 5
  lowpass filter design, 345
  reference guide, 6, 413
  signal processing toolbox, 29

  student edition, 6
  symbolic toolbox, 6
  user's guide, 6, 413
Matrix
  circulant, 177
  Toeplitz, 37
matrix-vector multiplication, 38,
  43
Merging formula, 166
Minimax approximation error,
  278
Minimax problem, development
  of, 278
Minimum-phase filter, 304
Minimum stopband attenuation,
  246
Mirror-image symmetry, 304
mod, 130
Modeling, system, 376
Modem, 371, 380
Moving average (MA) filters,
  34
μ-law, 385
Multiplier, 183

N-point sequence, 129
Narrowband interference,
  suppression of, 377
  project in, 379
Natural frequency, 30
Nonrecursive filters, 34
Number sequence, 7
Numerator polynomial, 82
Nyquist component, 143
Nyquist rate, 63, 384

Operations on sequences, 10
  folding, 12
  sample products, 13
  sample summation, 12
  scaling, 11
  shifting, 12
  signal addition, 10
  signal energy, 13
  signal multiplication, 11
  signal power, 13
Optimum filter, 374
Overlap-add method of
  convolution, 160
Overlap-save method of
  convolution, 158
  high-speed, 170
Overview of digital signal
  processing, 2
ovrlpadd, 178
ovrlpsav, 159

par2dir, 193
Parallel form, IIR filter structure,
  184, 190
parfiltr, 192
Parks-McClellan, 413
  algorithm, 284
Particular solution, 30, 105, 107
Passband cutoff frequency, 302
Passband ripple parameter, 302
Passband tolerance, 225
Peak side lobe magnitude, 246
Period, fundamental, 10
Periodic conjugate symmetry,
  142
Periodic sequences, 10, 117
Periodic shift, 146
Periodicity, M-fold, 173
Phase delay, constant, 228
Phase response, 54
  of analog prototype filters, 327
Pitch detection, 402–403
plot, 71
Poles in system function, 96
poly, 93
Polynomial
  denominator, 82
  numerator, 82
Practical D/A converters, 67
Proakis, J. G., 414
  (see also Ketchum, J. W.)
Projects
  adaptive channel equalization,
    381
  adaptive line enhancement,
    380
  ADPCM, 395
  binary data communications,
    408
  binary spread spectrum
    communications, 410
  DM and ADM, 399
  DTMF, 407
  LPC, 403
  PCM, 387
  suppression of sinusoidal
    interference, 379
  system identification, 376
Properties of
  DFT, 139
  DTFT, 41, 47
  linear convolution, 37
  linear-phase FIR filters, 228
  magnitude squared response,
    304
  ROC, 83
  z-transform, 84
Pulse code modulation (PCM),
  384
  A-law nonlinearity, 386

μ-law nonlinearity, 385–386
  project in, 387

Rabiner, L. R., 414
Radix-2 decimation-in-frequency
  FFT, 167
Radix-2 decimation-in-time FFT,
  165, 166
rand, 10
randn, 10
Random sequence, 10
real2dft, 176
Reconstruction formula in the
  z-domain, 127
Reconstruction of analog signals,
  66
Rectangular window, 125, 244,
  245
Recursive filters, 34
  (see also IIR filters)
Reflection coefficients, 208
Region of convergence (ROC), 81
  properties of, 83
Relationships between system
  representations, 102
Relative linear scale, 302
Relative specifications, 225
rem, 129
remez, 285
residuez, 91
Response, 20
  amplitude, 231
  to arbitrary sequences, 55
  to complex exponential, 54
  frequency, 54
  impulse, 21
  magnitude (or gain), 54
  phase, 54
  to sinusoidal sequences, 54
  steady-state, 55
  unbounded, 107
  zero-input, 33
  zero-state, 33
Ripple parameter, passband, 302
roots, 32, 96
Row vector, 7

Sampling, 61
  interval, 61, 123
  theorem, 63
Sampling and reconstruction of
  analog signals, 60
Sampling and reconstruction in
  the z-domain, 124
sdir2cas, 308
Second-order sections, 184, 185,
  190

Sequences
  causal, 22
  exponential, 9
  finite-duration, 7
  folded-and-shifted, 24
  infinite-duration, 8
  N-point, 129
  negative-time, 82
  number, 7
  operations on, 10
  periodic, 10, 117
  positive-time, 81
  random, 10
  sinusoidal, 9
  two-sided, 83
  types of, 8
  unit sample, 8
  unit step, 8
Shifts
  circular, 146
  periodic, 146
sigadd, 11
sigfold, 12
sigmult, 11
Signal
  analysis, 4
  band-limited, 62
  dilation, 36
  filtering, 4
  processing, 2
Signals
  analog, 2, 7
  digital, 2
  discrete-time, 7
  energy, 13
  power, 13
sigshift, 12
$sinc(x)$, 67
Sinusoidal sequence, 9
Solutions
  difference equation, 105
  homogeneous, 29, 105, 107
  particular, 30, 105, 107
  steady-state, 105, 107
  transient, 105, 107
  zero-input, 34, 105, 107
  zero-state, 34, 105, 107
Specifications
  absolute, 225
  filter, 224
  magnitude-only, 225
  relative, 225
  relative linear, 302
Spectral transformations
  (see Frequency-band
    transformations)
Spectrum
  analyzers, 2, 182
  energy density, 49, 153

high-density, 136
high-resolution, 136
power, 153
spline, 73
Spread spectrum
  communications, 411
  project in binary, 412
Stability, 22
  bounded-input bounded-output
    (BIBO), 22
  in z-domain, 102
Staircase filter, 290
stairs, 71
Steady-state response, 55, 105
Step invariance, 327, 365
stepseq, 9
Stopband attenuation parameter,
  302
Stopband tolerance, 225
Storey, J. R.
  (see George, D. A.)
stp_invr, 366
Structures, digital filter, 182
  all-pole lattice filter, 212
  all-zero lattice filter, 208
  basic elements, 183
  FIR filter, 197
  IIR filter, 183
  lattice-ladder, 214
Summable, absolutely, 22
Superposition summation, 21
Suppression of narrowband
  interference, 377
Synthesis
  even and odd, 17
  unit sample, 17
System function, 95
System identification, 376
  project in, 376
System modeling
  (see System identification)
System representation
  from difference equations,
    96
  relationships between, 102
  transfer function, 97
  in the z-domain, 95
Systems
  discrete 20
  linear, 20
  LTI, 21

Table
  amplitude response and
    β-values for linear-phase FIR
    filters, 278
  comparison of analog filters,
    350

frequency transformations for digital filters, 352
$Q(\omega)$, L and $P(\omega)$ for linear-phase FIR filters, 279
window function characteristics, 251
z-transform, 87
Telecommunications, 371, 381
  modems, 371, 380
Theorem
  alternation, 284
  frequency sampling, 125
  sampling, 63
  z-domain stability, 103
Time-varying impulse response, 21
Toeplitz matrix, 37
Tolerance
  passband, 225
  stopband, 225
  transition band, 225
Tone detection, 406
Touch Tone, 403–404
Transfer function representation, 97
Transformations
  bilinear, 327, 336
  filter, 327
  frequency-band, 350
  linear fractional, 337
  spectral, 350
Transient response, 105, 107
Transition band tolerance, 225
Transition bandwidth
  approximate, 246, 251
  exact, 247, 251
triang, 253

Triangular window
  (see Bartlett window)
Twiddle factor, 165
Two important categories of DSP, 4

U_buttap, 307
U_chb1ap, 316
U_chb2ap, 320
U_elipap, 324
Unbounded response, 107
Unit circle, 81
Unit sample sequence, 8
Unit sample synthesis, 17
Unit step sequence, 8

Vectors
  column, 43
  row, 7
Voice synthesis, 5

Widrow, B., 414
Window design techniques, 243
  basic idea, 245
Window function characteristics, 251
Windowing, 243
Windows
  Bartlett (triangular), 248
  Blackman, 250
  Hamming, 249
  Hanning, 249
  Kaiser, 250
  rectangular, 125, 244, 245

xcorr, 29

z-domain
  causal LTI stability theorem, 103
  LTI stability theorem, 103
  sampling and reconstruction in, 124
  stability and causality, 102
  system representation, 95
z-domain system function, 95
z-transform
  the bilateral, 80
  complex conjugation, 84
  convolution, 85
  differentiation in the z-domain, 85
  folding, 84
  frequency shifting, 84
  inverse, 81, 89
  linearity, 84
  multiplication, 85
  one-sided, 105
  reconstruction formula, 127
  sample shifting, 84
  table, 87
  z-transform properties, 84
Zero-input response, 33, 105, 107
Zero-order hold (ZOH) interpolation, 68
Zero-padding, 135
Zero-state response, 33, 105, 107
zeros, 135
Zeros in system function, 96
zmapping, 353
zplane, 96